COGNITIVE COMPUTING SYSTEMS

Applications and Technological Advancements

COGNITIVE COMPUTING SYSTEMS

Applications and Technological Advancements

Edited by

Vishal Jain, PhD

Akash Tayal, PhD

Jaspreet Singh, PhD

Arun Solanki, PhD

First edition published 2021

Apple Academic Press Inc.
1265 Goldenrod Circle, NE,
Palm Bay, FL 32905 USA

4164 Lakeshore Road, Burlington,
ON, L7L 1A4 Canada

CRC Press
6000 Broken Sound Parkway NW,
Suite 300, Boca Raton, FL 33487-2742 USA

4 Park Square, Milton Park,
Abingdon, Oxon OX14 4RN

First issued in paperback 2023

© 2021 Apple Academic Press, Inc.

Apple Academic Press exclusively co-publishes with CRC Press, an imprint of Taylor & Francis Group, LLC

Reasonable efforts have been made to publish reliable data and information, but the authors, editors, and publisher cannot assume responsibility for the validity of all materials or the consequences of their use. The authors, editors, and publishers have attempted to trace the copyright holders of all material reproduced in this publication and apologize to copyright holders if permission to publish in this form has not been obtained. If any copyright material has not been acknowledged, please write and let us know so we may rectify in any future reprint.

Except as permitted under U.S. Copyright Law, no part of this book may be reprinted, reproduced, transmitted, or utilized in any form by any electronic, mechanical, or other means, now known or hereafter invented, including photocopying, microfilming, and recording, or in any information storage or retrieval system, without written permission from the publishers.

For permission to photocopy or use material electronically from this work, access www.copyright.com or contact the Copyright Clearance Center, Inc. (CCC), 222 Rosewood Drive, Danvers, MA 01923, 978-750-8400. For works that are not available on CCC please contact mpkbookspermissions@tandf.co.uk

Trademark notice: Product or corporate names may be trademarks or registered trademarks and are used only for identification and explanation without intent to infringe.

Publisher's Note
The publisher has gone to great lengths to ensure the quality of this reprint but points out that some imperfections in the original copies may be apparent.

Library and Archives Canada Cataloguing in Publication

Title: Cognitive computing systems : applications and technological advancements / edited by Vishal Jain, PhD, Akash Tayal, PhD, Jaspreet Singh, PhD, Arun Solanki, PhD.

Names: Jain, Vishal, 1983- editor. | Tayal, Akash, 1973- editor. | Singh, Jaspreet, 1976- editor. | Solanki, Arun, 1985- editor.

Description: First edition. | Includes bibliographical references and index.

Identifiers: Canadiana (print) 20200380516 | Canadiana (ebook) 20200380737 | ISBN 9781771889315 (hardcover) | ISBN 9781003082033 (ebook)

Subjects: LCSH: Artificial intelligence. | LCSH: Computational intelligence. | LCSH: User interfaces (Computer systems)

Classification: LCC TA347.A78 C64 2021 | DDC 006.3—dc23

Library of Congress Cataloging-in-Publication Data

CIP data on file with US Library of Congress

ISBN: 978-1-77188-931-5 (hbk)
ISBN: 978-1-77463-763-0 (pbk)
ISBN: 978-1-00308-203-3 (ebk)

DOI: 10.1201/9781003082033

About the Editors

Vishal Jain, PhD, is Associate Professor at the Department of Computer Science and Engineering, School of Engineering and Technology, Sharda University, Greater Noida, UP, India. Before that, he worked for several years as Associate Professor at Bharati Vidyapeeth's Institute of Computer Applications and Management (BVICAM), New Delhi, India. He has more than 15 years of experience in academia. He obtained the degrees of PhD (CSE), MTech (CSE), MBA (HR), MCA, MCP, and CCNA. He has more than 370 research citation indices with Google Scholar (h-index score 9 and i-10 index 9). He has authored more than 70 research papers in reputed conferences and journals, including *Web of Science* and *Scopus*. He has authored and edited more than 10 books with various reputed publishers, including Apple Academic Press, CRC, and Taylor & Francis Group. His research areas include information retrieval, semantic web, ontology engineering, data mining, ad hoc networks, and sensor networks. He received a Young Active Member Award for the year 2012–13 from the Computer Society of India, Best Faculty Award for the year 2017 and Best Researcher Award for the year 2019 from BVICAM, New Delhi, India.

Akash Tayal, PhD, is Associate Professor in the Department of Electronics and Communication at the Indira Gandhi Delhi Technical University for Women, Delhi, India. He holds a PhD from IIT Delhi. His current research interests include machine learning, data science and data analytics, image processing, stochastic dynamic facility layout problem, optimization, and meta-heuristic and decision science. He has published articles in various international journals and conference proceedings, such as *ANOR, COI, IEEE, GJFSM, IJOR* and *IJBSR*. He is a life member of the GLOGIFT Society and the Indian Institution of Industrial Engineering.

Jaspreet Singh, PhD, is Associate Professor at the Department of Computer Science, School of Engineering, GD Goenka University, Gurgaon, India. Dr. Singh completed his PhD in Information Technology at Guru Gobind Singh Indraprastha University. He also cleared the prestigious UGC-NET and GATE examinations. He has over 17 years of teaching experience at postgraduate level. Dr. Singh has authored chapters for three books. He is a keen researcher and has published many research publications in reputed international journals and conferences, including SCI-indexed journals such as *Wireless Networks* and the *International Journal of Communication Systems*. Dr. Singh is on the editorial boards of several reputed journals and has also served as a reviewer and TPC member for many journals and conferences. He has guided over 100 dissertations at the master's level and is currently supervising three PhD scholars. His research interests include wireless networks, sensor networks, algorithms, and deep learning.

Arun Solanki, PhD, is Assistant Professor in the Department of Computer Science and Engineering, Gautam Buddha University, Greater Noida, India, where he has also held various additional roles over the years. He has supervised more than 60 MTech dissertations. His research interests span expert system, machine learning, and search engines. He has published many research articles in SCI/Scopus-indexed international journals and has participated at many international conferences. He has been a technical and advisory committee member of many conferences and has chaired and organized many sessions at international conferences, workshops, and seminars. Dr. Solanki is Associate Editor for the *International Journal of Web-Based Learning and Teaching Technologies* (IJWLTT) and has guest edited special issues of *Recent Patents on Computer Science*. Dr. Solanki is the editor of many books and is working as the reviewer for journals published by Springer, IGI Global, Elsevier, and others. He received MTech degree in computer engineering from YMCA University, Faridabad, Haryana, India. He has received his PhD in Computer Science and Engineering from Gautam Buddha University, Greater Noida, India.

Contents

Contributors

Dinesh Bhatia
Department of Biomedical Engineering, North Eastern Hill University, Shillong, Meghalaya 793022, India

Pratiyush Guleria
National Institute of Electronics and Information Technology Shimla, Himachal Pradesh 171001, India

Akanksha Gupta
Department of Computer Science, Swami Shraddhanand College, University of Delhi, New Delhi, Delhi 110036, India

Rebakah Jobdas
Faculty of Computer Science, Kadi Sarva Vishwavidyalaya, Gandhinagar, Gujarat 382015, India

Hemdev Karan
Department of Computer Engineering, Vivekanand Education Society's Institute of Technology, Chembur, Mumbai, Maharashtra 400074, India

Gurpreet Kaur
University Institute of Engineering and Technology, Punjab University, Chandigarh 160025, India

Pooja Khurana
Department of Mathematics, Manav Rachna International Institute of Research and Studies, Faridabad, Haryana 121004, India

Amod Kumar
National Institute of Technical Teachers Training and Research, Chandigarh 160026, India

Deepak Kumar
Department of Mathematics, Manav Rachna International Institute of Research and Studies, Faridabad, Haryana 121004, India

Naresh Kumar
Department of Mathematics, Indian Institute of Technology, Roorkee, Uttarakhand 247667, India

Ravinder Kumar
Skill Faculty of Engineering and Technology, Shri Vishwakarma Skill University, Gurugram, Haryana 122003, India

Vinod Kumar
Department of Mathematics, University College of Basic Science and Humanities, Guru Kashi University, Talwandi Sabo, Punjab 151302, India

Raghani Madhu
Department of Computer Engineering, Vivekanand Education Society's Institute of Technology, Chembur, Mumbai, Maharashtra 400074, India

Gautam Menghani
Department of Computer Engineering, Vivekanand Education Society's Institute of Technology, Chembur, Mumbai, Maharashtra 400074, India

Tarun Methwani
Department of Computer Engineering, Vivekanand Education Society's Institute of Technology, Chembur, Mumbai, Maharashtra 400074, India

Lino Murali
School of Engineering, Cochin University of Science and Technology, Kochi, Kerala 682022, India

Ramyashree
Department of Computer Science and Engineering, N.M.A.M. Institute of Technology, Nitte, Karnataka 574110, India

Lokesh Kumar Shrivastav
University School of Information, Communication and Technology, Guru Gobind Singh Indraprastha University, New Delhi, Delhi 110078, India

Priyanka Singh
Department of Civil Engineering, Amity School of Engineering and Technology, Amity University, Noida, Uttar Pradesh 201301, India

Umang Singh
Institute of Technology and Science, Mohan Nagar, Ghaziabad, Uttar Pradesh 201007, India

Manu Sood
Department of Computer Science, Himachal Pradesh University, Shimla, Himachal Pradesh 171005, India

Mohit Srivastava
Chandigarh Engineering College, Landran, Mohali 140307, India

Nitin Tejuja
Department of Computer Engineering, Vivekanand Education Society's Institute of Technology, Chembur, Mumbai, Maharashtra 400074, India

Shivani A. Trivedi
Faculty of Computer Science, Kadi Sarva Vishwavidyalaya, Gandhinagar, Gujarat 382015, India

S. Varadharajan
Department of Civil Engineering, Amity School of Engineering and Technology, Amity University, Noida, Uttar Pradesh 201301, India

Kirthanashri S. Vasanthan
Amity Institute of Molecular Medicine and Stem Cell Research, Amity University, Noida, Uttar Pradesh 201301, India

P. S. Venugopala
Department of Computer Science and Engineering, N.M.A.M. Institute of Technology, Nitte, Karnataka 574110, India

Shwetambara Verma
Professor, Department of Civil Engineering, School of Engineering and Technology, Kaziranga University, Jorhat, Assam.

V. Vishnuprabha
School of Engineering, Cochin University of Science and Technology, Kochi, Kerala 682022, India

Daleesha M. Viswanathan
School of Engineering, Cochin University of Science and Technology, Kochi, Kerala 682022, India

Abbreviations

AC	autocorrelation
ACF	autocorrelation function
ADC	analog-to-digital converter
AI	artificial intelligence
AN	affective neuroscience
ANFIS	adaptive network-based fuzzy inference system
ANN	artificial neural network
ANOVA	analysis of variance
AODB	agent-oriented database
AOR	agent–object relationship
API	application programming interface
ARIMA	autoregressive integrated moving average model
ARM	association rule mining
AUC	area under curve
BCI	brain–computer interface
BP	backpropagation
BP	blood pressure
BPNN	backpropagation neural network
BSN	body sensor network
CART	classification and regression tree
CBOW	continuous bag of words
CIoT	cognitive IoT
CNN	convolutional neural network
CoML	cognitive machine learning
CSR	continuous speech recognition
CWR	connected speech recognition
DBN	deep belief network
DM	digital marketing
DNN	deep neural network
DSS	decision support system
DT	decision tree
ECG	electrocardiogram
EEG	electroencephalography
EMG	electromyogram

ESM	exponential smoothing model
FA	finite automata
FA	fly ash
FFNN	feedforward neural network
fMRI	functional magnetic resonance imaging
GA	genetic algorithm
GB	Gaussian Bayes
GLM	generalized linear model
GNN	graph convolution network
GPS	global positioning system
GPU	graphics processing unit
GRU	gated recurrent unit
HMM	hidden Markov model
HMS	health monitoring system
HR	heart rate
ICA	independent component analysis
IDA	intelligent database agent
IDRS	Indian Diabetes Risk Score
iDSS	intelligent decision support system
IE	information extraction
IoT	Internet of things
ISSR	integrated speaker and speech recognition
IWSR	isolated speech recognition
KNN	*k*-nearest neighbors
KPCA	kernel-based principal component analysis
LPCC	linear predictive coding coefficient
LR	language recognition
MAS	multiagent system
MCQs	multiple choice questions
MCU	microcontroller unit
MDRF	Madras Diabetes Research Foundation
MFCC	Mel-frequency cepstral coefficient
MHMS	mobile health monitoring system
MIS	machine intelligence system
ML	machine learning
MLP	multilayer perception
MLR	multiple linear regression
MRI	magnetic resonance imaging
NB	Naïve Bayes

NLG	natural language generation
NLP	natural language processing
NN	neural network
NRC	nonrecurring cost
OCN	organizational cognitive neuroscience
OID	object identifier
PCA	principal component analysis
PCF	partial autocorrelation function
PFC	prefrontal cortex
PLP	perceptual linear prediction
POS	parts of speech
PP	polypropylene
PPG	photoplethysmography
QR	quick response
RASTA	relative spectral processing
RASTA-PLP	relative spectra perceptual linear predictive
RBF	radial basis function
RBFNN	radial basis function neural network
RC	reinforced concrete
RcNN	recursive neural network
ReLU	rectified linear unit
RF	random forest
RHMS	remote health monitoring system
RMSE	root-mean-square error
RNN	recurrent neural network
ROC	receiver operating characteristic
RR	respiratory rate
RST	rhetorical structure theory
SGD	stochastic gradient decent
SI	speaker identification
SN	social neuroscience
SOM	self-organizing map
SpO_2	oxygen saturation
SPR	speaker recognition
SR	speech recognition
SV	speaker verification
SVM	support vector machine
SVR	support vector regression
T2D	Type 2 diabetes

TD	text-dependent
TF-IDF	term frequency-inverse document frequency
TID	text-independent
TSVR	twin support vector regression
TWSVM	twin support vector machine
UDT	user-defined data type
WC	waist circumference
WHMS	wearable health monitoring system
WHO	World Health Organization
WSD	word-sense disambiguation
ZCR	zero-crossing rate

Preface

It gives us immense pleasure to introduce to you the first edition of the book entitled *Cognitive Computing Systems: Applications and Technological Advancements*. Cognitive computing systems are one of the emerging areas under the umbrella of artificial intelligence. The cognitive computing systems mimic the working and learning mechanism of human brains. The major challenge for cognitive systems is to devise models that are self-learning and can infer useful information from extensive ambiguous data. Many techniques are pivotal to build a successful cognition system, including natural language processing, speech recognition, neural networks, deep learning, sentiment analysis, etc. The objective of this book is to introduce cognitive computing systems, bring out their key applications, discuss different areas and technologies used in cognitive systems, and study existing works, underlying models, and architectures. Some unique features of this book that make it a useful resource for learning are the following:

- Simple and articulate language
- Supported with large examples
- Inclusion of recent and up to date work
- Discussion on open issues
- Present various scientific works on real-world application.

Each chapter presents the use of cognitive computing and machine learning in some application areas. This book will be asset for researchers and faculties to know in detail the fine print of cognitive computing. The work in this book has been contributed by well-qualified and eminent faculties. The authors are keen researchers in the area of artificial intelligence, cognitive computing, and allied areas.

All editors of this book are experts in this domain, each having excellent publications, research experience, and years of experience. A brief profile of the editors is listed as follows. Dr. Vishal Jain is an Associate Professor at Department of Computer Science and Engineering, School of Engineering and Technology, Sharda University, Greater Noida, UP, India. Dr. Akash Tayal is an Associate Professor with the Department of Electronics and Communication Engineering, Indira Gandhi Delhi Technical for Women, Delhi, India. Dr. Jaspreet Singh is an Associate Professor with the Department of Computer Science and Engineering, School of Engineering, G. D. Goenka University, Gurgaon, India, and Dr. Arun Solanki is an Assistant Professor with Gautam Buddha University, Greater Noida, India.

The idea behind this book is to simplify the journey of aspiring data scientists and machine learning enthusiasts across the world. Through this book, they will enable us to work on machine learning problems and gain from experience. This book will provide a high-level understanding of various machine learning algorithms along with tools and techniques.

This book would not have been possible without the motivation and valuable contribution of many people. First of all, we would like to thank the almighty God for providing the inspiration and the motivation to keep doing good work. We are grateful to our mentors, who are our inspirational models. This book would not have been possible without the support of our families, including the kids, who cooperated on many counts including sparing us the precious time to complete this book effectively, from their schedule of already tight periods, that otherwise would have been spent on their upbringing.

We hope that the readers make the most of this volume and enjoy reading this book. Suggestions and feedback are always welcome.

PART I

Using Assistive Learning to Solve Computationally Intense Problems

CHAPTER 1

High-Frequency Stochastic Data Analysis Using a Machine Learning Framework: A Comparative Study

LOKESH KUMAR SHRIVASTAV[1] and RAVINDER KUMAR[2*]

[1]*University School of Information, Communication and Technology, Guru Gobind Singh Indraprastha University, New Delhi, Delhi 110078, India*

[2]*Skill Faculty of Engineering and Technology, Shri Vishwakarma Skill University, Gurugram, Haryana 122003, India*

[*]*Corresponding author. E-mail: ravinder.kumar@svsu.ac.in*

ABSTRACT

High-frequency stochastic data analysis and prediction are challenging and exciting problems if we aim to maintain high level of accuracy. The stock market dataset is selected randomly for the experimental investigation of the study. Historical datasets of a few stock markets have been collected and used for this purpose. The model is trained, and the results are compared with the real data. In the past few years, specialists have tried to build computationally efficient techniques and algorithms, which predict and capture the nature of the stock market accurately. This chapter presents a comparative analysis of the literature on applications of machine learning tools on the financial market dataset. This chapter provides a comparative and brief study of some relevant existing tools and techniques used in financial market analysis. The main objective of this chapter is to provide a comparative study of novel and appropriate methods of stock market prediction. A brief explanation of advanced and recent tools and techniques available for the analysis and a generalized and fundamental model in R-language for the stock market analysis and prediction are also provided in this chapter. In addition, this chapter presents a review of significant challenges and futuristic challenges of the field.

1.1 INTRODUCTION

In recent years, high-speed data acquisition technology has demanded an appropriate and advanced analysis and prediction mechanism. With easy access and advancement of the storage system, a massive amount of data are generated every second. It forced the demand for a high-speed data analytics processing system. The development of information communication technology and complex computational algorithms and the collection, analysis, and prediction using high frequency are possible [1]. In recent years, machine learning frameworks play an important role in providing an excellent forecast and fast computation on a massive amount of high-frequency data. High-frequency data mean that datasets are collected in a fine regular or irregular interval of time and are referred to as high-frequency stochastic time-series datasets [2]. Problem-solving technology may be classified into two parts: hard computational method and soft computational method. The hard computational method can deal with precise models to achieve the solution accurately and quickly. Prof. L. Zadeh introduced the soft computational method in 1994. It is a mechanism that deals with uncertainty, robustness, and better tolerance objectives of the model. The soft computing method is a hybrid mechanism that is a combination of fuzzy logic, evolutionary computation, probabilistic reasoning, and machine learning. The fundamental constituent of soft computing is machine learning that was introduced by A. Samuel in 1959, which explores the study and development of appropriate technologies and tools that can learn and adopt the nature from the data and predict its futuristic nature. The application of machine learning has gradually grown and now reached on its optimum and very advanced level, where it can examine and forecast the futuristic nature of the dataset. In the past few decades, machine learning has produced a compelling result in low-frequency datasets. A low-frequency dataset means a dataset that will be collected in an hour or a day. Nowadays, we are living in the era of nanotechnology, where data can be produced in the fine interval of minutes or second or nanoseconds. These kinds of datasets are beneficial for machine learning tools and techniques, where chances of deep learning can increase more, and the prediction capability will produce a better result. The uses of machine learning techniques over high-frequency data are applied and used from social networks to medical science and nanoscience to rocket science. The existing frameworks for analysis and prediction of these high-frequency datasets can be classified into the following two categories.

- *Statistical model:* This is a traditional mathematical method, in which advanced statistical models and procedures can analyze the high-frequency dataset. It is good to predict in terms of stock market, but it demands some assumption that decides the accuracy of the model [3]. Therefore, it cannot be utilized as an intelligent system.
- *Soft-computing-based model:* Machine learning frameworks are compelling and are capable of capturing the dynamics of stock price and predicting the futuristic trend by using the time-series mechanism.

Recent research has shown that both models have their own significance to capture the high-frequency datasets [4,5]. Analysis of the stock price trend has been challenging for both investors and researchers. Financial time series is a significant source of information for stock market prediction. Finding hidden patterns is the requirement of analysis and prediction for price fluctuations. The fast development of computing capacity, as well as modern and smart machine learning frameworks, makes it possible to produce new solutions for data analysis. Because of the slow and complicated process of manual and fundamental statistical methods, its prediction has less use because a model has no use if streaming of the dataset is faster than the production of its result. A process that demands longer time to arrive for the forecast has no use if the objective is to predict the result in time. In this work, we will try to produce the generalized machine learning model to make an intelligent system, which will capable of producing the futuristic nature of the stock market.

The last decade was the era of the single model, where one model is taken for the analysis and prediction. The experimental results proved that these models are good, but for the betterment of the model, a hybrid model was applied and analyzed. It is found that the performance estimation of the hybrid framework is better than that of a single framework. However, much space is available to produce a better model and the best model [6]. In this study, we have taken stock market datasets that are collected on the fine interval of the single minute, so these are high-frequency and nonlinear datasets. The availability of the high-frequency dataset is infrequent in the field of stock market. Therefore, it was rarely utilized in the recently reviewed experimental setup. In this study, a base, generalized, and very famous statistical model, autoregressive integrated moving average (ARIMA) model, will be applied on this high-frequency dataset of the stock market on the time of the experiment, and it will be compared with a fundamental machine learning model generalized linear model (GLM) to define these two models very carefully. The GLM is a basic regression model of machine learning with multidistribution capabilities. If the ARIMA model will able to compete for the GLM, then it will be compared with other advanced machine learning models in a later stage of this study. The ARIMA model is the fundamental and open model that gives liberty to its developer to understand the nature of the datasets at each stage of its processing. This is the reason why it is called a base model. This model carries its own simplicity to understand the nature of nonlinear datasets and saturates it, as desired. This approach can be fruitful to guide the student and investors to build a more reliable and optimized intelligent financial forecasting system. The primary use of this work is to explore and provide fundamental obstacles and futuristic dimension and guidelines in directions of particular research. These simulated results have shown deviation from the actual result. Therefore, a deeper comparative ensemble and advanced analysis and simulation are required to build a more optimized intelligent system to predict the stock market behaviors more precisely and correctly. The machine learning framework is therefore required for a more optimized result. For this purpose, we will compare this result with the artificial neural network (ANN) model or a twin support vector machine (TWSVM) or more complicated support vector machine (SVM) by using different hybrid kernels to precise the behavior of high-frequency datasets in a better manner.

The rest of this chapter is organized as follows. Section 1.2 presents a detailed review of recent literature on prediction. Section 1.3 presents the shortcoming of various prediction methods. Section 1.4 presents the proposed prediction method. In Section 1.5, detailed results and the experimental setup are presented. Section 1.6 presents the conclusion and future direction.

1.2 LITERATURE REVIEW

Available relevant pieces of literature from 2007 to 2018 prove that stock market behaves nonlinearly. Different researchers used different techniques and tools that are applied to get more reliable and improved results to optimize predictions. Recent studies show that hybrid techniques can produce better results than the single analysis model.

Khemchandani and Chandra [7] proposed a TWSVM, which is a set of two smaller SVM classifiers and is capable of solving nonlinear problems. It is four times faster and generalized type of model than the traditional SVM. Khemchandani and Chandra found that the proposed TWSVM gave better results and is capable of solving smaller size quadratic problems.

As compared with the single hyperplane, which is present in the traditional SVM, the TWSVM has two hyperplanes, where either of the planes is close to one of two datasets and far from the other datasets. In addition, Khemchandani and Chandra observed that the TWSVM is more favorable than a single SVM because a nonlinear kernel can easily be applied to precise the result.

Lu et al. [8] suggested a two-staged hybrid framework that is a mixture of support vector regression (SVR) and independent component analysis (ICA), which was applied on financial prediction. They found the individual modeling framework as SVR is good enough to precise the prediction result of the financial time series, but it is not capable of handling the inherited high noise. This problem degrades the prediction capabilities. It forced to shift in the hybrid framework so that one can handle inherited noise of the dataset and others can optimize the result of prediction and create the overfitting and underfitting problem. They proposed a hybrid model that is a combination of SVR and ICA, wherein the first-stage ICA takes care of influenced noise, and the second-stage SVR will be used for the prediction of financial time series. ICA has a perfect type of mechanism to separate the potential source signal from any mixed datasets without any advanced knowledge of separation mechanism. In the proposed model, at the first stage, an independent component (IC) is produced using ICA.

Moreover, the particular ICs that had higher noise were removed. The ICs that had less noise was used as input for the SVR model. In the first segment, the ICA was utilized for predicting variables to generate the ICs. In the next segment, noisy ICs were identified and removed, and the rest of the ICs without noise served as input for the SVR predicting model. The result obtained from the experiment suggests that the proposed model outperforms the SVR model. This model was applied to two different datasets as Nikkei 225 opening price and the TAIEX closing price, and the proposed model was compared with the traditional model. It was observed that the proposed model provides a precise and better result as compared to the traditional model. The experiment suggests that in the future, some other hybrid mechanisms can be developed for better prediction result.

Peng [9] proposed a different type of hybridization that is twin support regression because the speed of learning in terms of classical SVR is deficient. The classical SVR is constructed based on a convex quadratic function and linear inequalities. These problems forced to build a twin support vector regression (TSVR). This fragmented SVR is smaller than the classical SVR. The help of two nonparallel planes can develop it. The experimental results suggest that the used TSVR are much faster than the classical SVR, and it is capable of solving the twin quadratic programming problem efficiently without the use of equality constraints. It was also observed that the TSVR is not capable to handle the hybridization problem. It was also observed that the TSVR loses sparsely because a more in-depth study is required to handle it.

Wang and Zhu [10] studied many different types of machine learning algorithm and found that an SVM is the most promising tool in the field of financial market. The SVM is a kernel-based algorithm that uses input space X and Hilbert space F by the use of kernel function space X. Finally, they propose a two-stage kernel mechanism applied in SVR in time-series forecasting in the financial market. Herein, they took some candidate kernels and then produced a mixture kernel, which is capable of producing the better result of the prediction. The regularization parameters were selected from all the available combinations to optimize the result. The experimental setup worked in two steps. In the first step, a standard SVR was applied by using all available candidate kernels to find a linear combination of sparse regularization parameters. In the second step, the regularization parameter will automatically be solved by using a solution path algorithm. The model was applied to the S&P500 and NASDAQ market and gave a promising result. The model continuously outperforms the financial market, and the over return or profit is statistically significant, which is the best part of this study.

Wang et al. [11] developed a hybrid model, which is a combination of ARIMA model, exponential smoothing model (ESM), and the backpropagation neural network (BPNN), to predict the stock market price based on time series. They applied the best of all three models to optimize the result. The threshold factor of the hybrid model was decided by genetic algorithm (GA). It is a linear mixture of the ARIMA model and the ESM and a nonlinear mixture of the BPNN on the two original and available datasets. In the experiment, the opening of the Dow Jones Industrial Average Index and the closing of the Shenzhen Integrated Index are used. The experimental result proves that the proposed hybrid model works better than any individual model. Thus, this study proves that the hybrid combination of tools is a powerful technique in prediction, and it is dominant in the field of management science, which can be applied in other generalized fields. However, this model has its own limitation; the authors recommend some more powerful hybrid combinations as multivariate adaptive regression splines and SVR can be developed to precise and improve the time series and high-frequency forecasting.

Devi et al. [12] suggested a fundamental and generalized approach for the stock market prediction in terms of time series. In this experimental setup, the authors use the last five years, from 2007 to 2011, historical data of NSE—Nifty Midcap50 companies (top four companies that have maximum Midcap value) were taken for time-series prediction. The actual dataset was collected and trained by using the ARIMA model with different criteria. The Akaike information criterion and the Bayesian information criterion are used to observe the correctness of the model. The Box–Jenkins mechanism is used to analyze the model. To forecast the nature of the dataset, mean absolute error, and mean absolute deviation are used to analyze the fluctuation in the actual historical data. It is realized that more modified and advanced approaches will be applied to find the information hidden in the stock market.

Enke and Mehdiyev [13] developed a model that is based on hybrid prediction. It works based on combined differential evolution with a fuzzy clustering and fuzzy inference neural network to produce indexing result. First, the input can be generated with the help of stepwise regression analysis. They pick the sets of input that has the most robust prediction capability. Second, a differential-evolution-based fuzzy clustering method can be applied for the extraction of rules to produce a result. Finally, fuzzy inference in a neural network is developed to predict the final result. The model developed was used for stock market forecasting. The experimental results, simulation, and lower root-mean-square error (RMSE) suggest that linear regression models, probabilistic neural models, a regression neural network, and a multilayered feedforward neural network (FFNN) may produce better results in terms of this type of regression. This study suggests that by allowing the fuzzy models as augment and by using type-2 fuzzy sets, the computational and expressive power can be improved, and the modified model can produce better results and is able to capture the unpredictive nature of the stock market shortly.

Patel et al. [14] suggested the two-stage fusion technique, in which in the first stage, they applied SVR. At the second stage, they applied the mixture of ANN, random forest, and SVR, resulting in SVR–ANN, SVR–RF, and SVR–SVR to predict the model. The potential of this mixture model is compared with the single-stage modeling techniques, in which RF, ANN, and SVR are used single-modeled techniques. The particular outcomes suggest that two-stage hybrid or mixture models are superior to that of the single-stage prediction modeling techniques. They suggested a hybridizing method to get better and more accurate results.

Sheta et al. [15] suggested a model, which is a combination of SVMs and ANNs, to build hybrid prediction models. The hybrid prediction models were compared on the basis of the various evaluation criteria. The produced model was compared with the various evaluation criteria. Twenty-seven potentially useful variables were selected that may affect the stock movement and its analysis. The analyzed SVM model with the capability radial basis function (RBF) kernel model gives better prediction result compared to ANN and regression techniques. The results were analyzed under the evaluation criteria. This study recommends applying other hybrid soft computing tools to precise the stock market prediction result.

Chiang et al. [16] developed a model, which is a combination of the ANN and particle swarm optimization, to act as an adaptive intelligent stock trading decision support system and predict the futuristic nature of the stock market. The particular system has its own limitation because it demands technical indicators and particular patterns for the input pattern.

Tkáč and Verner [17] investigated and analyzed Generalized AutoRegressive Conditional Heteroskedasticity, linear regression, discriminant analysis, where the nature of neural networks is capable of finding better results without using a statistical assumption. This study identified linear regression, logit, discriminant analysis, and the ARIMA model as benchmark methods. This study also observed that the ANN has better potential than any other statistical and soft computing methods in terms of determination coefficient, mean square error, or classification or prediction accuracy. The advantages of conventional models are their transparency, simplicity, generalized nature, and ability to comprehensibly analyze the received output. It was realized that due to parallel and complex nature of the neural network, there is no authentic and recognized value of synaptic weight in the hidden layer, which makes it impossible to establish the relation between input and output datasets. This problem established the need for the hybrid model, which is a combination of the ANN and the traditional approach. This study observed that proposed the hybrid network is better than the conventional feedforward network supported by gradient-based tools and techniques. The particular hybridization is suitable for a particular type of task. Therefore, this study suggests the need for a metaheuristic method to optimize the result performance. The essential problems such as lack of analytical abilities and formal background can be addressed to improve the performance of the ANN. Therefore, general methodology universal guidelines for the selection of hidden layers, control variables, and overall design of the topology to improve the performance of the ANN are needed. This chapter established that the ANN is an undisputable better method than any traditional method due to user-friendliness of software packages and general availability of data analysis technique.

Shrivastav and Kumar [18] suggested the deep neural network (DNN) strategy that can trend data in real time. They defined three parameters: log return, pseudo log return, and trade indicator. They formulated and calculated all these terms in their paper. By using the DNN, they predict the next 1-min pseudo log return. The used architecture was chosen arbitrarily, which has one input layer, five hidden layers, and one output layer. The DNN was trained after every 50 epochs.

Chourmouziadis and Chatzoglou [19] suggested a twofold system, in which they applied the fuzzy system. In the first stage, they used a fuzzy system in the short-term trading that discards the overflowed confidence of classical data and uses the detailed assessment. In the second stage, they applied a novel trading technique and an "amalgam" between compromised sets of mainly picked unrequited technical indicators to produce alarming signals and then supplied these signals to required design and required fuzzy system, which produces the part of the portfolio that is to be invested. That short-term fuzzy system is tested for the ASE general index for a longer time. This particular model has its own limitations such as weights of the fuzzy rules. Therefore, it is very difficult task because the success rate of the model depends on the capability of selecting the required technical indications. The proposed strategy analyzes the nature of prediction for the short interval of time. It is a good strategy for small-size datasets with 66% accuracy. In the testing period, it provides 81% accuracy for the traders for the prediction. The DNN is a simple model; therefore, the author recommends the other model as deep recurrent neural network, deep belief network, convolution DNN, deep coding network, and other network to get a more precise and accurate result for the vast datasets.

Qiu et al. [20] suggested a model that uses the ANN, GA, and simulated annealing. It produces satisfactory results in the proposed 18 input sets that can successfully predict the stock market returns. It can be applied to minimize the dimension of the available input variables. They recommended applying the ANN and other models to predict the stock market.

Zhong and Enke [21] suggested three-dimensional reduction techniques, which are fuzzy robust principal component analysis (PCA), PCA, and kernel-based principal component analysis (KPCA). These techniques are used to simplify and rearrange the original data structure through the use of ANN and dimension reduction. Proper selection of kernel for the excellent performance of KPCA is very essential. Zhong and Enke [21] suggested the mechanism for selecting automatic kernel functions to get a better result. The simulated results suggested that combining the ANNs with the PCA gives little better classification accuracy than the rest of other two combinations.

Barak et al. [22] suggested a fusion model, which is based on multiple diverse base classifiers that handle standard input and a meta-classifier that precise the prediction. The combination of diverse methods such as Ada-Boost, Bagging, and Boosting is used to produce diversity in the sets of the classifier. The experimental result produces that Bagging performance is superior infusion with 83.6% of accuracy when mixed with Decision Tree,

Rep Tree, and logical analysis of data (LAD) Tree producing result accuracy of 88.2% with bloom filter Tree, LAD Tree, and Decision Tree and Naive Bayes in terms of risk prediction. This study helps select the prominent individual classifier and produce a mixture model for the stock market prediction. The fusion model was compared with a wrapper GA model to produce a benchmark. The fusion model was applied on a particular dataset of the Tehran Stock Exchange, and the Bagging and Decision mixture model performed better than the other two algorithms. This study recommends optimizing the classification parameter, predicting other significant responses and textual information, and applying technical features, and customizing the proposed approach will optimize the prediction capabilities of the mixture model.

Chonga et al. [23] suggested the seven-set feature application on three significant approaches such as autoencoder, restricted Boltzmann machine, and PCA to construct three-layer DNNs to predict the futuristic stock returns by using data representation and the DNN. It is applied on the Korean stock market index and found that the DNN produces slightly better results than the linear autoregressive machine learning model in the training set. However, these advantages were mostly disappeared in the testing phase. It works better in the limited resource, but in the case of high-frequency stochastic datasets, its performance may be doubtful.

Despite enormous research in the area, no researchers can produce a single established model that can give an optimized and precise model of computationally intelligent system for the stock market. Moreover, high-frequency datasets are rarely used in this area. These problems demand new or hybrid tools, and techniques should be applied to these high-frequency datasets, which justifies the significance of this study. This study is the first step toward the model of a computationally intelligent system for the stock market prediction. For this purpose, different tools and techniques should be analyzed and realized to create a benchmark. Therefore, in this study, the most famous statistical ARIMA model will be utilized to maintain the quality of the system.

1.3 SHORTCOMING OF VARIOUS PREDICTION METHODS

1.3.1 HYBRID MECHANISM

The individual model has drawbacks such as local optima, overfitting, and difficulties to pick other parameters that directly decline the accuracy of prediction. These problems can be solved by combining two or more models in an individual approach to get a more precise and accurate result. It is a combination of parametric and nonparametric approaches. First, the parametric approach is applied to model and implement the trend of price movement. Second, a nonparametric approach is applied to the result of the parametric approach to enhance and improve the final performance. The parametric approach is applied in the case of binomial tree, finite-difference method, and Monte Carlo method, whereas the nonparametric method is applied in the linear neural network, multilayer perception (MLP), and SVM. In the next part of our study, some new or advanced machine learning tools (TWSVM) will we applied to get a more precise result.

Al-Hnaity and Abbod [24] suggested a hybrid mechanism to optimize the results. It is a certain linear combination of the models with individual weight factor. To maintain this criterion, a new and modified hybrid model is proposed to address the nonlinear characteristics time-series pattern of the stock market. The proposed model is a set of BPNN, SVR, and SVM, where the GA decides the weight of these models. By the proposed model, another required model can also be combined to optimize the result with the help of GA.

1.3.2 EXISTING FORECASTING MODELS

Financial time-series forecasting is the main challenge of machine learning techniques. Some commonly known soft computing techniques are discussed in the following [25].

1.3.2.1 REGRESSION ANALYSIS MODEL

Regression analysis is a very fundamental and generalized technique that has already proved its efficiency in the field of high-frequency data analysis, where problem variables are moving according to the time or time series [26]. Stock market prediction is one example of high-frequency data analysis, where the attributes of the stock market move or fluctuate according to the time series. The ARIMA model has already shown the ideal model to settle a benchmark. The hybrid ARIMA model is a better option that finds a correlation for autocorrelated errors. For the comparative analysis with the nonparametric machine learning framework, this study proposes a GLM.

This model contains a collection of regression techniques such as Gaussian, Poisson, Binomial, Gamma, Ordinal, and Negative regression. All these regression techniques have their own significance so that they can be utilized according to the demand of the problems.

1.3.2.2 ARTIFICIAL NEURAL NETWORK

The ANN [27,28] is a fundamental mechanism in the analysis and prediction of high-frequency polynomial datasets due to its capability to adapt discontinuities and nonlinearities. It is an autodriver type of mechanism where nonlinear behavior datasets can be easily analyzed and predicted with the help of a time series. However, the experimental result proves that it cannot handle nonstationary, multidimensional, and excessive noisy datasets. In these conditions, the FFNN can provide a better result than the ANN due to feedback and past simulations. Still, the ANN got popularity because of its simplicity, and it can provide a benchmark for further development, and it is already used in the area of stock market prediction and analysis. It is a mathematical mechanism, which is inspired by the biological neural network present inside the human's brain, and can easily correlate the input stream with the output result. It can handle the uncertainty and irregularities of the datasets.

1.3.2.3 SUPPORT VECTOR MACHINE

Vladimir Vapnik introduced it in AT&T Bell Laboratories. The experimental result has already shown that it is a far better mechanism than the ANN and the FFNN [29,30]. It is a competent and powerful mechanism that can solve the problem of classification and regression in terms of both analysis and prediction. It creates some support points that correlate in vector in terms of a two-dimensional system and provide a hyperplane in terms of a three-dimensional environment. It takes training datasets from the classified datasets and represents it into a higher dimensional space in terms of classification and isolates it from the dissimilarities by using linear regression techniques. It utilizes the predetermined kernel function according to the requirement of the datasets, where the optimal hyperplane can be produced, which is called support vector with the maximum margin.

1.3.2.4 KERNEL FUNCTIONS FOR MACHINE LEARNING FRAMEWORK

Kernel is a mechanism of finding the similarity index between the two data points. It can be utilized without knowing to feature space of the problem domain because it contains all information about its input datasets as a relative position in feature space. As described by Howley and Madden [31], it is a process that reveals regularities in the dataset in the representation of new feature space that is not detectable in the real representation. The major merit of the method is that it allows the use of a process that is based on linear functions in the processed feature space that is computationally efficient as well as expressible [32]. The kernel trick is a process in which the no explicit representation can be done in the new feature space. By using polynomial computational cost, it is capable of expressing the feature spaces whose expressibility is greater than the polynomial. This type of combined approach of the kernel can be easily used for the purpose of solving problems of machine learning, such as unsupervised classification (clustering), semisupervised learning, and supervised classification or regression. The SVM classifier is the classic and most eminent example of the classic kernel approach found under the supervised learning umbrella. The SVM is the most famous model where the dataset can be taken in terms of the dot product of pairs to solve the decision function and the optimization problem. Due to this capability, the SVM handles the problem of nonlinearity. By using a kernel function $K(x, z) = \varphi(x), \varphi(z)$, where $\varphi(x)$ is used as mapping to the particular feature space and the replacement of the dot product can find it, is utilized to calculate the dot product of the two sampled datasets available in the feature space. The kernel is also utilized to observe the maximum margin and provide the maximum separating hyperplane in the feature space of the SVM that creates the optimal decision boundary. Finally, in the SVM, a kernel function is capable of calculating the optimal separating hyperplane without the explicit use of the mapping function in the feature space.

1.3.2.5 CHOOSING THE RIGHT KERNEL

The performance of the SVM depends upon the selection of the kernel, which is a very challenging task and depends upon the datasets, problem, and its nature [33]. Manual selection of the kernel is the very complex work, which can be resolved by the automatic kernel selection procedure with the prior knowledge of the datasets. The earlier experiment suggests that mixture kernel can enhance the capability of the SVM single kernel. The kernel can be mixed in an inappropriate way to enhance the performance of the system by $K(x, z) = K_1(x, z) + K_2 (x, z)$ by the basic kernel building process. It is always not possible to construct the right mixture of the kernel; therefore, an automatic kernel mixture procedure is required. First, we will try to understand the types of the kernel; then, we will understand the fundamental mixture kernel building process.

1.3.2.6 TYPES OF KERNEL FUNCTIONS

The particular list of few popular kernel functions taken from the existing and recent literature works is described as follows.

1.3.2.6.1 Linear Kernel

It is the most fundamental and simple kernel function, which can be formed by the inner product $\langle x,y \rangle$ and adding an optional constant c in it

$$k(x,y) = x^T y + c. \tag{1.1}$$

1.3.2.6.2 Polynomial Kernel

It is a nonstationary kernel function, where first all the training data will be normalized for the particular set of the problem, in which x and y are input vectors, c is a constant term (adjustable parameter), α is a slope, and d is a polynomial degree that can be selected by the users

$$k(x,y) = \left(x^T y + c\right)^d. \tag{1.2}$$

1.3.2.6.3 Gaussian Kernel

This is an essential kernel function that can be applied in different learning algorithms. It is well known for supervised learning classification as an SVM and SVR. It is also known as a radial basis kernel, where x is the center and σ is the radius that can be given by the user. Here, the two samples x and x' express feature vectors in the input domain, $\|x - y\|^2$ represents squared Euclidian distance between the two features domain, and σ is a free parameter

$$k(x,y) = \exp\left(\frac{\|x-y\|^2}{2\sigma^2}\right). \tag{1.3}$$

1.3.2.6.4 Analysis of Variance (ANOVA) Kernel

It is a very particular type of kernel, as convolution kernels are ANOVA and Gaussians. To design and develop an ANOVA kernel, we have to consider $X = S^N$ for some set S and kernels $k^{(i)}$ on $S \times S$, where $i = 1, \ldots, N$. For $P = 1, \ldots, N$, the ANOVA kernel of order P is explained as follows:

$$K_p(x,x') = \sum_{1<i_1<\ldots<i_p\leq N} \prod_{p=1}^{P} K^{ip}(x_{ip}, x'_{ip}). \tag{1.4}$$

1.3.2.6.5 Hyperbolic Tangent (Sigmoid) Kernel

It is also a particular type of kernel applied in the field of many great machine learning ANNs, FFNNs, and the rest of the fields where we can use a and r, where $a > 0$. It is taken as a scaling parameter of the input dataset, and r can be used as a shifting parameter, which controls threshold mapping. If $a < 0$, it will reverse and scale the dot product of the input dataset

$$k(x,y) = \tanh\left(2x^T y + c\right). \tag{1.5}$$

1.3.2.6.6 B-Spline (RBF) Kernel

It is a specialized type of kernel defined in the interval of [−1, 1]. It can be explained with the help of a recursive formula

$$k(x,y) = B_{2p+1}(x-y). \tag{1.6}$$

where $p \in N$ with $B_{i+1} = B_i B_0$.

1.3.2.7 HYBRID KERNEL METHOD

A hybrid or mixture kernel can enhance the capability of analysis and prediction, so it can be used widely. For this purpose, we can take the number of candidate kernels and try to implement a method to find the sparse linear mixture of these kernels to enhance its productivity of prediction and analysis of a future dataset. The SVM is a probable example of this type of kernel. By using the particular mixture of the kernel, we can produce the customized SVM.

Two or more than two kernels can be combined by using the simple kernels, which can develop a mixture kernel or new kernel as follows:

Let K_1 and K_2 be kernels over XxX, $X \subseteq R^n$, $a \in R^+$, and let $f(\cdot)$ be a real-valued function on X

$$\varnothing : X \rightarrow R^m \tag{1.7}$$

where K_3 is a kernel over $R^m \times R^m$ and B is a symmetric positive-semidefinite $n \times n$ matrix. Then, the following functions are kernels:

- $K(x, y) = K_1 (x, y) + K_2(x, y)$
- $K(x, y) = K_1 (x, y) + K_2(x, y)$
- $K(x, y) = cK_1 (x, y)$
- $K(x, y) = K_3 (\varnothing(x), \varnothing(y))$
- $K(x, y) = f(x) + f(z)$
- $K(x,y) = x' By$

These are the fundamental procedures to develop the mixture kernel, which is also a valid kernel.

Different types of kernels can be used for different subsets of x, and we can mix the information taken from different sources; each kernel will start analyzing similarity as provided by the domain. If the two inputs from two different sources will be taken, we obtain

$$\begin{aligned} K_A(x_A, x_B) + K_B(x_A, x_B) &= \varnothing_A(x_A)^T \varnothing_A(y_A) + \varnothing_A(x_A)^T \varnothing_A(y_A) \\ &= \varnothing(x)^T \varnothing(y) \\ &= \varnothing(x, y) \end{aligned} \tag{1.8}$$

where $x = [x_A, x_B]$ is the concatenation of two different representations of the different sources.

This method will provide the linear mixture of kernels. Different types of kernel can be used to predict different types of datasets. However, all kernels cannot produce appropriate results. Therefore, finding appropriate kernels and its appropriate mixture (linear, quadratic, conditional, etc.) is a big challenge.

1.3.2.8 HYBRID MODELING MECHANISM

As we discussed earlier, the individual model has own drawbacks such as local optima, overfitting, and difficulties to pick other parameters that directly decline the accuracy of prediction [34,35]. These problems can be solved by combining two or more models in an individual approach to get a more precise and accurate result. It is a combination of parametric and nonparametric approaches. First, the parametric method is applied to design the model and get it trained by its price movement. Second, a nonparametric method is used on the result of the parametric method to enhance and improve the final performance. The parametric approach is applied in the case of binomial tree, finite-difference method, and Monte Carlo method, whereas the nonparametric method is applied in the linear neural network, MLP, and SVM. Different researchers provide hybridization, which is a combination of two or more individual models.

1.3.2.9 TWIN SUPPORT VECTOR MACHINE

The TWSVM is an advanced and new kind of framework that gives eminent analysis and prediction results in the case of regression as well as classification [36]. It is based on the mechanics of the generalized eigenvalues proximal SVM and produces the twin nonparallel plane by correlating twin of quadratic programming problems. The experimental results suggest that it is better than the conventional SVM. Earlier, the TWSVM was constructed to solve the problem of binary classification. However, in the later stage of the modification, it was started to apply on multiclass classification problems. Due to promising and empirical result, the application of TWSVM has gradually increased and provides better analysis and prediction capability in the field of all types of classification and regression problems.

1.3.2.10 FINANCIAL TRADING FRAMEWORK WITH FORECASTING

Cavalcante et al. [37] gave a model to forecast the stock market by using a machine learning framework. Producing a computational intelligent system that can be applied in time-series forecasting demands the following functional steps: data preparation, algorithm definition, training, and forecasting evaluation. In the first step of data preparation, all abnormalities and noise should be detected and normalized before further processing. At the next stage, it requires algorithm definition and determination of soft tools and techniques, which can be used to model and forecast the data where the architectural method can be defined and selected. In order to create a model, it requires selecting an appropriate training process or method with the adjustment of training parameters and training procedure. In the last phase, the evaluation metrics analysis and accuracy measurement results can be obtained by simulation of the trained dataset. Two additional phases are required to be added for the betterment of the intelligent prediction system: the trading strategy and the money evaluation. Stock market forecasting techniques are used to get the deviation between the real and forecasted datasets to minimize the loss. The last added stage, which is the money evaluation, is analyzing the capability of the financial forecast system rather than generating profits. The significant phase is to maximize the profit with the help of prediction of the stock market.

1.4 EXPERIMENTAL SETUP AND RESULTS

Adhikari et al. [38] explained that the ARIMA model was introduced by Box and Jenkins in 1970. This method is the composition of many activities with a time-series mechanism. It prominently results in short-term forecasting. The ARIMA (p,q,d) model is applied on the dataset of Coca Cola, which was taken from June 22, 1999 to December 16, 1999 in a fine interval of a single minute. At the first glance, it was observed that the original dataset, which is in high volume, is not stationary. For this purpose, first, the standard dataset is converted into a log(dataset). However, it was found that log (dataset) is also not stationary. Therefore, the dataset is differentiated to make it saturated. Then, the autocorrelation function (ACF) and the partial autocorrelation function (PCF) will be calculated, and finally, the auto ARIMA model is applied for the prediction with p, q, d assumptions.

1.4.1 DATA COLLECTION AND PREPARATION

High-frequency data are collected from the stock market based on the first interval. These were collected from June 22, 1999 to December 16, 1999 in a fine interval of a minute. The volume of data is very high, that is, 49,058. It has Index, Date, Time, Open, High, Low, and Close attributes. However, we will use the only Index as Minute and Close for this analysis and prediction.

1.4.2 DATA PREPROCESSING AND EXTRACTION

Data preprocessing and extraction are essential steps for precise and accurate prediction. Data for holidays will not be available. Therefore, we need some preprocessing algorithms to minimize them. For smooth preprocessing, a feature extraction mechanism should be applied to minimize the irrelevant or blank data to produce a more accurate and precise result. It will automatically find new data by R-Studio by using an inbuilt function: anyNA(dataset).

1.4.3 SUMMARY OF THE DATASET

A summary is a general-purpose function in R-language that completely analyzes central tendencies of the datasets as min, first quartile, median, mean, third quartile, and max of both attributes Minute and Close represents in Table 1.1.

TABLE 1.1 Summary of the Dataset

Minute	Close
Min: 1	Min: 1242
First quartile :12,277	First quartile: 1332
Median: 24,507	Median: 1366
Mean: 24,528	Mean: 1363
Third quartile: 36,758	Third quartile: 1404
Max: 49,068	Max: 1452

1.4.4 ALGORITHM DEFINITION AND PARTITION

This is a critical step for the prediction of the ARIMA model that has been realized according to the available datasets, as shown in Figure 1.1.

First, we will divide the dataset into two parts, in which one part will be used for training and the other will be used in the testing phase.

```
valid in <- createDataPartition(y$Close, p=0.80, list=FALSE)
valid <- dataset[-validindex,]
dataset <- dataset[validin,]
dim(dataset)
[1] 39,256 2
dim(valid)
[1] 9812 2
```

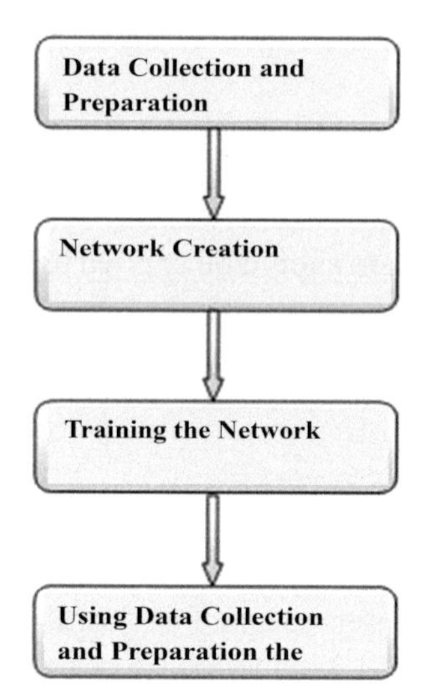

FIGURE 1.1 Data flow in the ARIMA model.

1.4.5 FORECASTING EVALUATION AND IMPLEMENTATION

The nature of a real high-frequency stochastic dataset is dispersed in nature. The real fluctuation and fluctuation from its mean value of the close are shown in Figures 1.2 and 1.3, respectively.

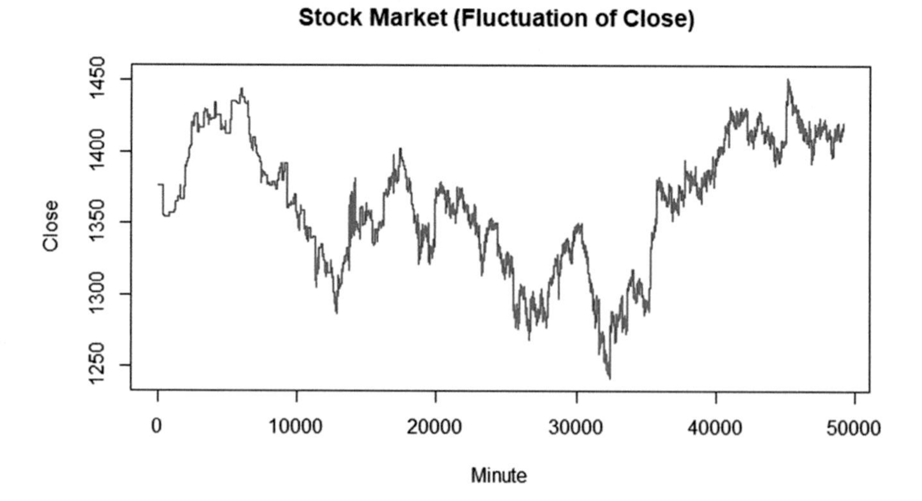

FIGURE 1.2 Stock market close fluctuation.

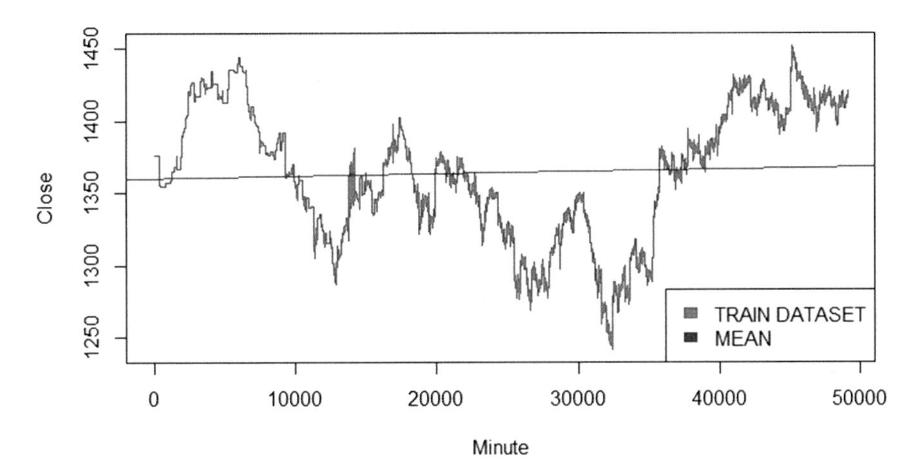

FIGURE 1.3 Stock market close deviation from the mean.

1.4.6 MODEL IMPLEMENTATION BY THE ARIMA MODEL

Implementation of the ARIMA model requires appropriate steps of the algorithm. This algorithm is the best way of exploring the model behavior as well as the dataset behavior. The algorithm is represented as follows:

- Find the log(Close) value.
- Obtain the diff(log(Close)), as shown in Figure 1.4.
- Get the ACF of diff(log(Close)).
- Find the PCF of diff(log(Close)), as shown in Figure 1.5.
- Obtain the time series of diff(log(Close)).
- Train the model with time series or diff(log(Close)) results by auto ARIMA.
- Forecast the result up to the desired period, as shown in Figure 1.6.
- Test the model with some hypotheses.

First, we have taken Close attribute for training and established its graph, as shown in Figure 1.2. To find its deviation, we plotted another graph, as in Figure 1.3. To reduce its fluctuation, we took log(Close) and plotted. Still, we need to reduce the fluctuation; therefore, we took diff(log(Close)). Now, we calculated the ACF and the PACF and tested the dataset for saturation, which is required the further procedure. Now, we applied time series and auto ARIMA to predict the model up to the desired year based on previous training.

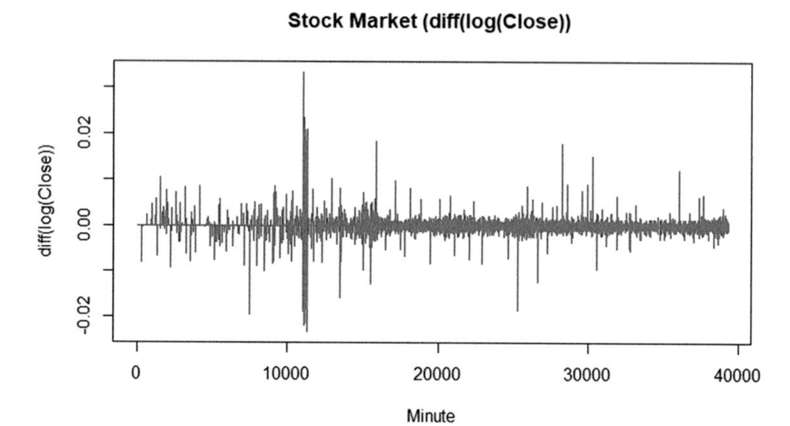

FIGURE 1.4 Stock market diff(log(close)).

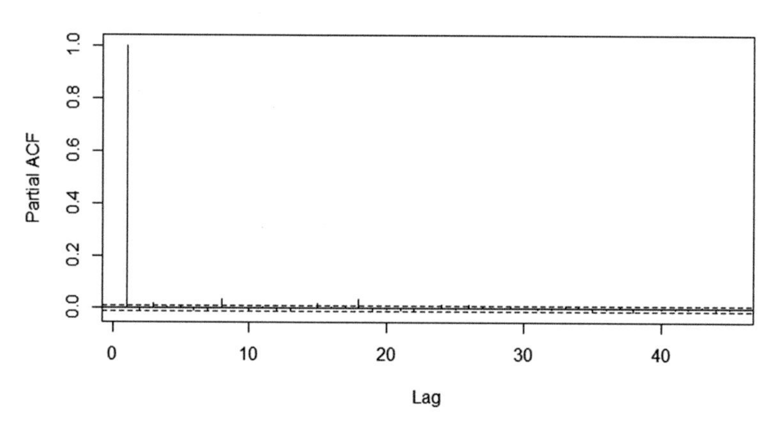

FIGURE 1.5 Stock market close PCAF.

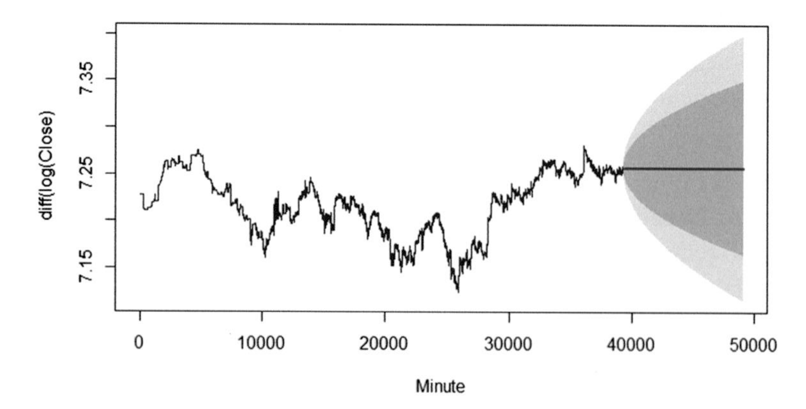

FIGURE 1.6 Forecasting in the ARIMA model.

```
adf.test(z)
adf.test(diffz)
closearima <- ts(z, start = c(1999,06), end=c(1999,06), frequency = 1)
pclose<-auto.arima(z)
closeforval=forecast(pclose, h=9812).
```

1.4.7 TEST OF THE ARIMA MODEL

Finally, it is tested by the Dickey–Fuller test, checks the null hypotheses, and searches for unit root that is available in the autoregressive model. The experimental test performed on the datasets of stock market concludes that in all steps, the time series was stationary that was done by adf.test(). The test results of both log(Close) and diff(log(Close)) are equal in terms of Lag Order, where its p value is below 1 and its alternate hypothesis is also stationary. Therefore, both results satisfied that the ARIMA model was successfully implemented with a bonafide dataset, as shown in Table 1.2.

TABLE 1.2 Test Result of ARIMA (0,1,2)

Test Name	Dickey–Fuller	Lag Order	*p*-Value	Alternative Hypothesis
Augmented Dickey–Fuller Test for log(Close)	–1.7325	33	0.6923	Stationary
Augmented Dickey–Fuller Test for diff(log (Close))	36.551	33	0.01	Stationary

1.4.8 MODEL IMPLEMENTATION BY THE GLM

GLM general-purpose machine learning tools can be used in regression and classification [39,40]. In the case of classification, such tools perform as binary classifiers. The selection of classification and regression depends on the nature of the dataset. In the case of factor type of dataset, the classification will be applied, and in the case of numeric data, regression type can be utilized. This study proposes the GLP with Gaussian regression because the available dataset is numeric. The model will be developed by using very advanced, memory efficient, and speedy package H_2O in R-language. This model will follow all preprocessing steps as provided in the ARIMA model. The experimental result is shown in Figure 1.7.

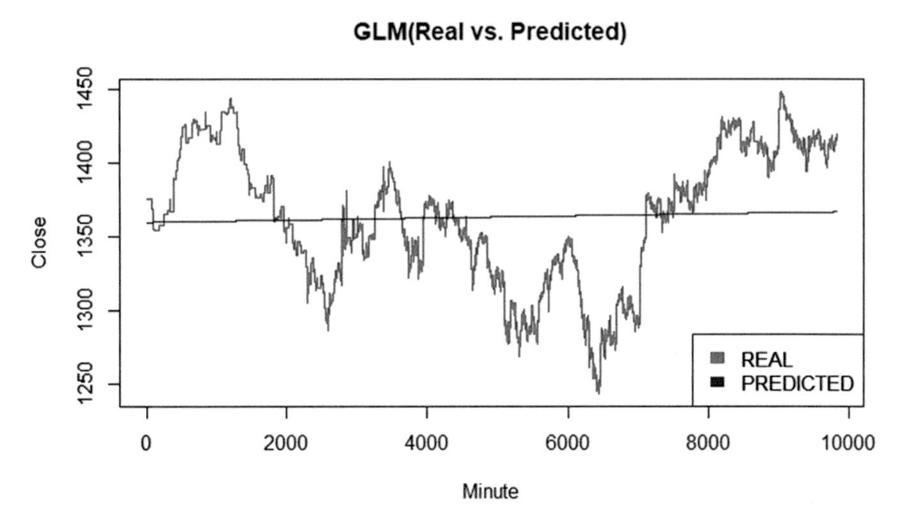

FIGURE 1.7 Forecasting in the GLM.

1.4.9 COMPARATIVE ANALYSIS

The experimental results describe that both the ARIMA model and the GLM are unable to capture the nonlinear and Brownian nature of the big high-frequency dataset. Figure 1.8 and Table 1.3 show the comparative results, respectively.

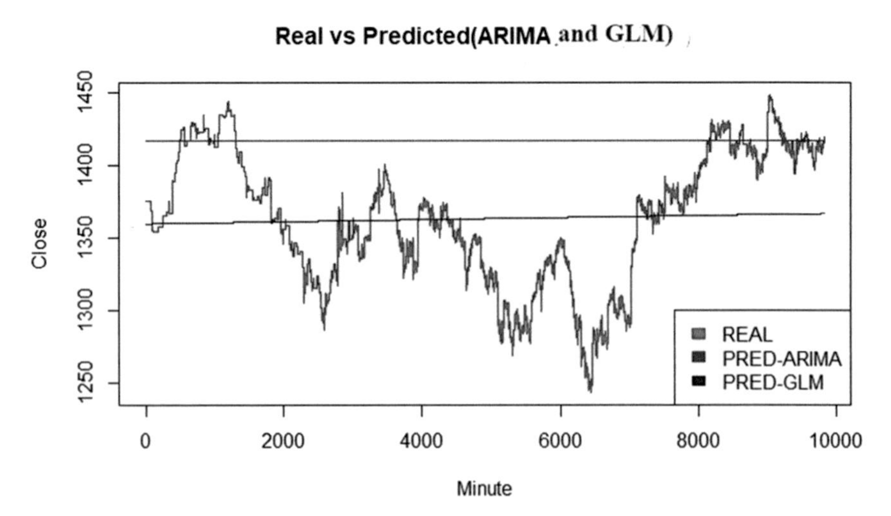

FIGURE 1.8 Comparative Forecasting by the ARIMA model and the GLM, where the comparative numerical value of the twenty days results are shown in Table 1.3.

TABLE 1.3 Analysis of 20-Day Data

Index	Minute	Real Close	ARIMA Close	GLM Close
1	5	1376	1417.51	1359.875
2	8	1376	1417.51	1359.875
3	14	1376	1417.51	1359.876
4	16	1376	1417.51	1359.876
5	17	1376	1417.51	1359.876
6	20	1376	1417.51	1359.877
7	21	1376	1417.51	1359.877
8	22	1376	1417.51	1359.877
9	24	1376	1417.51	1359.877
10	26	1376	1417.51	1359.877
11	30	1376	1417.51	1359.878
12	31	1376	1417.51	1359.878
13	32	1376	1417.51	1359.878
14	35	1376	1417.51	1359.879
15	37	1376	1417.51	1359.879
16	39	1376	1417.51	1359.879
17	44	1376	1417.51	1359.88
18	47	1376	1417.51	1359.88
19	53	1376	1417.51	1359.881
20	60	1376	1417.51	1359.882

1.4.10 COMPARATIVE PERFORMANCE ESTIMATION

The root mean square defines the behavior of the model. It also defines that the lower RMSE is better for the model

$$\text{RMSE} = \sqrt{\frac{1}{n}\sum_{i=1}^{n}(r-p)^2} \tag{1.9}$$

n: total number of samples

r: real sample value

p: predicted value

RMSE of the ARIMA model: 70.59261

RMSE of the GLM: 45.23671.

This performance estimation explores that both the ARIMA model and the GLM are not able to capture the nature of high-frequency stochastic big data in a good manner. However, the GLM provides a better comparative result than the ARIMA model. The use of other available regression techniques can improve GLM performance.

1.5 CONCLUSION AND FUTURE DIRECTION

This chapter applied the most straightforward ARIMA model and GLM to realize the approach and to understand the behavior of the stochastic nature of the stock market. The ARIMA model is the best available statistical model for exploring the nonlinear and Brownian behavior of the stock market. The experimental results suggest that both these models are unable to capture the nature of the high-frequency dataset of the stock market. However, the GLM is slightly better than the ARIMA model. The discussed techniques and approach can be fruitful to guide the student and investors to build the more reliable and optimized intelligent financial forecasting model. The primary use of this work is to explore and provide fundamental obstacles and futuristic dimension and guidelines in directions of the particular research. The simulated result showed the deviation from the actual result. Therefore, this study and analysis recommend a more in-depth, comparative, and ensemble analysis, and simulation is required to build a more optimized intelligent system to predict the stock market behaviors more precisely and accurately. This study also recommends shifting from statistical modeling to machine learning frameworks for more precise, automated, and timeliness result prediction. For this purpose, the futuristic studies and research will need advanced machine learning nonparametric models for the betterment of prediction results in terms of a high-frequency stochastic dataset of the stock market.

KEYWORDS

- **high-frequency data**
- **stock market prediction**
- **machine learning**
- **ARIMA model**
- **artificial neural network**
- **support vector machine**

REFERENCES

1. G. S. Atsalakis and K. P. Valavanis, "Surveying stock market forecasting techniques—Part II: Soft computing methods," *Expert Syst. Appl.*, vol. 36, no. 3 PART 2, pp. 5932–5941, 2009.
2. S. Shunrong, H. Jiang, and T. Zhang, "Stock market forecasting using machine learning algorithms," Dept. Elect. Eng., Stanford Univ., Stanford, CA, USA, 2012, pp. 1–5.
3. R. Kumar, P. Chandra, and M. Hanmandlu, "Local directional pattern (LDP) based fingerprint matching using SLFNN," in *Proc. IEEE 2nd Int. Conf. Image Inf. Process.*, 2013. pp. 493–498.
4. R. Kumar, "Fingerprint matching using rotational invariant orientation local binary pattern descriptor and machine learning techniques," *Int. J. Comput. Vis. Image Process.*, vol. 7, no. 4, pp. 51–67, 2017.
5. M. Tkáč and R. Verner, "Artificial neural networks in business: Two decades of research," *Appl. Soft Comput.*, vol. 38, pp. 788–804, 2016.
6. J.-J. Wang, J.-Z. Wang, Z.-G. Zhang, and S.-P. Guo, "Stock index forecasting based on a hybrid model," *Omega*, vol. 40, no. 6, pp. 758–766, 2012.
7. R. Khemchandani and S. Chandra, "Twin support vector machines for pattern classification," *IEEE Trans. Pattern Anal. Mach. Intell.*, vol. 29, no. 5, pp. 905–910, 2007.
8. C.-J. Lu, T.-S. Lee, and C.-C. Chiu, "Financial time series forecasting using independent component analysis and support vector regression," *Decis. Support Syst.*, vol. 47, no. 2, pp. 115–125, 2009.
9. X. Peng, "TSVR: An efficient twin support vector machine for regression," *Neural Netw.*, vol. 23, no. 3, pp. 365–372, 2010.
10. L. Wang and J. Zhu, "Financial market forecasting using a two-step kernel learning method for the support vector regression," *Ann. Oper. Res.*, vol. 174, no. 1, pp. 103–120, 2010.
11. J. J. Wang, J. Z. Wang, Z. G. Zhang, and S. P. Guo, "Stock index forecasting based on a hybrid model," *Omega*, vol. 40, no. 6, pp. 758–766, 2012.
12. B. U. Devi, D. Sundar, and P. Alli, "An effective time series analysis for stock trend prediction using ARIMA model for nifty midcap-50," *Int. J. Data Min. Knowl. Manag. Process*, vol. 3, no. 1, pp. 65–78, 2013.
13. D. Enke and N. Mehdiyev, "Stock market prediction using a combination of stepwise regression analysis, differential evolution-based fuzzy clustering, and a fuzzy inference neural network," *Intell. Autom. Soft Comput.*, vol. 19, no. 4, pp. 636–648, 2013.
14. J. Patel, S. Shah, P. Thakkar, and K. Kotecha, "Predicting stock market index using fusion of machine learning techniques," *Expert Syst. Appl.*, vol. 42, no. 4, pp. 2162–2172, 2015.
15. A. F. Sheta, S. E. M. Ahmed, and H. Faris, "A comparison between regression, artificial neural networks and support vector machines for predicting stock market index," *Soft Comput.*, vol. 7, no. 8, pp. 55 63, 2015.
16. W. C. Chiang, D. Enke, T. Wu, and R. Wang, "An adaptive stock index trading decision support system," *Expert Syst. Appl.*, vol. 59, pp. 195–207, 2016.
17. M. Tkáč and R. Verner, "Artificial neural networks in business: Two decades of research," *Appl. Soft Comput.*, vol. 38, pp. 788–804, 2016.
18. L. K. Shrivastav and R. Kumar, "A novel approach towards the analysis of stochastic high frequency data analysis using ARIMA model," *Int. J. Inf. Syst. Manage. Sci.*, vol. 2, no. 2, pp. 326–331, 2019.

19. K. Chourmouziadis and P. D. Chatzoglou, "An intelligent short term stock trading fuzzy system for assisting investors in portfolio management," *Expert Syst. Appl.*, vol. 43, pp. 298–311, 2016.
20. M. Qiu, Y. Song, and F. Akagi, "Application of artificial neural network for the prediction of stock market returns: The case of the Japanese stock market," *Chaos, Solitons Fractals*, vol. 85, pp. 1–7, 2016.
21. X. Zhong and D. Enke, "Forecasting daily stock market return using dimensionality reduction," *Expert Syst. Appl.*, vol. 67, pp. 126–139, 2017.
22. S. Barak, A. Arjmand, and S. Ortobelli, "Fusion of multiple diverse predictors in stock market," *Inf. Fusion*, vol. 36, pp. 90–102, 2017.
23. E. Chong, C. Han, and F. C. Park, "Deep learning networks for stock market analysis and prediction: Methodology, data representations, and case studies," *Expert Syst. Appl.*, vol. 83, pp. 187–205, 2017.
24. B. Al-Hnaity and M. Abbod, "A novel hybrid ensemble model to predict FTSE100 index by combining neural network and EEMD," in *Proc. Eur. Control Conf.*, 2015, pp. 3021–3028.
25. R. Aguilar-Rivera, M. Valenzuela-Rendón, and J. J. Rodríguez-Ortiz, "Genetic algorithms and Darwinian approaches in financial applications: A survey," *Expert Syst. Appl.*, vol. 42, no. 21, pp. 7684–7697, 2015.
26. E. Alpaydin, *Introduction to Machine Learning*, 2nd ed. Cambridge, MA, USA: MIT Press, 2010.
27. E. Guresen, G. Kayakutlu, and T. U. Daim, "Using artificial neural network models in stock market index prediction," *Expert Syst. Appl.*, vol. 38, no. 8, pp. 10389–10397, 2011.
28. A. A. Adebiyi, A. O. Adewumi, and C. K. Ayo, "Comparison of ARIMA and artificial neural networks models for stock price prediction," *J. Appl. Math.*, vol. 2014, 2014, Art. no. 614342.
29. A. Kazem, E. Sharifi, F. K. Hussain, M. Saberi, and O. K. Hussain, "Support vector regression with chaos-based firefly algorithm for stock market price forecasting," *Appl. Soft Comput.*, vol. 13, no. 2, pp. 947–958, 2013.
30. B. M. Henrique, V. A. Sobreiro, and H. Kimura, "Stock price prediction using support vector regression on daily and up to the minute prices," *J. Finance Data Sci.*, vol. 4, no. 3, pp. 183–201, 2018.
31. T. Howley and M. G. Madden, "An evolutionary approach to automatic kernel construction," in *Proc. Int. Conf. Artif. Neural Netw.*, 2006, pp. 417–426.
32. T. Hofmann, B. Schölkopf, and A. J. Smola, "A review of kernel methods in machine learning," Max Planck Inst., Nijmegen, Germany, Tech. Rep. 156, 2006.
33. N. Cristianini and J. Shawe-Taylor, *An Introduction to Support Vector Machines and Other Kernel-Based Learning Methods*. Cambridge, UK: Cambridge Univ. Press, 2000.
34. J.-J. Wang, J.-Z. Wang, Z.-G. Zhang, and S.-P. Guo, "Stock index forecasting based on a hybrid model," *Omega*, vol. 40, no. 6, pp. 758–766, 2012.
35. M. Ouahilal, M. E. Mohajir, M. Chahhou, and B. E. E. Mohajir, "A novel hybrid model based on Hodrick–Prescott filter and support vector regression algorithm for optimizing stock market price prediction," *J. Big Data*, vol. 4, no. 1, 2017, Art. no. 31.
36. R. Khemchandani, P. Saigal, and S. Chandra, "Improvements on ν-twin support vector machine," *Neural Netw.*, vol. 79, pp. 97–107, 2016.

37. R. C. Cavalcante, R. C. Brasileiro, V. L. F. Souza, J. P. Nobrega, and A. L. I. Oliveira, "Computational intelligence and financial markets: A survey and future directions," *Expert Syst. Appl.*, vol. 55, pp. 194–211, 2016.
38. R. Adhikari and R. K. Agrawal, "An introductory study on time series modeling and forecasting," 2013, *arXiv:1302.6613*.
39. [Online.] Available: http://docs.h2o.ai/h2o/latest-stable/h2o-docs/data-science/glm.html
40. A. Elliot and C. H. Hsu, "Time series prediction: Predicting stock price," 2017, *arXiv: 1710.05751*.

CHAPTER 2

Cognitive Machine Learning Framework to Evaluate Competency during the Course Endorsement for Computer Science Postgraduates

SHIVANI A. TRIVEDI* and REBAKAH JOBDAS

Faculty of Computer Science, Kadi Sarva Vishwavidyalaya, Gandhinagar, Gujarat 382015, India

**Corresponding author. E-mail: satrivedi@gmail.com, drshivaniat@gmail.com*

ABSTRACT

The approach of content designing along with content delivery by the facilitator changes in accordance to the environment undergoing a dramatic shift. Career-oriented learners look forward for the endorsement in the course which persuades them in career advancement. In India looking at the current scenario having more than 789 universities and 20,000 colleges offering postgraduate courses, the challenging task is to find better course endorsement options. The research framework is based on neuro-education, the consequence of using brain imaging methods is to serve as analytical tools for measuring the educational interferences of the learner for shaping a better arrangement for increase learning ability. The neuro signals collected through the EEG device are analyzed with the machine learning algorithms resulting to a cognitive machine learning model to foresee the competency of student in picking a course labeling a output to Attention Span endeavor to choose the course. The research work will be organized in four phases. Phase 1 is discretize cognitive skill of a learner. Phase 2 is procuring neuro signals of a learner. Phase 3 is applying active learning methods to facilitate the content at the time of the course endorsement. Phase 4 is developing cognitive machine learning model to generate neuro feedback. The neuro feedback is generated, beneficial to facilitator as well as the learner. The learner can endorse the course to uplift knowledge domain. In this study, a sample size of 60 undergraduate program and postgraduate program learners are taken into consideration while choosing elective course for the higher semester.

2.1 INTRODUCTION

This chapter amalgamates cognitive machine learning (CoML) and the brain–computer interface (BCI) framework to elate competency during the course endorsement for computer science postgraduates. The main aim of this chapter is to represent the development of the CoML framework to predict the competency of a learner as observer, performer, listener, and reader. This chapter also aims to identify learner's skill in learning a course, which is benefiting from gaining confidence after successful course completion. The other reason is to impact the effective teaching–learning method's applicability on the learners. The CoML model could assist the facilitator in predicting learner's performance and the effectiveness of an adapted approach in content designing [1]. These CoML techniques could provide guidelines to the facilitator for developing better teaching strategies to improve learner's performance [2]. This study can be further carried out on a prototype for a web-based assessment tool at a global level. The research work is organized in four phases. Phase 1 determines the cognitive skill of a learner. Various cognitive skills are responsible for critical thinking, appraised to the learner counting on their learning ability. The learning ability categorizes the learner categories as performer, learner, reader, and observer. Phase 2 procures neurosignals of a learner. The electroencephalography (EEG) signals in the form of alpha, beta, gamma, and theta waves are acquired through an EEG device. The raw signals are filtered and converted from analog to digital values ranging from 0 to 100 Hz. Phase 3 applies active learning methods to facilitate the content at the time of the course endorsement. Active learning methods are used to gain the attention of the learners. Active learning methods using an educational game with the audiovisual content can be adopted. To take more involvement of learners in the teaching–learning process, Phase 4 develops the CoML model to generate neurofeedback. The CoML model is the reverse engineering process to evaluate the cognitive ability of the learner by predicting the outcomes after performing the experiments based on active learning methods [3]. The neurofeedback, when generated, is beneficial to the facilitator as well as the learner. The learner can endorse the course to the uplift knowledge domain. The facilitator can explore new pedagogies to amuse the learner's ability. In this chapter, for experiment purpose, a sample size of 40 postgraduate learners in computer science was taken into consideration. While choosing an elective course such as artificial intelligence (AI) and digital marketing (DM). AI is human intelligence in a simulation environment for information acquisition. DM is a marketing trend using digital technologies. These two subjects offered as elective for computer science postgraduates; the following section helps in course endorsement.

2.2 COURSE ENDORSEMENT OPTIONS

The approach of content designing along with content delivery by the facilitator changes by following the environment undergoing a dramatic shift. The endorsement into a particular course equips the learner for high-quality ability along with the academic rigor expected [4]. The exploring of learning theory, strategies for communication between facilitator–learner and learner–learner, engagement techniques between facilitator–learner, learner–learner, and learner–content contingent upon the cognitive skill applied, and the knowledge retention [5] can be achieved. Career-oriented learners take the endorsement in the course, which persuades them in career advancement. In India, looking at the current scenario having more than 789 universities and 20,000 colleges offering postgraduate courses, the challenging task is to find better course endorsement options. Many prominent institutions in the world such as Digital Vidya and Coursera provide demo sessions online. Based on those demo sessions, students decide to endorse either AI or DM.

2.3 COGNITIVE NEUROSCIENCE

The prospective of neuroscience training packages is to surge the inclusive brain utility affected by the neurocognitive functions, which abode in the brain circuits while performing core tasks such as observing, experimenting, listening, and reading. This paves a way to augment academic skill acquisition [6]. The consequence of using brain imaging methods is to serve as analytical tools for measuring the educational interferences of the learner. To read through the brain and to detect signals, noninvasive methods such as functional magnetic resonance imaging (fMRI) and EEG are used. These neuroimaging methods have a latent ability to enhance the cognitive subprocess at a profound level than reliance on only behavioral methods [7].

Neuroeducation is a promising area spanning the gap to progress bidirectional activity in the brain of learners and facilitators impacting the learning process [8]. Neural science and cognitive bases of academic skills for shaping a better arrangement for learning environments optimally achieve these capabilities according to the learning ability [9]. The neurosignals collected through the EEG device are analyzed, and the CoML model is developed to foresee the competency of the student in picking a course [10]. Machine learning (ML) aims toward the development of computer programs that can access and utilize data for personal upgradation. In this chapter, the algorithm builds a CoML model from a set of EEG data, which contain input data from the brain, and labels the output as attention span or concentration [11]. Therefore, unsupervised learning algorithms are used to find structure in the data, such as grouping or clustering of data points based on maximum attention, attention span, and duration of time spent by the learner in the two-step course endorsement activities. These outcomes have tossed a light for comprehending the root of disarrangements looked by learners. To endeavor and choose the course and its substance to complete the speculation of time in discovering that course is better for professional success or not can be obtained. At the facilitator side, the insight is to revise the content and assure the learner to boost in course endorsement.

BRAIN SIGNALS FREQUENCY

Frequency Range	Name of waves	Description
>40 Hz	Gamma	High mental activity, Fear
13–39 Hz	Beta	Active, Cognition Mode
7–13 Hz	Aplha	Relaxation, pre-sleep, Pre-wake state
8–12 Hz	Mu	Sensorimotor Rhythm
4–7 Hz	Theta	Deep Meditation, NREM Seep
<4 Hz	Delta	Deep Dreamless Sleep

FIGURE 2.1 Description of brain frequency signals.

The essence of the thought process, sensations, and performances is the passing of flow of charge between the neurons within the brain. Harmonized electrical rhythms form brainwaves from many types of neurons passing information to each other, as shown in Figure 2.1. The delta band frequency records below 4 Hz are observed in children below the age of one year and decline as they grow. In grown-up individuals, delta rhythms are customarily known in deep, dreamless sleep and trifling in adults when awake. If these bands recorded in colossal number signify deformity then it is a link to a neurological disorder. Sometimes, these bands are taken as an artifact, which is a noise signal generated by the muscular movement of the neck or jaw. Theta band frequency ranges from 4 to 7 Hz. An insignificant quantity of theta waves chronicled if an individual is receptive and energetic. When the individuals are in drowsy or meditative, theta waves are recorded at the upper level. Compared to delta waves, a huge amount of theta activity in receptive individuals is associated with a neurological disorder. The theta band has been concurrent with meditation or concentration and a wide series of cognitive processes such as reasoning or sensible state [12]. Alpha band frequency is between 8 and 12 Hz. Their magnitude upsurges when their own eyes shut and the body relaxes, and they incapacitate when the eyes are open and rational thinking is on. Alpha activity is associated with logical reasoning. Cumulative mental effort and reasoning cause a conquest of alpha activity, particularly from the frontal areas [13]. Subsequently, these rhythms might be applicable signals to measure cognitive skills. Beta band frequency ranges between 13 and 30 Hz [14] related to wakening mindfulness or responsiveness. Attention or concentration is measured when the brain signals are detected with beta waves. Beta brainwaves are classified into three bands depending upon the frequency: low beta (12–15 Hz) is associated with a "fast idle" or absorbed thought process, beta (15–22 Hz) is associated with logical thinking or dynamically problem-solving ability, and high beta (22–38 Hz) is associated with intricate thinking ability, incorporating new capabilities, and high apprehension. Persistent frequency of brain signal processing is not a proficient way to invoke the brain, as it takes remarkable energy. Gamma band frequency signals range from 30 to 100 Hz. The existence of gamma waves in the brain activity of a healthy individual connects to certain muscular movements. Certain research trials have publicized an association in individuals between motor actions, and gamma waves have been traced during muscle contraction [13]. The EEG signal recording system consisted of electrodes with conductive media, amplifiers with filters, analog-to-digital converter, and the recording device. BCI readings are detected when electrodes are in use. They are basically of the following types: single-use electrodes (gel-based), reusable electrodes (gold-plated, silver-plated, stainless steel, or tin), headset and EEG cap, saline-based electrodes, and needle electrodes [15]. There are some companies hosted BCI games in the market. A company named Emotiv [13,16] has technologically developed EPOC Neuroheadset. It introduced several sets of BCI-based games, such as Cortex Arcade and Spirit Mountain Demo Game. Besides, the company owns a highly efficient BCI, called EPOC Neuroheadset, with 14 electrodes at a very reasonable price along with a friendly interface for programmers. They set a fast development strategy to BCI-based applications. NeuroSky, another company, also arcades the headset named Mind Wave. Neuroheadset with the inbuilt interface for developing software applications can accumulate brainwaves and derivate into various mental states. Many other software companies such as Microsoft have revealed curiosity to research BCI, challenging the expansion of preliminary original applications that used BCI [17]. The Mind Wave Mobile Headset consists of a measuring device and a sensor that contacts with the forehead. The orientation points are positioned on the ear pad. The microcontroller chip processes the signals and delivers data to applications in digital form. This instrument is connected to the analyzer via Bluetooth. The EEG electrode is positioned on the own forehead (on the frontal cortex) during an experiment. The headset securely processes the signals and delivers outputs in the form of EEG power scales (alpha waves and beta waves), which rate to the level of attention, meditation, and measure of an eye blink. Ranges of attention and meditation are specified and conveyed to a meter with a virtual e-Sense range of 1–100. Values 20–40 are abridged ranges, which are not suitable for consideration, and values ranging from 1 to 20 reflect strongly sunk e-Sense values. The values falling in the range of 40–60 are reflected to be unbiased. The resulting values greater than 60 are considered to be elevated values. Values in the range of 80–100 are considered to be high levels of e-Sense.

2.4 COGNITIVE MACHINE LEARNING

Cognitive computing is used to explain AI systems that intend to pretend human cognition [13,18]. Many artificial intelligent technologies require a computer system. To build cognitive models that imitate social cognition processes, including ML, deep learning, neural networks (NNs), neurolinguistic programming, and sentiment analysis. This chapter represents the CoML framework for the learner. Using the CoML framework, the learners can elate the competency to predict their scope to embrace course. According to bloom's technology, the cognitive skills of a learner are diversified into creating, evaluating analyzing, applying, understanding, and remembering [25]. ML techniques have been extensively investigated and enriched. In supervised learning, the past experienced data with labels are processed scientifically. The perceptron, logistic regression, and the nearest-neighbor rule are new representative ML methods. Based on the perceptron, NNs and support vector machine (SVM) were proposed later. The ML models, Naïve Bayes (NB), and Random Forest (RF), have been utilized more extended time and have shown more significant achievement beyond the diversity of errands. Unlike supervised ML, unsupervised learning is based on training data without labels. In unsupervised learning, data are taken as input and generate a group of data or cluster with the data records. The cluster analysis method is used for elementary data analysis to create trends and patterns in the group of data.

2.5 STATE OF THE ART

The literature survey was carried out with the published work in *Cognitive Neuroscience*, *Cognitive Computing*, and *CoML* research domains. Table 2.1 presents a set of 25 relevant research papers. Colt et al. [26] have worked on a method for computing the scores of the key performance indicators for IT professionals to find the competency assessment process. They have estimated vital performance indicators by considering four performance levels. Limitation of this research is a small size of the sample as they have cited. They have implemented no computerized techniques. They have focused on problem-solving skills. The authors have identified this skill as essential for promoting graduates in the job. The reason is that during training, the skills that are acquired by the graduates and that employers look into candidates are problem-solving skills. The authors have presented a model to evaluate graduates with their problem-solving ability to meet with industry job competency. To train the proposed model, they have used ML techniques to predict suitable jobs for graduates. The authors have addressed challenges, which are faced by graduates and industry in the recruitment process. In their work, they have highlighted trends, methods, and gaps in the recruitment process for skill assessment and expectation [27]. Kart [25] worked on learning-style detection-based cognitive skills. He focused on three objectives: the first is identifying the learning style based on the cognitive abilities of a novice vigorously, the second is a mapping between cognitive skills and Bloom's taxonomy with learning object, and the third is deriving the knowledge competency level for improvement by way of keeping track and feedback mechanism through the reinforcement model. The implemented system classifies learners into two groups based on memory and concentration. Serby et al. [28] have worked to determine the efficiency in a postgraduate thoracoscopy program. In evaluating a single group using the test, the model comprises multiple-choice questions (MCQs) and psychomotor skill measures. They have taken pre- and postcourse training assessment data of the trainer to evaluate them. This work is adopted by other areas of medical educators to do procedural training based on the competency-based paradigm. Dlamini and Leung [29] worked to predict sales performance while hiring using PMaps scientific sale assessments. In their assessment tools, all participants who scored high in PMaps Sales Assessment™ have performed better at insurance sales job. They are going to extend their work to develop a predictive algorithm for the client for predicting likely sales performance of a candidate at the time of the prehiring stage. Van Lieshout et al. [30] have implemented ML techniques such as NB classifier, logical reasoning, SVM, decision tree method RF, and NNs. They have mentioned that the study of ML applied to email spam classification for a long time requires a tremendous amount of records for immense precision [31]. Their models accomplished superior performance with small and biased samples in comparison with other representative ML methods.

Currently, machine learning algorithms are used for predicting academic achievements. The ML-based app is used for self-evaluation of teacher-specific instructional methods and tools [31,32]. Most of the researchers have used ML techniques to predict student's academic performance. Navalyal and Gavas [27] have compared classification techniques such as decision tree, LASSO, K-NN, and SVM to evaluate student's performance. Authors of [34] have also used decision tree methods such as ID3, C4.5, and classification and regression tree (CART) techniques. To access the academic achievements of undergraduate students with qualitative attributes such as economic status, resource utilization, and living location. The authors of [35] have used CART decision tree with cross-validation and RF to evaluate student's performance for engineering students. Delorme et al. [24] have developed intelligent tutorial systems, which promised to deliver an adaptive learning experience to improve student learning outcomes. The prospective of neuroscience training packages is to surge the inclusive brain utility affected by the neurocognitive functions. Taking abode in the brain circuits while performing core tasks such as reading. The mathematical solution was paving the way to augment the academic skill acquisition [25]. The consequence of using brain imaging methods is to serve as analytical tools for measuring the educational interferences of the learner. To read through the brain and to detect signals, noninvasive methods such as fMRI and EEG are used [26]. These neuroimaging methods have latent ability to enhance the cognitive subprocess at a profound level than reliance on only behavioral methods. The academic competence fosters better if neural and cognitive bases are considered. There is a strong connotation of imaging data with verbal strategy reports than with the problem size, validating the capture of brain activation in the process of mental tasks [27]. These insights can be applied to educational settings to measure the skills taught in the classroom, such as responding to reasoning and problem solving. Neuroeducation is a promising area spanning the gap to progress bidirectional activity in the brain of learners and facilitators impacting the learning process. The causal cognitive functions designed by learning environment confine the flexibility of brain circuits rendering to poor performance in academics. The neural science and cognitive bases of academic skills should be organized for shaping a better arrangement for learning environments that optimally achieve these capabilities according to the learning ability [28]. BCIs have not attained much potential proven outside the laboratory because of their robustness. The authors have followed the instructional design principles at several levels to identify the flaws in BCI training protocols. Here, the instructions are provided to the user in the tasks to execute and feedback provided to them. Poor performance is due to signal processing algorithms for analysis of EEG signals [34].

2.6 PROBLEM IDENTIFICATION

Based on the literature review, it has been found that in higher education having a competitive and a vast domain of knowledge attainment process, when the learner has to endorse into a specified course there is much ambiguity about the prerequisites ought to be acquired. The learner accommodates to the class as per the individual interest. Learner incompetency to obtain the competition results in end of the journey. The requirement, the cognitive data of the student, is to acquire in counting with the performance indicator to build a CoML model. Cognizance factors of shedding light for self-assessment are not *a priori* considered in addition to the course endorsement competency factors according to the existing state of the art. The research objective is to investigate whether there is a significant effect on students in activity-based teaching. To clarify the concept and their academic performance, as well as in the evaluation of the effectiveness of teaching methods, three supervised ML techniques are compared to find the accurate analytical result. The significance of this study provides better insight into the teaching community for the use of an effective teaching method used by them with ML techniques.

2.7 SETTING BENCHMARK

The proposed model is based on three areas of the existing system. The first area is cognitive theory, where the researchers have marked different angles of cognition, such as creativity, analyzing, comprehension, evaluation, and understanding. These functions take place in the brain. Many of the researchers have worked to associate the actions taken place in the mind to the cognition tasks. For reading a brain signal, a hardware device that can read the analog signals and convert them to the computer-readable format is required. This task is to be performed by the BCI. Reading the signs to prototype a model and then establish an ML algorithm to prove the accuracy of correctly classified instances is the domain of ML. This chapter compares the three areas, as mentioned in Table 2.1, which describes the work done by different authors implementing the technologies related to cognitive theory, BCI, and ML techniques applied to cognition. Many of the authors have involved one or more technologies. Competency assessment using key performance is applied for measuring IT professional competency but has not implemented the BCI. The CRESST model is used for learning cognitive theory for training and evaluation. The Kraiger model used AI and SVM for predictive mapping of graduate skills to industry roles (Mwakondo, 2016). Student's academic performance prediction has been done using SVM [7]; competency-based metrics are used to determine the effectiveness of a postgraduate thoracoscopy course. Assessment is based on orientation style, personality, attitude, interest stream, and subject selection; career planning made EEG oscillations for memory retention [19]. The interplay of interests between natural and artificial cognitive systems for the spectacular medium. A tool to make disabled people communicate was developed using the EMOTIV instrument to measure EEG signals [20].

2.8 COML FRAMEWORK

This chapter proposes a method for acquiring the competency level of learners considering the performance indicators resulting from the competency algorithm, as shown in Figure 2.2. The performance markers are

TABLE 2.1 Comparative Study of the Existing Work

Author	Technology Used			Remarks
	Cognitive Theory	**Brain–Computer Interface**	**CoML**	
(Alexandra & Brad, 2017)	Competency Assessment using Key Performance	NA	NA	Applied for measuring IT professional Competency
(Mwakondo, 2016)	CRESST Mode for Learning, Cognitive Theory For Training, Evaluation Kraiger's model	NA	AI, Neural Networks, Statistical techniques, SVMs	Predictive Mapping of Graduate's Skills to Industry Roles
(Akangah et al., 2018)	NA	NA	CART Decision Tree, Random Forest,	Predicting Academic Achievement in Fundamentals of Thermodynamics
(Oloruntoba, 2017)	NA	NA	SVM	Student Academic Performance Prediction Using SVM
(Colt et al., 2010)	competency-based Metrics	NA	NA	Use to Determine the Effectiveness of a Postgraduate Thoracoscopy Course
(Bhargava, 2015)	Assessment based on Orientation Style, Personality, Attitude, Interest	NA	NA	Stream and subject Selection, Career Planning
Klimesch, Wolfgang,1999	NA	EEG oscillations in alpha and beta	NA	Memory performance concerning age
Foresight, 2003	The interplay of interests between natural and artificial cognitive systems	NA	NA	Cognitive systems

TABLE 2.1 *(Continued)*

Author	Technology Used			Remarks
	Cognitive Theory	Brain–Computer Interface	CoML	
Anupama, H.S. Cauvery, N.K. Lingaraju, G.M.	NA	Emotiv Instrument used for signals	NA	A tool to make disabled people communicate
Sunday, Joseph Henry, Nwani	NA	BCI uses in clinical applications	NA	Recording brain wave with an action performed
Proposed CoML framework	Cognizance Factors for self-assessment	YES	Supervised Learning: Decision Tree Algorithms, SVM, Naive Bayes Unsupervised Learning: *K*-Means Clustering, Density-Based Cluster	Competency in course Endorsement

estimated considering the cognitive levels of a learner in the IT profession while endorsing a course. Following the cognitive ability of the learner, the course content is designed to identify the competency level of a learner.

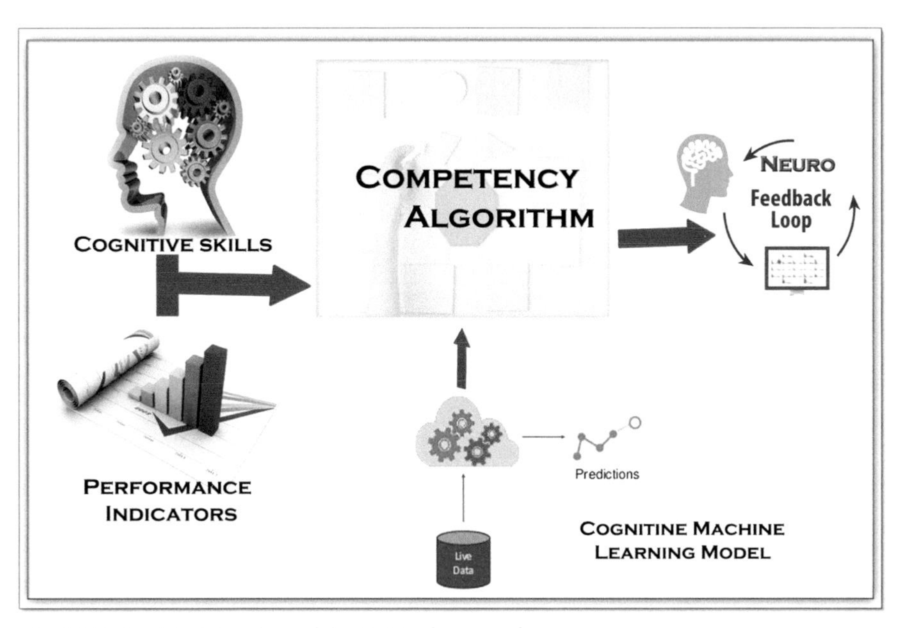

FIGURE 2.2 Modular view of the CoML framework.

The proposed framework is based on four-layered architecture, as shown in Figure 2.3. The first layer is data acquisition. This layer acquires cognitive data from the BCI device and the performance evaluation interface. The acquired data are stored in excel file format, that is, CSV. The second layer is data processing. This layer performs preprocessing data tasks. The preprocessing tasks are data transformation, normalization, cleaning, encoding, and feature engineering. Feature engineering skills play a vital role in preparing data for further execution. The third layer is ML experimentation. This layer evaluates and executes supervised and unsupervised ML algorithms. For the implementation of the ML algorithm, dynamic programming skill is the requirement. Python, Java, and R-language support dynamic programming. The output of ML experimentation is used to generate a predictive analytical report of the learner in the last layer of the architecture.

The performance indicator difference is very minimal. The learners had similar features and characteristics irrespective of their performance levels. Based on this analysis, the computed performance indicators can be integrated

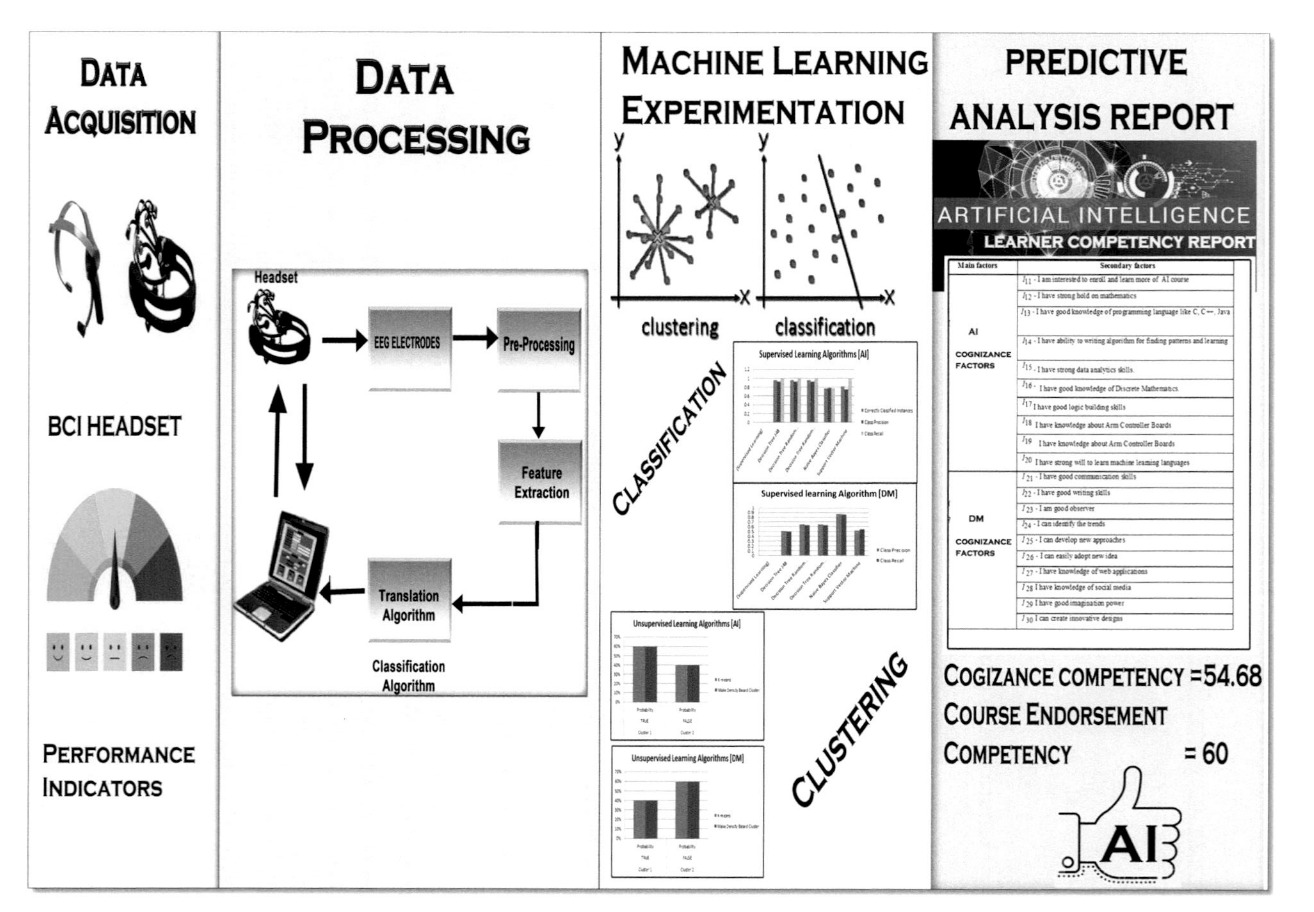

FIGURE 2.3 Architectural view of the CoML framework.

into an endorsement of a course facilitating in measuring the performance of their technical subjects. The result is analyzed by the facilitator as well as the learner in the form of neurofeedback.

This model is to design the framework in classification and computation of the performance indicators. These indicators were *a priori* defined, adapted by the competency model. The self-cognizance factors and course endorsement competency factors are considered for two courses, namely, AI and DM, as shown in Table 2.2.

TABLE 2.2 Cognizance Factors for AI and DM

Main Factors	Secondary Factors
I_1 (AI-Cognizance Factors)	I_{11}—I am interested in enrolling and learning more of AI course
	I_{12}—I have a stronghold on mathematics
	I_{13}—I have good knowledge of programming language like C, C++, Java
	I_{14}—I have ability to writing algorithm for finding patterns and learning
	I_{15}—I have strong data analytics skills.
	I_{16}— I have good knowledge of Discrete Mathematics.
	I_{17}—I have excellent logic building skills
	I_{18}—I understand Arm Controller Boards
	I_{19}—I work on Arm Controller Boards
	I_{20}—I have a strong will to learn ML languages
I_2 (DM- Cognizance Factors)	I_{21}—I have excellent communication skills
	I_{22}—I have excellent writing skills
	I_{23}—I am a reasonable observer.
	I_{24}—I can identify the trends
	I_{25}—I can develop new approaches.
	I_{26}—I can quickly adopt the new idea.
	I_{27}—I know web applications.
	I_{28}—I work on social media.
	I_{29}—I have good imagination power.
	I_{30}—I can create innovative designs

The course endorsement competency factors have four categories based on learning ability, knowledge applicability, hardware, and software compatibility, as shown in Table 2.3.

TABLE 2.3 Course Endorsement Competency Factors for AI and DM

Course Endorsement Competency—AI	
Main Factors	Secondary Factors (P)
CEC	P1. Learning Ability
	P2. Hardware Compatibility
	P3. Software Compatibility
	P4. Applicability
Course Endorsement Competency—DM	
Main Factors	Secondary Factors
CEC	P1. Learning Ability
	P2. Exemplary Study
	P3. Case Studies
	P4. Applicability

2.9 EXPERIMENTAL PROCEDURE

The experiment was conducted for computer graduate learners with a sample size of 40. The learners are subjected to register for the course endorsement session. During the class, the cognitive values are taken from the BCI machine. In the form of digital signals procured for a sign for every second, learning is subjected to answer online questionnaire on the particular subject. Brain signals are accumulated in hertz. The average value of the attention span and the score attained at the end of the session. The brain wave visualizes as shown in Figure 2.4 by the learner, while the cognition action is performed, but for the researcher, data have to be converted into the digital format. The brain wave visualizes are represented in different frequencies as alpha, beta, gamma, and theta waves in e-sense of attention and meditation meters from 0 to 100 Hz.

Attention and meditation are resolved and documented on an interface measuring scale with a relative e-sense scale ranging from 1 to 100. The values between 20 and 40 are diminished levels, and values from 1 to 20 are treated as weakly lowered signals. A neutral value is considered in the range of 40–60. Values above 60 are considered to be values higher than the standard. The frequency is to interpret to an algorithm depending upon the alpha, beta, gamma, and theta waves; we can determine the concentration depending upon the spatial memory of the cognitive skills. The data as per Table 2.4 are the attention span for one single learner having a duration of 26 s. Twenty-six frequency signals are recorded in terms of attention and

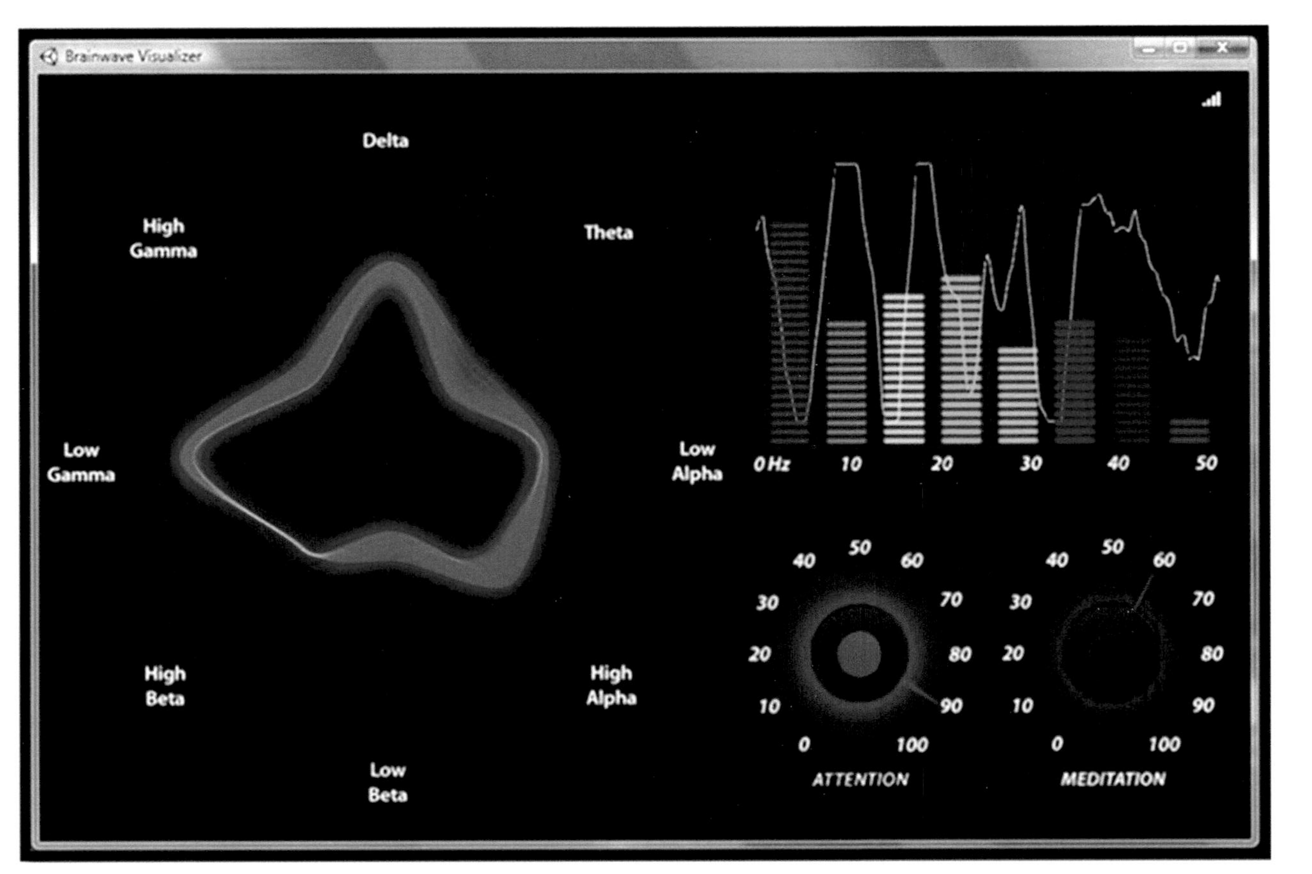

FIGURE 2.4 Brain wave visualizer.

meditation. As the scale ranges from 0 to 100, each value is multiplied with 100 to get attention span. It is organized to be ahead for operating systems such as Windows and MAC OS and statistical analysis by the WEKA tool.

TABLE 2.4 Attention Span of a Single Learner

Name: LEARNER 1				
Time stamp: 2019/04/09 3:21:56 PM				
Time	**Meditation Frequency (Hz)**	**Attention Frequency (Hz)**	**Conversion**	**Attention Span**
1:09:19	0.64	0.44	0.44×100	44
1:09:20	0.11	0.44	0.44×100	44
1:09:21	0.22	0.53	0.53×100	53
1:09:22	0.81	0.48	0.48×100	48
1:09:23	0.68	0.41	0.41×100	41
1:09:24	0.33	0.30	0.30×100	30
1:09:25	0.33	0.26	0.26×100	26
1:09:26	0.43	0.38	0.38×100	38
1:09:27	0.32	0.44	0.44×100	44
1:09:28	0.54	0.54	0.54×100	54
1:09:29	0.35	0.67	0.67×100	67
1:09:30	0.82	0.61	0.61×100	61
1:09:31	0.85	0.53	0.53×100	53
1:09:32	0.83	0.56	0.56×100	56
1:09:33	0.59	0.56	0.56×100	56
1:09:34	0.75	0.67	0.67×100	67
1:09:35	0.76	0.74	0.74×100	74
1:09:36	0.68	0.78	0.78×100	78
1:09:37	0.74	0.69	0.69×100	69
1:09:38	0.55	0.61	0.61×100	61
1:09:39	0.59	0.56	0.56×100	56
1:09:40	0.56	0.54	0.54×100	54
1:09:41	0.66	0.75	0.75×100	75
1:09:42	0.59	0.84	0.84×100	84
1:09:43	0.59	0.83	0.83×100	83

2.9.1 RESULTS AND DISCUSSION

The results for the cognizance factors for AI and DM following Likert scale ranging from strong competent to not competent are given in Tables 2.5–2.14. These cognizance factors are performance indicators for self-assessment of the learner to identify the prerequisites owned by a learner before a course endorsement, which is summarized in Table 2.5. The performance is analyzed in the form of a Likert scale ranging from 1 to 5:1 signifies strong competent and 5 signifies not competent. The proprietary algorithm for the BCI is used to get the data saved for analytical purposes.

2.9.2 COGNIZANCE FACTOR FOR AI

The cognizance factors ensuring the willingness of the learner to endorse a new course such as AI are shown in Table 2.5. These cognizance factors are performance indicators for self-assessment of the learner to identify the prerequisites owned by a learner before a course endorsement, which is summarized in Table 2.5. The performance is analyzed in the form of a Likert scale ranging from 1 to 5:1 signifies strong competent and 5 signifies not competent.

The cognizance factor concerning each student is obtained by classifying into the scale of strong competitors to no competency, as shown in Table 2.6. The learners are given a suffix of L1–L40 with a prefix of the course code AI denotes artificial intelligence.

2.9.3 COURSE ENDORSEMENT COMPETENCY FOR AI

The EEG values obtained from the BCI device termed as "Cognitive (COG)." The scores are obtained by the evaluation of 50 MCQs; each phase has 10 questions from the AI concept. The first 10 questions assess the learning ability through an audio–video credential, followed by 10 MCQs to answer. The second phases of indicators procure from the existing knowledge domain of the learner under the hardware and software compatibility, as shown in Table 2.7. The performance indicator values are shown in Table 2.8. The last 10 questions on the applicability have to be answered.

TABLE 2.5 Cognizance Factors for AI

Cognizance Factors	Strong Competent	Moderate Competent	Average Competent	Less Competent	Not Competent
I am interested in enrolling and learning more of AI course	12	5	3	3	0
I have a stronghold on mathematics	9	1	6	5	1
I have good knowledge of programming language like C, C++, Java	10	3	2	8	1
I have ability to writing algorithm for finding patterns and learning	8	3	5	6	2
I have strong data analytics skills.	6	4	1	9	4
I have good knowledge of Discrete Mathematics	7	4	7	4	1
I have a strong will to learn ML languages	12	1	5	2	3
I have excellent logic building skills	8	3	3	8	2
I know arm Controller Boards	9	2	4	5	3
I have working experience with sensors and actuators	7	2	6	6	3

TABLE 2.6 Result for the Learner Competency Final Outcome for AI

Learner Competency	AI_{L1}	AI_{L2}	AI_{L3}	AI_{L4}	AI_{L5}	AI_{L6}	AI_{L7}	AI_{L8}
Strong competent	8	4	3	1	4	2	0	0
Moderate competent	2	5	1	0	2	2	1	1
Average competent	1	0	4	1	1	2	3	2
Low competent	0	3	2	4	2	2	4	5
Not competent	0	1	0	1	2	0	0	0

TABLE 2.7 Result for the Learner Competency Based on Parameters for AI

Learner Tag	AI_{L1}		AI_{L2}		AI_{L3}		AI_{L4}		AI_{L5}	
Parameters	CAI	S	CAI	S	CAI	S	CAI	S	CAI	S
P1. Learning ability	53.86	6	67.71	6	45.04	6	62.88	6	53.4	8
P2. Hardware compatibility	62.74	3	60.62	3	44.44	7	63.12	6	59.98	6
P3. Software compatibility	62.79	3	52.41	3	48.86	3	43.23	3	56.22	2
P4. Applicability	56.47	9	254.9	9	54.14	7	61.87	6	75.35	9

2.9.4 COGNIZANCE FACTOR FOR DM

The cognizance factors for DM include good communication, reading, and writing skills. These cognizance factors show flexibility for learning. The performance is analyzed in the form of a Likert scale ranging from 1 to 5:1 signifies strong competent and 5 signifies not competent.

The analyzed data shown in Table 2.8 throw light on the competency of each learner while answering the 10 cognizance questions. Table 2.8 describes the learner's competency level based on performance indicators. The learners are given a suffix of L1–L40 with a prefix of the course code DM denotes digital marketing.

2.9.5 COURSE ENDORSEMENT COMPETENCY FOR DM

The EEG values are obtained from the BCI device. The scores are obtained by the evaluation of 50 MCQs; each phase has 10 questions from the DM concepts, as shown in Table 2.8. The first 10 questions assess the learning ability through an audio–video credential, followed by 10 MCQs. The second phase of indicators procures from the existing knowledge domain of the learner following the hardware and software compatibility. The last 10 questions on the applicability have to be answered.

TABLE 2.8 Cognizance Factors for DM

Cognizance Factors	Strong Competent	Moderate Competent	Average Competent	Less Competent	Not Competent
I have good communication skills	8	4	3	2	8
I have good writing skills	3	4	1	10	10
I am a good observer	2	3	3	7	7
I can identify the trends	2	3	5	7	7
I can develop new approaches	1	3	1	9	9
I can quickly adopt a new idea	3	1	3	8	8
I know web applications	4	3	0	3	3
I work on social media	4	5	0	9	9
I have good imagination power	1	3	7	4	4
I can create innovative designs	1	0	0	0	0

TABLE 2.9 Result for the Learner Competency Outcome for DM

Learner Tag	DM_{L1}	DM_{L2}	DM_{L3}	DM_{L4}	DM_{L5}	DM_{L6}	DM_{L7}	DM_{L8}
Strong competent	4	0	0	0	0	7	1	1
Moderate competent	6	2	0	0	1	3	0	0
Average competent	0	1	0	2	1	0	0	1
Low competent	0	4	4	2	4	0	0	5
Not competent	0	1	2	3	2	0	6	1

TABLE 2.10 Result for the Learner Competency Based on Parameters for DM

Learner Tag	DM_{L1}		DM_{L2}		DM_{L3}		DM_{L4}		DM_{L5}	
Parameters	CDM	S	CDM	S	CDM	S	CDM	S	CDM	S
P1. Learning ability	43	8	59	7	64	7	67	7	60	7
P2. Exemplary study	55	7	49	5	66	7	73	6	60	7
P3. Case studies	49	7	52	4	61	6	45	3	65	5
P4. Applicability	46	3	49	5	13	5	56	7	76	4

2.9.6 RESULT ANALYSIS THROUGH ML TECHNIQUES

The data received from the proposed model are input to the ML algorithm based on supervised learning decision tree algorithms: J48, RF, random tree, NB classifier, and SVM-implemented cognitive EEG data and performance indicators of learners based on four parameters for the course endorsement. Before implementing data to the ML algorithm based on digital data retrieved from the EEG signals, these have to be converted into numeric data based on the numerosity reduction method of data discretization. The converted numeric and nominal data of cognitive EEG data and performance indicators of the learner through WEKA for further processing based on a supervised and unsupervised learning algorithm. Decision tree techniques, J48 and random tree, identified that the cognizance factor variable is the first decision variable to decide about the competency of the learner for the course, as shown in Figure 2.5.

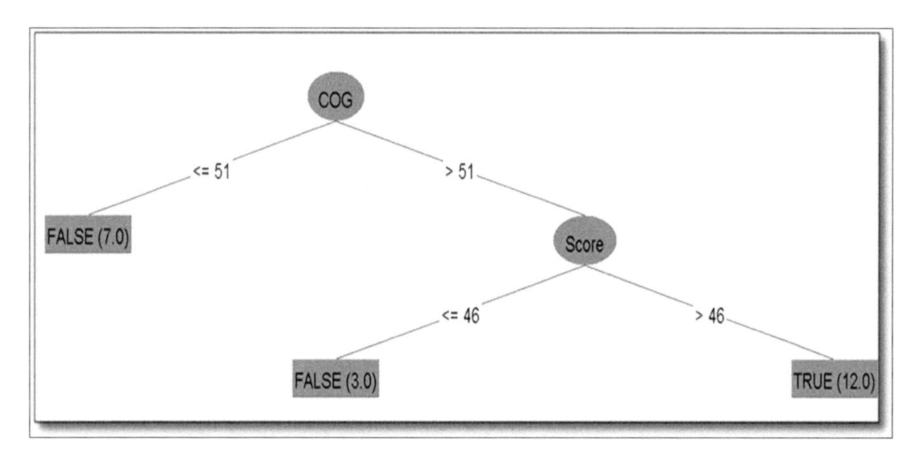

FIGURE 2.5 Decision tree J48.

Correctly classified instances, predicted value, and sensitivity calculation generated from implementing supervised learning techniques are mentioned in Table 2.11 and Figure 2.6. These statistics represent that the 100% recall value is interpreted as the completeness of the result. The consistent appraisals of class precision 92.3% for decision tree techniques are relevant values of prediction, and 77.3% of NB classifier and 75% of SVM class precision values are appropriate values of the forecast for data acquired from the experiment of AI course endorsement.

TABLE 2.11 Class Precision and Class Recall for AI

ML Techniques for AI Competency (Supervised Learning)	Class Precision (%)	Class Recall (%)
Decision Tree J48	92.3	100
Decision Tree Random Forest	92.3	100
Decision Tree Random Tree	92.3	100
Naïve Bayes Classifier	79	77
Support Vector Machine	75	100

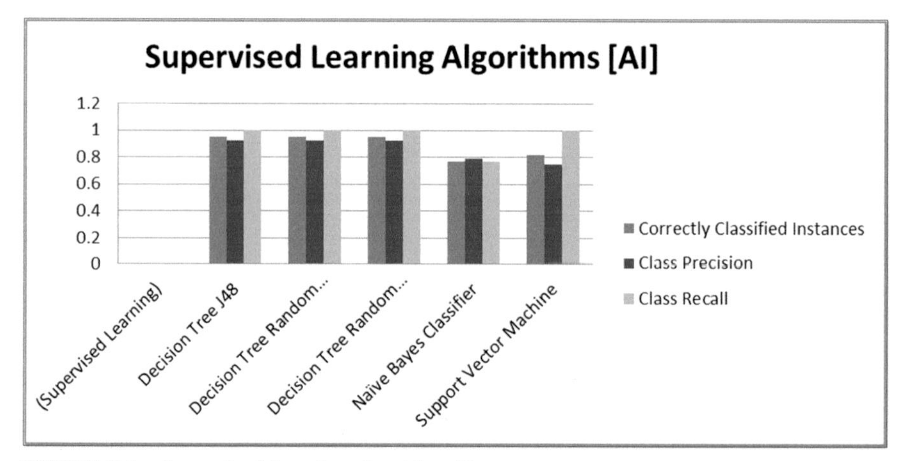

FIGURE 2.6 Supervised learning algorithm AI.

Unsupervised ML algorithms *k*-means and density-based clustering are applied, and results are mentioned in Table 2.12 and Figure 2.7. Two clusters are represented in Table 2.13 showing the probability from the training data sets and the probability of course competency predicted to be 60% and course incompetency predicted to be 40% for data acquired from the experiment of AI course endorsement.

TABLE 2.12 ML Techniques AI for Unsupervised Learning

ML Techniques AI Competency (Unsupervised Learning)	Cluster 1 TRUE Probability (%)	Cluster 2 FALSE Probability (%)
K-means	60	40
Make Density-Based Cluster	60	40

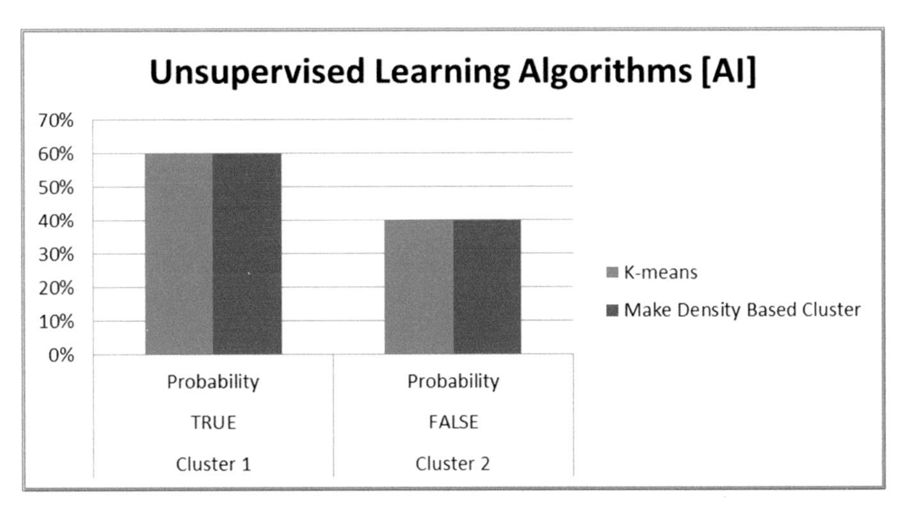

FIGURE 2.7 Unsupervised learning algorithm AI.

Correctly classified instances predicted the value and sensitivity calculation generated from implementing supervised learning techniques mentioned in Table 2.13 and Figure 2.8. These statistics represent that the 50%–86% of the recall value interprets the completeness of the result. The consistent appraisals of class precision of 65% for decision tree techniques are relevant values of prediction. Therefore, 87% of NB classifier, 53% of SVM, and class precision values are appropriate values of a forecast for data acquired from the experiment of DM course endorsement.

TABLE 2.13 Class Precision and Class Recall for DM

ML Techniques for DM Competency (Supervised Learning)	Class Precision (%)	Class Recall (%)
Decision Tree J48	52	50
Decision Tree Random Forest	65	64
Decision Tree Random Tree	65	64
Naïve Bayes Classifier	87	86
Support Vector Machine	53	55

Unsupervised ML algorithms *k*-means and density-based clustering are applied, and results are mentioned in Table 2.14 and Figure 2.9. Table 2.14 represents two clusters' probability from the training data sets and the probability of course competency predicted to be 40% and course incompetency predicted to be 60% for data acquired from the experiment of DM course endorsement.

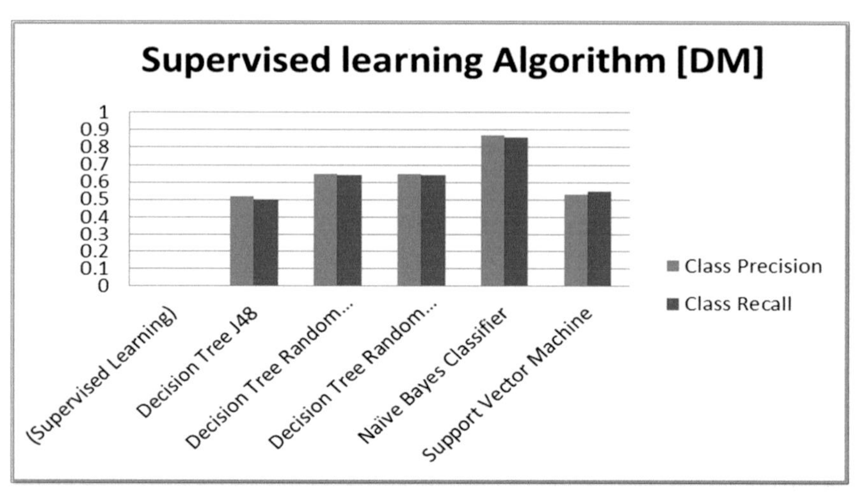

FIGURE 2.8 Supervised learning algorithm DM.

TABLE 2.14 ML Techniques DM for Unsupervised Learning

ML Techniques DM Competency (Unsupervised Learning)	Cluster 1 TRUE Probability (%)	Cluster 2 FALSE Probability (%)
K-means	40	60
Make Density-Based Cluster	40	60

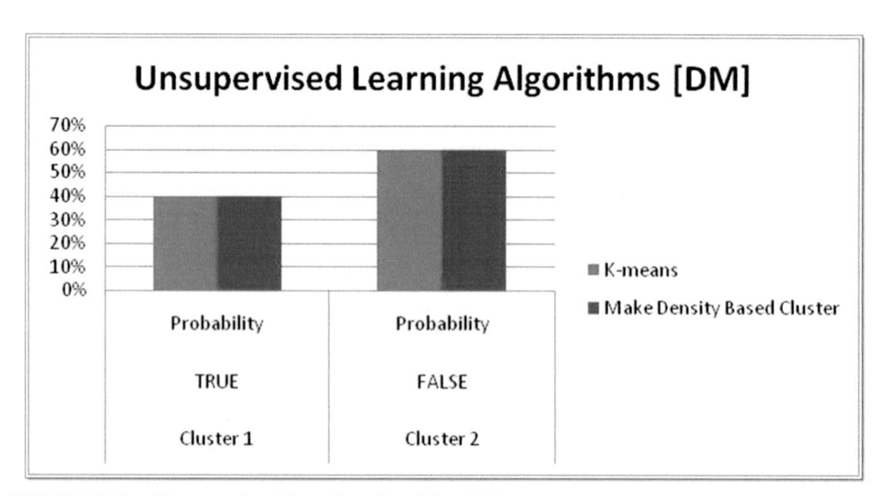

FIGURE 2.9 Unsupervised learning algorithm DM.

2.10 STRENGTH AND LAPSE OF THE FRAMEWORK

The proposed model in the CoML framework has an advantage over the other models as the cognition values are derived from brain, which speaks the tongue of the mind, unlike the traditional way of questionnaire or survey. The cognizance factors are drawn on to self-assess the learners by reading the EEG signals. The validity of the analysis of the CoML model utilizes as pre- and postevaluation of a course fulfillment. ML algorithms are implemented on the model to check the accuracy and correctly classified instances.

The model mainly requires a BCI headset, without which the cognitive signals cannot be procured, and has a battery life of 8 h. If the accurate selections of the frequency are not accumulated in a precise manner, then there is a loss of data. Data are also lost if the headset gets disconnected during the experiment. The model is implemented on bigger sample size and more courses for more clustering and classification. The automated interface is generated, which reduces the time duration of data procurement.

2.11 CONCLUSION

The proposed model has conceded out the experimental procedure appropriated with cognizance factors, BCI, and ML techniques to classify the competency values for course endorsement. The learners are classified into five competency categories. This chapter also focuses on considering the cognitive ability of the learner with the conceptual capacity of understanding the concepts, hardware, and software compatibility putting forth the applicability. The analytical results from WEKA have confirmed the model accuracy of 92.3% under the ML algorithms and class precision of 87% under the NB classifier. The CoML framework based on neuroeducation, which is the science, deals with the learner's unique cognitive strengths and their appropriate learning style fostering successful intensification of academic skills and confidence to proceed for the selected course.

2.12 FUTURE SCOPE

Future work can be extended by the automation process of data procurement with the help of user-friendly language such as Phyton. At present, the model is implemented for computer science course, which can be applicable for discrete courses such as management, commerce, etc. Uplifting the course content as per the learning ability of the learners can help gain more knowledge. The automated system implemented on static IP can acquiesce the virtual users having the headset to measure their cognitive ability for endorsing a course.

KEYWORDS

- **cognitive skills**
- **neuro science**
- **neuro education**
- **EEG signals**
- **neuro-feedback**
- **machine learning**
- **active teaching learning**
- **cognitive machine learning (CoML)**

REFERENCES

1. T. O. K. Ichihara, "Span of attention, backward masking, and reaction time," *Percept. Psychophys.*, vol. 29, no. 2, pp. 106–112, 1981.
2. R. Verbrugge, "Logic and social cognition," *J. Philos. Logic*, vol. 38, no. 6, pp. 649–680, 2009.
3. D. Ansari, "Neuroeducation—A critical overview of an emerging field," *Neuroethics*, vol. 5, no. 2, pp. 105–117, 2011.
4. N. Bousbia, I. Rebaï, J. M. Labat, and A. Balla, "Analysing the relationship between learning styles and navigation behavior in the web-based educational system," *Knowl., Manage. E-Learn.: Int. J.*, vol. 2, no. 4, pp. 400–421, 2010.
5. T. Chamillard and D. Karolick, "Using learning style data in an introductory computer science course," *ACM SIGCSE Bull.*, vol. 31, no. 1, pp. 291–295, 1999.
6. M. Nakayama, H. Yamamoto, and R. Santiago, "The impact of learner characteristics on learning performance in hybrid courses among japanese students," *Electron. J. e-Learn.*, vol. 5, no. 3, pp. 195–206, 2006.
7. G. Schalk, D. McFarland, T. Hinterberger, N. R. Birbaumer, and J. Wolpaw, BCI2000: A general-purpose brain-computer interface (BCI) system," *IEEE Trans. Biomed. Eng.*, vol. 7, no. 4, pp. 223–228, 2015.
8. K. Crowley, A. Sliney, I. Pitt, and D. Murphy, "Evaluating a brain-computer interface to categorize human emotional response," in *Proc. 10th IEEE Int. Conf. Adv. Learn. Technol.*, 2010, pp. 276–278.

9. M. Duvinage et al., "A P300-based quantitative comparison between the Emotive Epoc headset and a medical EEG device," in *Proc. 9th IASTED Int. Conf. Biomed. Eng. BioMed.*, 2012, pp. 37–42.
10. K. D. E. Mezghani and R. B. Halima, "DRAAS: Dynamically reconfigurable architecture for autonomic services," in *Web Services Foundations*. New York, NY, USA: Springer, 2014, pp. 483–505.
11. M. J. Khan and K.-S. Hong, "Passive BCI based on drowsiness detection: An fNIRS study," *Biomed. Opt. Express*, vol. 6, no. 10, pp. 4063–4078, 2015.
12. J. Lagopoulos et al., "Increased theta and alpha EEG activity during nondirective meditation," *J. Alternative Complementary Med.*, vol. 15, no. 11. pp. 1187–1192, 2009.
13. A. L. Crowell et al., "Oscillations in sensorimotor cortex in movement disorders: An electrocorticography study," *Brain: J. Neurol.*, vol. 135, no. 2, pp. 615–630, 2012.
14. M. Ungureanu, C. Bigan, R. Strungaru, and V. Laarescu, "Independent component analysis applied in biomedical signal processing politehnica," *J. Inst. Meas. Sci. SAS*, vol. 4, no. 2, pp. 1–8, 2004.
15. T. Brown, M. C. Thompson, J. Herron, A. Ko, H. Chizeck, and S. Goering, "Controlling our brains—A case study on the implications of brain-computer interface-triggered deep brain stimulation for essential tremor," *Brain-Comput. Interfaces*, vol. 3, no. 4, pp. 165–170, 2016.
16. L. Botrel, E. M. Holz, and A. Kübler, "Brain Painting V2: Evaluation of P300-based brain-computer interface for creative expression by an end-user following the user-centered design," *Brain-Comput. Interfaces*, vol. 2, no. 3, pp. 135–149, 2015.
17. B. H. Green and G. Manager, *The Internet of Things in the Cognitive Era*, Dec. 16, 2015.
18. E. Mezghani, R. B. Halima, and K. Drira, "DRAAS: Dynamically reconfigurable architecture for autonomic services," in *Web Services Foundations*. New York, NY, USA: Springer, 2014, pp. 483–505.
19. R. Mehar and A. Kaur, "Career choice preferences among rural and urban adolescents in relation to their intelligence," *Educ. Quest—Int. J. Educ. Appl. Social Sci.*, vol. 6, no. 3, pp. 197–206, 2015.
20. W. Klimesch, "EEG alpha and theta oscillations reflect cognitive and memory performance: A review and analysis," *Brain Res. Rev.*, vol. 29, nos. 2/3, pp. 169–195, 1999.
21. M. Dunleavy, C. Dede, and R. Mitchell, "Affordances and limitations of immersive, participatory augmented reality simulations for teaching and learning," *J. Sci., Educ. Technol.*, vol. 18, no. 1, pp. 17–22, 2009.
22. H. S. Anupama, N. K. Cauvery, and G. M. Lingaraju, "Brain-computer interface and its types—A study," *Int. J. Adv. Eng. Technol.*, vol. 3, no. 2, pp. 739–745, 2012.
23. B. He, B. Baxter, B. J. Edelman, C. C. Cline, and W. W. Ye, "Noninvasive brain-computer interfaces based on sensorimotor rhythms," *Proc. IEEE*, vol. 103, no. 6, pp. 907–925, 2015.
24. A. Delorme, C. Kothe, N. Bigdely-Shamlo, A. Vankov, R. Oostenveld, and S. Makeig, "MATLAB-based tools for BCI research," in *Human–Computer Interaction and Brain–Computer Interfaces*. New York, NY, USA: Springer, 2010, pp. 241–259.
25. M. Kart," Effects of learning styles on student outcomes in a general education computing course," *Consortium Comput. Scencesin Colleges*, vol. 30, no. 4, pp. 37–43, 2015.
26. H. G. Colt, M. Davoudi, S. Quadrelli, and N. Z. Rohani, "Use of competency-based metrics to determine the effectiveness of a postgraduate thoracoscpy course," *Respiration*, vol. 80, no. 6, pp. 553–559, 2010.

27. G. U. Navalyal and R. D. Gavas, "A dynamic attention assessment and enhancement tool using computer graphics," *Human-Centric Comput. Inf. Sci.*, vol. 4, no. 1, 2014, Art. no. 11.
28. H. Serby, E. Yom-Tov, and G. F. Inbar, "An improved P300-based brain-computer interface," *IEEE Trans. Neural Syst. Rehabil. Eng.*, vol. 13, no. 1, pp. 89–98, 2005.
29. M. Dlamini and W. S. Leung, "Evaluating machine learning techniques for improved adaptive pedagogy," in *Proc. IST-Africa Week Conf.*, 2018, pp. 1–10.
30. M. van Lieshout et al. Neurocognitive predictors of ADHD outcome: A 6-year follow-up study," *J. Abnormal Child Psychol.*, vol. 45, no. 2, pp. 261–272, 2017.
31. T. D. Sunny, T. Aparna, P. Neethu, J. Venkateswaran, V. Vishnupriya, and P. S. Vyas," Robotic arm with brain—Computer interfacing," *Procedia Technol.*, vol. 24, pp. 1089–1096, 2016.
32. F. Pestilli, M. Carrasco, D. J. Heeger, and J. L. Gardner," Attentional enhancement via selection and pooling of early sensory responses in human visual cortex," *Neuron*, vol. 72, no. 5, pp. 832–846, 2011.

CHAPTER 3

Cognitive Decision-Making through an Intelligent Database Agent for Predictive Analysis

SHIVANI A. TRIVEDI* and REBAKAH JOBDAS

Faculty of Computer Science, Kadi Sarva Vishwavidyalaya, Gandhinagar, Gujarat 382024, India

Corresponding author. E-mail: satrivedi@gmail.com, drshivaniat@gmail.com

ABSTRACT

Cognitive decision-making is an intellectual process aimed at the selection of a course of action among various choices. Information technology applications that support decision-making processes had an evolutionary transformation of spreadsheet software to the complex decision-support systems. The cognitive process of decision-making is being practiced through various intelligent information systems. The intelligent decision support system is being used by decision makers to make smart decisions in real-time dynamic environment based on intelligent data. To achieve intelligent data, it is needed to transform existing relational data and big data into agent-oriented data and subsequent design of data model to be in tune to the intelligent decision support system. For such system, the intelligent database agent (IDA) is designed to generate the analytical data for cognitive decision-making. The intelligent database agent is based on predictive data analytics. In any organization, predictive models utilize the sample which is found in historical and real-time transactional data to identify threats and prospects. IDA confines associations between many aspects to permit evaluation of threat or prospective associated with a particular set of conditions, steering cognitive decision-making. The development phases for the intelligent database agent is demonstrated and implemented to retrieve the information for cognitive decision-making in the proposed framework. The proposed framework is evaluated based on the complexity of verities of data, data query complexity, i-DSS query Execution Time, memory block used, volume of data, level of aggregation, reusability, flexibility, scalability and manageability are taken into the consideration for justification.

3.1 INTRODUCTION

In this chapter, the intelligent database agent (IDA) is designed to generate analytical data for cognitive decision-making. The IDA is based on predictive data analytics. The predictive analysis consists of a number of statistical techniques from data mining, predictive modeling, and machine learning, which analyze current and historical facts to make predictions about future or uncertain events. In any organization, predictive models utilize the sample that is found in historical and real-time transactional data to identify threats and prospects. The IDA confines associations among many aspects to permit evaluation of threat or prospective associated with a particular set of conditions, steering cognitive decision-making. The defining pragmatic effect of these technical methods is that predictive analytics provides a probability for each individual decision variables, for example, client, worker, healthcare patient, merchandise, automobile, part, appliance, or other executive unit in order to decide, notify, or influence executive processes that can be relevant across a large number of individuals, such as in digital marketing, marketing, credit risk assessment, fraud detection, manufacturing, healthcare sector, and administrative operations. An intelligent database architecture is proposed in this chapter. The proposed intelligent database architecture is based on three layers. The first layer is user directive and intelligent acquisition. This layer acquires decision-making directives from the user. Decision makers give input to the in-data intelligence, which specifies the future needs for decision-making. A user directive also specifies the processing methods of that data. Data aggregation, data filtration, data integration, and other processing methods can be identified in general as a resource to generate meaningful information from data. The second layer is inherit intelligence transmission to database. This layer transforms the user directive data, the method in the specified methods defined by the user directive intelligent acquisition. In this layer, the database environment can be agent oriented, agent and operation database environment, multidimensional database environment, and distributed unstructured data environment to set up decision-making. The third layer is the in-data intelligent database. This layer embeds the bits of intelligence as in-data intelligence, that is, agent in the database. This in-data intelligence "Agent" includes base data, the method to analyze that data as needful for decision-making by user's direction, and this in-data defined by two previous layers as an agent, in which data, method, and resource are defined for decision-making. Furthermore, in the proposed framework, development phases for the IDA are demonstrated and implemented to retrieve the information for cognitive decision-making. The proposed framework is evaluated based on the complexity of varieties of data; data query complexity, intelligent decision support system (iDSS) query execution time, memory block used, the volume of data, level of aggregation, reusability, flexibility, scalability, and manageability are taken into consideration for justification. Cognitive is the ability to retain knowledge, memory storing, conscious thinking, problem solving, and decision-making.

3.1.1 COGNITIVE DECISION-MAKING AND IDSS

Cognitive decision-making is an intellectual process aimed at selecting a course of action among various choices. Information technology applications that support decision-making processes, along with problem-solving activities, have increased and progressed over the past few decades. During the era ranging from the 1970s to 1990s, these applications had an evolutionary transformation of spreadsheet software, which was supporting simple applications to the complex decision support systems (DSSs) incorporating the use of statistical and optimization models. These systems further enriched with components of artificial intelligence to thread intelligence up to a certain level until the current era. The cognitive process of decision-making practiced through various intelligent information systems such as cognitive informatics, intelligent agent systems, expert systems, and iDSSs. Regardless of whether there is an action or opinion, every decision-making process produces an ultimate selection. This growth led to the development of varieties of DSSs with different tags such as management information systems, intelligent information systems, expert systems, management support systems, and knowledge-based systems. The iDSS is being used by decision makers to make smart decisions in a real-time dynamic environment. The data realized as a precious asset for intelligent decision-making. However, these raw data are rarely beneficial as their value depends on a user"s ability to extract knowledge useful for decision support. The common challenge faced by the user in the current era of cognitive decision-making is collecting diversified rapid generation of data from various sources such as social media, sensors, computer systems, and devices. Such diversified sources generate data with characteristics such as velocity, volume, value, variety, and veracity leading data to the big data. Apart from the management concern, the efficiency and deficiency of the cognitive decision-making process depend on the techniques of data analytics.

3.1.2 INTELLIGENT DATABASES SOLUTIONS FOR COGNITIVE DECISION-MAKING USING DSS

An intelligent information retrieval depends upon the collection methods of data, storage methods, and data analytics process. To deal with the above challenges, there is a need of rigorous analytical processing on the data to support timely, intelligent decision-making through the processing technologies such as data mining, analytical tools, MapReduce, Hadoop, data warehousing, online analytical processing technologies, web-based solution, ontology-based system, data warehousing and data mining techniques, OLAP, MOLAP, etc. The iDSS is an interactive software-based system intended to help decision makers to compile useful information from a combination of raw data, documents, and personal knowledge to identify and solve problems and make decisions. The architecture of the iDSS has three fundamental components: the database (or knowledge base or big data platform), the model (i.e., decision context and user criteria), and the user (i.e., user interface). The iDSS is to enhance its capability for data management and subsequent access for the growing need of improved framework, intended to focus on modeling of data using the agent paradigm to enhance the characteristics of data as "intelligent data" in the database for accelerating information retrieval, intelligence, adaptability, system integration, and scalability of the DSS. It is needed to transform existing relational data and big data into agent-oriented data and subsequent design of the data model to be in tune with the iDSS.

3.2 LITERATURE REVIEW

Literature survey is conducted in two areas to know the trends of intelligent DSS framework development. One is cognitive decision-making, iDSS, and database solutions for an iDSS.

It has acknowledged that iDSSs are the need for cognitive decision-making environment in an area such as business, health, agricultural, applied science, and biology, as well as in education [1–4]. DSSs are based on IT infrastructures to translate a wealth of information into tangible and lucrative results by way of collecting, maintaining, processing, and analyzing a large amount of complex data [6–11].

The multiagent-oriented paradigm provides cognitive decision-making in the area such as the stock market to manage portfolio and the risk management system based on preferences in the opinion of group decision-making [7,11]. A grouped DSS with an agent-oriented paradigm using fuzzy clustering and the analytical hierarchy process is for determination of important data used for asynchronous processing of data for multicriteria decision. A theory of managing data allocation in a distributed database system through mobile intelligent agents is implemented in many autonomous systems. As it is indicated that intelligent agents can learn from the system and provide some statistics about the mean value of the query occurrence on specific site, representing a percentage x of the cost can change the values to a variable to improve the response. Their research work is targeted to be implemented to improve the response of different databases on the different platform using Java Database Connectivity. An integration model for Global Resource Information Database (GRID) and multiagent system (MAS) using graphical description language called Agent Grid Integration Language. In this, the contributors have highlighted combining with the bottom-up vision of service GRID and top-down approach MAS. To enhance the concept of service into the perspective of dynamic service generation [12,13], an agent-assisted DSS is suggested to be useful for a high degree of cooperative problem-solving capability [14]. Using agent-oriented system development techniques, one can achieve intelligent acquisition, modeling, and delivery to enhance the information system characteristics such as intelligence, adaptability, system integration, and scalability [15,16]. Additionally, it enhances the two characteristics in the DSS, that is, interoperability and reuse of underlying data files in a heterogeneous database system using OMT and MaSE [17]. To generate learning the weighted value in the form of attributes, methods, and control mechanism in the database of case-based reasoning in agent-oriented DSS [18]. The database management system is a primary component to carry up to date and comprehensive data for the iDSS [19]. The JADE platform and the resource allocation problem in the multiagent environment can be solved using an integrated system to present the role of the database [16,20,21]. In the overall performance of the database, read optimized Relational Database Management System (RDBMS) that contract with other systems, which are write optimized. In this concept, query execution using column representation of data rather than row as column-oriented Database Management System (DBMS) to enhance the transaction performance is demonstrated in [22]. With the use of agent technology, the database can be made self-managing, maintainable, and at optimum performance. This technology replaces human database administrators (DBAs) with intelligent agents for DBA activities. For this, the contributors have implemented the prototype using JADE and Oracle 8.1.7. It further implemented agent-based cloud services to enhance the capability, persistence, and efficiency in the cloud computing environment [20–23]. Data warehouse, knowledgebase system, and other database systems are used as backend for a DSS for this database community that used columnar database SciDB and its query language SciQL [10]. According to [10] and [24], performance of the DSS with DW and online analytical processing (OLAP) does not work effectively as traditional online transactional processing (OLTP). There are three basic components: rows and columns, predicate logic, and fixed schemes. If anyone is missing or not appropriate, then selection of traditional RDBMS is not a good choice. As specified in [25] and [26], the RDBMS has limitation in handling complex structured data, and there is need to provide a framework to design a robust database for data collection and manipulation in the DSS. Furthermore, the database framework also encompasses the integration and interoperability with the existing system [27]. In the DSS framework, the use of data warehouse, data marts, and OLAP provides the highest level of functionality and decision support with historical data to meet with business intelligence, competitive intelligence, and knowledge management along with challenges such as unstructured data, business metadata, and metadata [10,14]. The database management system plays a vital role in the DSS [26]. Complex database queries for the DSS accelerated with affordable hardware architecture with large-scale multiprocessors such as GPUs, CUDA, CellBE, and Cell SDK [26]. One of the most important features of database technology is "Trigger," which is used to implement reactivity in database applications [28,29]. The column store database is used as an information retrieval tool with the custom index structure and the query evaluation algorithm [30]. The architecture of the traditional DSS [31] is closed and causes various problems such as lack of compatibility, adaptability, scalability, cost ineffectiveness, and accuracy of outcomes. The analytical study findings lead to adopting the agent–object relationship (AOR) model to achieve the stated characteristics in DSSs. Software modeling techniques such as entity relationship diagram, data flow diagram, and Unified Modeling Language are used for DSS framework modeling. Faizal et al. [32] and Taveter and Wagner [33] have worked for developing an enterprise model based on the AOR. They have defined business rules at the business level as integrity constraint, derivation rule, reaction rule, and deontic assignment. Furthermore, the business process is redefined as a social interaction process to do business. The need for deontic logic is also addressed to give importance to define in the form of the agent. In the formulating agent, the important term "RESOURCE" is introduced as they are subsumed, together with the material, information, and product. The challenges in this research are how to implement a broader view and more cognitive stance, by proposing to consider not only constrained, derivation, and reaction, but also deontic assignment rules. The agent-oriented paradigm adopted for high-level abstraction facilitates the conceptual and technical integration of communication and interaction with established information system technology [34,35]. It is mentioned in [34] that today's information system technology is based on the metaphors of data management and data flow. Moreover, it is under pressure to adopt concepts and techniques from the highly successful object-oriented paradigm, agent orientation. It is a need to transfer data from the relational database to the object-oriented database, from the object-oriented database to the XML database, and from the object-oriented database to the agent-oriented database (AODB). For global integrity with reduction of run-time overhead provides a mapping between tuples, object, and XML schema [15,17,25,37–39]. Many researchers are working on the transformation of the relational database to the object-oriented schema. Moreover, they are working on databases to get the advantages of object-oriented databases for managing complex data structure, better storage, and flexible and expandable data design [18,40–44]. Extension of query rewriting in OODB is the solution to overcome relational model limitation [32,42,45]. Study of the current market and historical review suggests the use of NoSQL database for various types of data structure and complex data [46]. The query caching algorithm is available in query optimization techniques such as pipelining, parallel execution, partitioning, indexes, and materialized view to increasing query performance, and hits are not containing the cache mechanism in OODB [47].

3.3 PROBLEM IDENTIFICATION

Based on the literature review findings, it can be identified that the existing DSS framework of cognitive decision-making environment has the following challenges.

- To design, develop, and implement the DSS framework, in which the following aspects are identified [31,32,48–50]:
  - To adopt user's constraint.
  - To draft resources and content priorities.
  - To determine the unstructured requirement engineering approach.
  - To manage ambiguity of data.
  - To determine important data from a large group of data.
  - To achieve asynchronous distributed and asynchronous processing.
- To develop automated human knowledge decision-making tasks with less intervention of human [5,51].
- To develop a framework for the database-driven DSS; this is useful for customized mortality prediction [10].
- To organize DSS as services, continuous analytics as service, and affordable analytics for masses (i.e., Big Data) [13,52].
- To develop a hypothetical database for derived data and fuzzy databases to handle uncertain information in intelligent decision-making [9].
- To develop hypothetical databases and "what-if" analysis-based algorithm to handle uncertain information in a decision-making environment [9].
- To achieve a database paradigm shift from relational to object-oriented data and object-oriented data to AODB to meet dynamic, flexible, scalable, expandable, and managing complex data model [36,38,39,40–42,46].

The objective of the undertaken work in this chapter is hereunder:

To add cognitive decision-making mental model and in-data intelligence in the development of IDA-oriented iDSS framework by way of enhancement of the system development methodology of the traditional DSS using the MAS development paradigm from the database perspective for predictive analysis.

3.4 COMPARATIVE STUDY OF THE EXISTING DDS FRAMEWORK AND DATABASE MODELS

The DSS architecture defines three modules, as shown in Figure 3.1: interface module, process module, and knowledge module. The first module contains contractor interface agent and client interface agents with the responsibility for receiving user specifications and delivering results. The processing module includes the coordinator agent with the responsibility for coordinating the various tasks that need to perform in cooperative problem solving. The last module contains a database agent, which carries the responsibility for keeping track of what data stored in the database. It provides predefined and ad hoc retrieval capabilities.

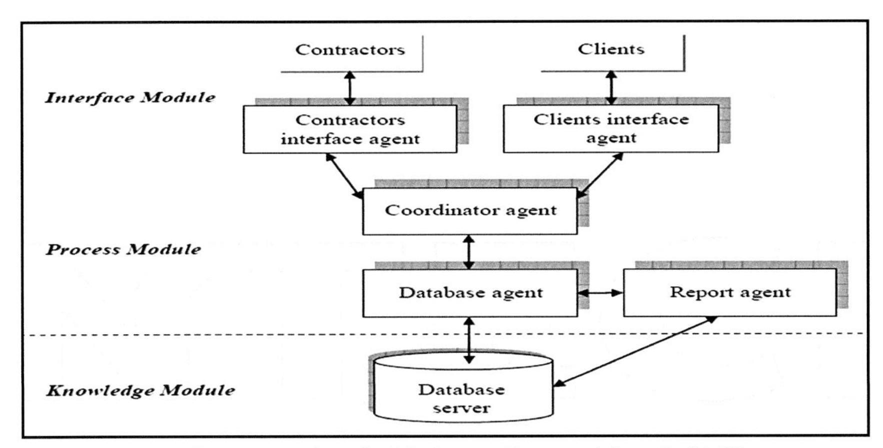

FIGURE 3.1 Intelligent agent-based DSS. Source: Recreated from Ref. [49].

The first databases were repositories of knowledge about data. That simplified the task of writing special-purpose programs for data manipulation to such an extent that in many cases, they could directly specify as queries by a naive user. Agent technology is used to develop enterprise and business modeling using agent-oriented business rules for the deontic assignment [54]. Table 3.1 shows the trends in the database solution used in research by other researchers, and it also depicts the proposed IDA data model.

3.5 PROPOSED SOLUTION

The proposed IDA for the cognitive decision-making iDSS framework is based on three-layer architecture shown in Figure 3.2. The proposed framework is expected to be capable to add cognitive decision-making mental model and in-data intelligence in the development of the iDSS framework by way of enhancement of the system development methodology of the traditional DSS to provide predictive analysis in cognitive decision-making using IDA. The first layer marked as "user directive intelligent acquisition" is acquiring the user constraints and can be supportive of retrieving autogenerated strategic information for decision-making from the agent. The second layer marked as "inherit-intelligence transmission to database" is dedicated to working for transmission of user directives with the related data

TABLE 3.1 Thematic Analysis of Database Researchers

Work Done by Author	Data Model Used	Database System Used	Future Directions and Challenges	Benefit Achieved
Object-oriented database from relational database [41]	Relational and object-oriented	Conceptual mapping	Implementing in ODBMS	Easy mapping with C#, Java
Data parallel acceleration of decision support queried [26]	Columnar Data Data-parallel model	Relational, Rapid mind	Multicore processor	Complex DSS query can be accelerated using the hardware platform cell/BE and GPU
Agent-oriented enterprise modeling [54]	Agent-oriented data model	Relational database	Agent communication	Business rule implementation
Practical application of triggers [28]	Relational	SQL-99 standard	Performance, uniformity, subtle behavior of trigger	Action event
Extending trigger by example to implement reactive agents in active databases [55]	Relational model, active rules using triggers	PostgreSQL	To generate Agent by example for the resource allocation problem solving	Resource allocation
Column store as an IR prototyping [30]	Column-oriented relational data model	MonetDB, ClueWeb1 2	Performance	Improved performance in search engine
Query optimization in OODB [42]	OOD	OODBMS	Complexity of data	Improved performance in individual subqueries
RDB to OODB [25]	Relational and OOD	Conceptual mapping	Implementation	Support to OOP
Query Optimization [42]	Object Oriented	OODBMS	Diversified Data	Data retrieval
Migration from RDB to OODB [43]	Relational and Object Oriented	OODBMS	Core issues are Consider	Direct and fast application pertinent object access
Xtriggers: XML database triggers [29]	Extended	XML, JAVA	Optimizing context path detection	Application of trigger on document instances and schema
Proposed IDA	Agent-oriented data modeling	OODBMS, nonrelational databases	Selection of nonrelational database	Easily migrate relation schema into an agent-oriented schema for cognitive decision-making in real-time scenario

in the operational database as well as AODB. These mainly take care of the creation of new data and the unique identification of decision-making in the dimension of existing data. The third layer marked as "in-data intelligence database" is responsible for reading and writing an "in-data intelligent," that is, data as an agent into the AODB.

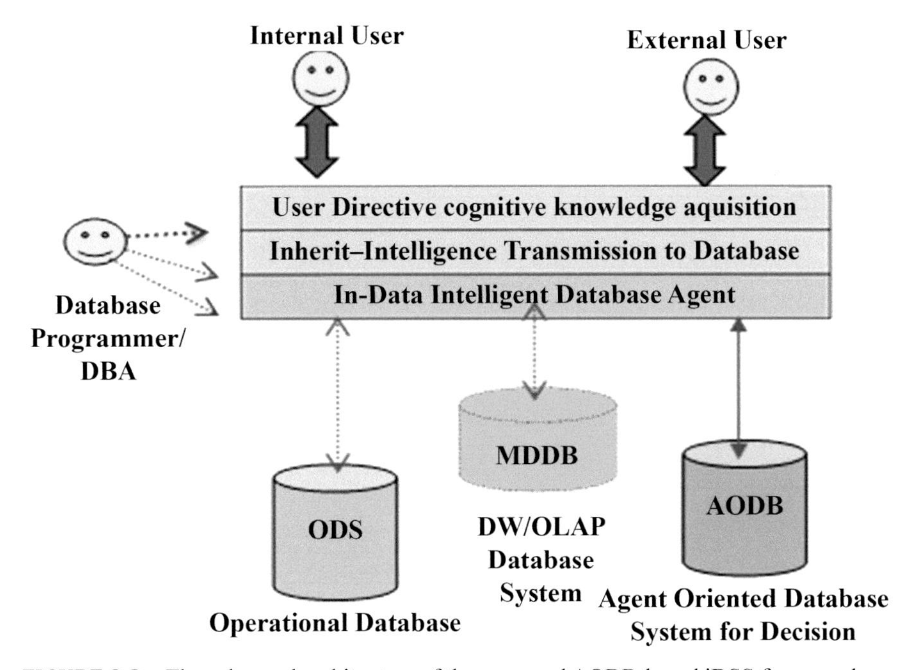

FIGURE 3.2 Three-layered architecture of the proposed AODB-based iDSS framework.

In user directive, cognitive knowledge acquisition layer acquires the cognition directives from the user. Decision makers give the input to the in-data intelligence, which specifies the future needs for decision-making. A decision maker's cognition directive also specifies the processing methods of that data. Data aggregation, data filtration, data integration, and other processing methods are identified in general as a resource to generate meaningful information from data for prediction.

Inherit-intelligence transmission to database layer transforms the user cognition directives data into the strategic rules for decision-making [56–58] to the in-data intelligence database layer to upgrade the intelligent agent of the database. The decision-making model is presented in the next section of this chapter.

In-data IDA layer embeds the cognitions as in-data intelligence, that is, agent in the database. This in-data intelligence agent includes base data, the method to analyze and predict that data as needful for decision-making by using direction, and this in-data defined by the previous two layers as an IDA. Figure 3.3 shows the three-view architecture of the smart database agent model for cognitive decision-making using iDSS for prediction.

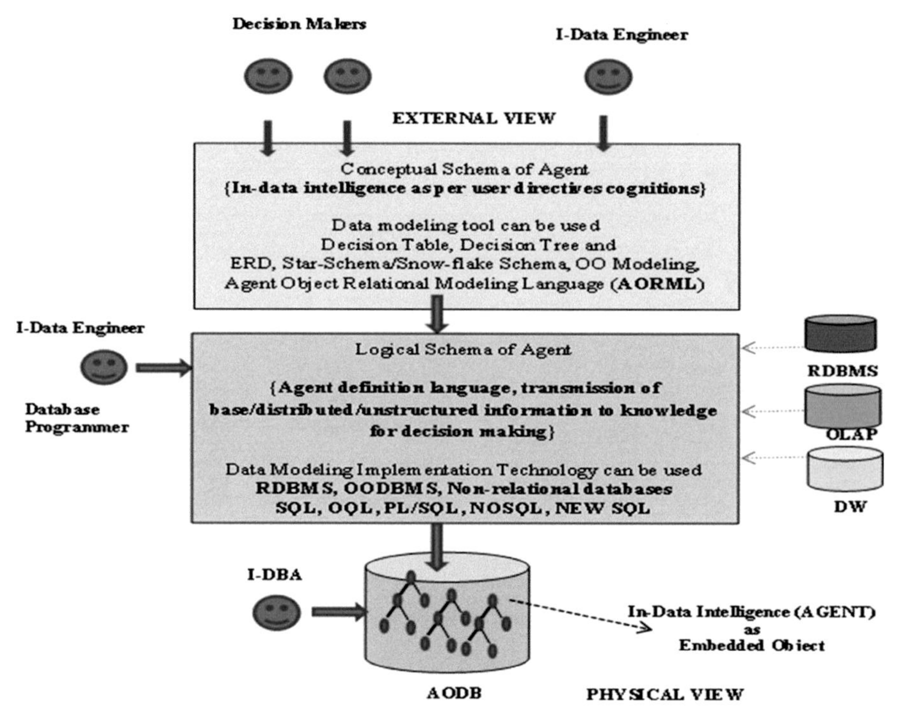

FIGURE 3.3 Three-view architecture of IDA for cognitive decision-making using i-DSS for prediction.

Decision Maker and I-Data Engineer together define the external view. Both users are vital users to generate the conceptual schema of the IDA. Decision Maker creates the need for various decision-making situation and essential piece(s) of information to make the right decision. For example, the educational institutional head needs to decide for the admission process based on currently admitted seats. Based on the decision criteria such as number of applicants, seats vacant, and historical data, other eligibility-criteria-based result declaration by other educational institution, and the current market trends for opting the courses are considered here as decision criteria. Based on these information needs of decision-making, the user I-Data Engineer identifies the current availability of data structure and unstructured information. Then, they prepare the agent-oriented AOR modeling language to structure the decision-making information in the agent-oriented data model for decision-making. These define the conceptual schema of the AODB system.

i-Data Engineer and Database Programmer together represent logical view. This group of the users then generates a logical schema from the agent models. The logical schema defines the in-data intelligence (agent) as an embedded object using object definition language. This view also describes the acquisition methods, link between two agents, and the link between the agent and relational schema, agent, and multidimensional schema as required. The agent, in this view, is defined as an embedded object, which contains data as pieces of data for decision-making. These include decision criteria, methods to link and gather the data from different resources, and actual resources from where the base data are stored.

Database Administrator defines the physical view. This user takes care of the defined agent in the object-oriented database. Figure 3.4 shows in-data intelligence, that is, agent-oriented data model. This agent contains an object and resource, as shown in Figure 3.4. This user also looks after the storage of the embedded object in the database and manages and controls the user accessibility of data from the database.

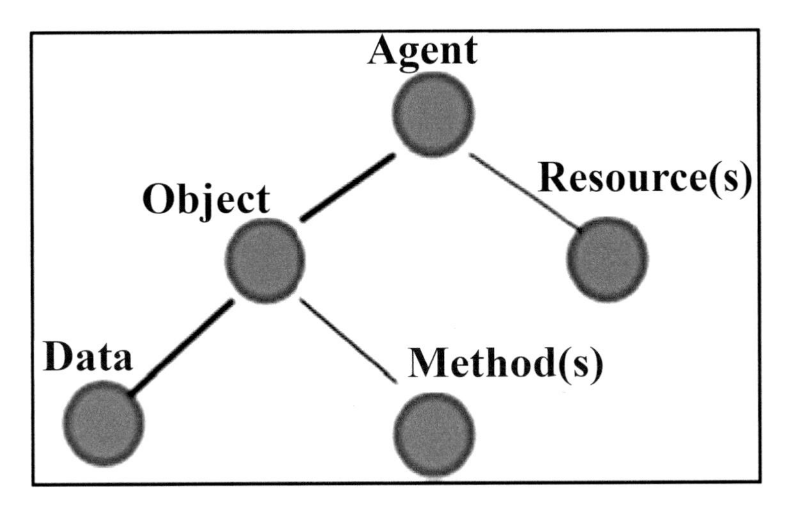

FIGURE 3.4 Agent representation.

3.5.1 DATA AS AGENT (IN-DATA INTELLIGENCE)

In the proposed framework, "Agent: in-data Intelligence" defines the knowledge as a resource for decision support in the data itself. Available data modeling techniques such as on the relational model, entity-relationship model, an object-oriented model, and not even in the multidimensional models such as star schema and snowflake schema are not fulfilling these needs. The proposed model considers the data object along with another object as a resource. This linked embedment of data object along with resource object becomes capable of satisfying the enhanced need of decision makers. The base-data-contained data object is subject to aggregation and analytics. These objects determine the agent as in-data intelligence. In addition, this represented data as an agent in Figure 3.5. An agent is defined with the object(s) and resources (s) required to the decision maker. Object(s) defines base-data object(s) that contains the base data and methods to manipulate it. The resource(s) represents the intelligent criteria and intelligent retrieval methods for decision makers.

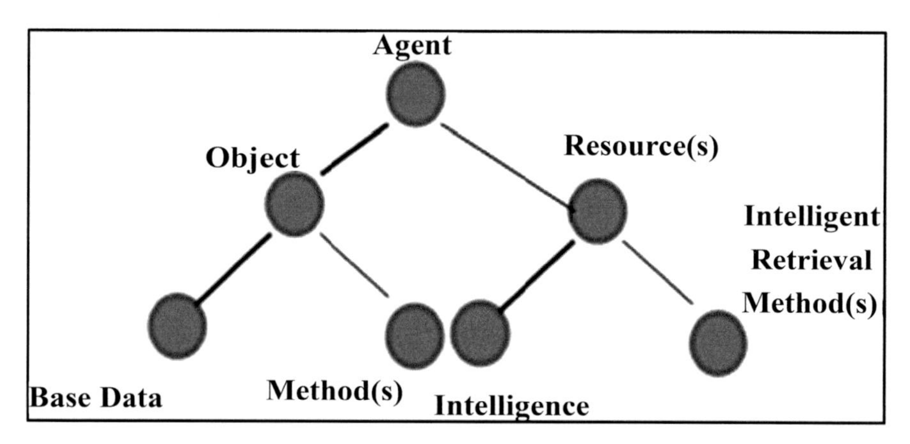

FIGURE 3.5 In-data intelligence (data as agent).

In general, a sketch of the agent is required to define database data as agent from the artificial intelligence perspective. The architectural components are agent type, performance measure, environment, actuators, sensors, percepts, and action. An agent is defined as rationale agent for each possible percept sequence. A rational agent should select an activity that is expected to maximize its performance measure, based on the evidence provided by the percept sequence and whatever built-in knowledge that the agent has for predictive decision-making. The conceptual architecture is based on PEAS task description defined as a performance measure, which is an objective criterion for the success of an agent's behavior. For example, a performance measure of a vacuum cleaner agent could be the amount of dirt cleaned up, the amount of time taken, the amount of electricity consumed, and the amount of noise generated. The second element is an "Environment," in which the agent operates with the following main properties such as accessible/inaccessible, deterministic/nondeterministic, episodic/sequential, static/dynamic, discrete/continuous, and single-agent/multiagent. The third is "Actuators," which are the set of devices that the agent can use to act. For a computer, it can be a printer or a screen. The fourth is "Sensors," which allow the agent to collect the percept sequence that uses for deliberating on the next action. Finally, the percepts refer to the agent's perceptual inputs at any given instant. An agent's percept sequence is the complete history of everything the agent has ever perceived. Now, the definition of *data* can be proposed that data are fact and figures with intelligence by suggesting data as agent = object+ resources. In this, object defines the basic data facts and figures and its methods to manipulate. Therefore, resources define another embedded object, which describes the environmental components of the agent with the environment, performance measures, actuators, sensors, and its methods to act and percepts. Table 3.2 represents in-data intelligence according to PEAS description. Figure 3.6 depicts an in-data-intelligence of object student.

TABLE 3.2 PEAS Description of Task Environment for an In-Data Intelligence

Agent Type	Performance Measure	Environment	Actuators	Sensors
Student	Attendance, results, fees Payments, etc.	Student Database Management System	Faculty, Director, Student, display, Mobile	GPS, Cameras, Keyboard, Mobile

3.5.2 PROPOSED IDA FOR IDSS DATABASE DEVELOPMENT METHODOLOGY

In this chapter, data modeling is proposed based on an agent-oriented software engineering approach to embed intelligence into the data object. Few methodologies that are in use for agent-based system developments are, namely, MaSE (1999), GAIA (2000), MESSAGE (2001), PROMETHEUS (2002), PASSI (2002), and SOAD (2002). In this work, the AOR model is used to design the proposed IDA model (AODB model). This model is based on two submodels. An external AOR model corresponds to environment analysis to communicate among agents, that is, as a database agent object interface, and an internal AOR model corresponds to a base-object and resource-object-data design model, as shown in Figure 3.6.

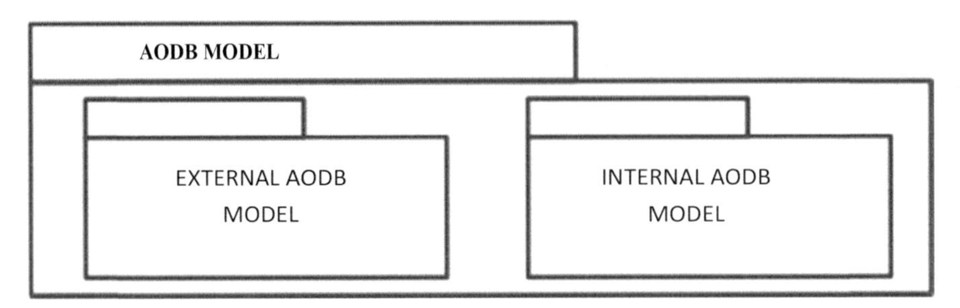

FIGURE 3.6 Agent Object Relational Database (AORDB) model consists of an external AOR model corresponding to environment analysis to communicate among agents.

An internal AOR model corresponds to a base-object and resource-object-data design model. An external AODB model is the view of an external user, that is, data engineer(s)/knowledge engineer(s)/decision maker(s), who observes the agents and their interactions in the problem environment under consideration. In this work, AODB modeling is implemented using the following database development phases.

3.5.3 PHASE-I: DECISION SUPPORT DATABASE, ENVIRONMENT ANALYSIS

In this phase, the database designer has to follow the following four phases to perform and clarify the data requirement at decision support level to define the intelligence in data and its automation through object methods.

To develop a conceptual schema of agent requirement, gathering is done based on the agent architecture, as discussed in the above section. Agent–object relational modeling is used to create a general view of the agent model, which is flexible, manageable, and reusable. It is required to identify BASE OBJECT/TABLES to store and manage necessary data based on PEAS. Then, after identification of RESOURCE object, it is essential to store and automate intelligence of the base object. Determining the requirements to define agent is based on the answer of the following questions to identify architectural components of the agent as suggested in PEAS description.

- Identify the actuators (decision makers) are involved in that environment, who has an actual need for the decision-making based on that data object?
- Which performance measures are required based on that data object to take the decision?
- What is the environment of that data object performance measures are getting affected?
- What are the data objects with its attributes and the methods?
- What is the intelligence required to define as a resource with data object?
- What is the different set of rules based on that data objects and the corresponding set of actions, which are required to perform automatically based on the perception of decision makers?
- Which are the events of the system so that decision-supported information such as performance measure and the action report can produce for actuators, that is, decision makers? Fact-finding methods, OBSERVATION, and DOCUMENT REVIEW techniques are used to analyze day-to-day routine analytical needs. Decision-making needs cognitive information collected from university authority. Moreover, identified users are Director, Dean, Registrar, Head of Institute, HoD, faculties, and students. Higher authorities at university need decision-making information for planning, as shown in Table 3.3. This decision-making requirement is highlighted according to different stakeholders of the university. This requirement is for capacity, quality, and efficiency expansion planning for the overall development of the university. While documenting the need, it is categorized in three general organizational, managerial decision-making types, namely, high level, middle level, and lower level. All these three types belong to internal stakeholder's category.

Next step in PHASE-I is followed to identify architectural components based on PEAS description. Table 3.4 shows the requirement determination of STUDENT agent for the university system along with its description and conceptual-type definition.

3.5.4 PHASE-II: IDA MODEL DESIGN

In this phase, as a database designer's activity, we have to develop the AORDB model, as suggested in the proposed AODB-based database development method [59]. Figure 3.7 represents that an agent can be defined through internal object type and external agents. Domestic object type contains base-data and resource-data objects.

TABLE 3.3 Strategic Requirement for Decision-making

High-Level Requirements for Decision-making	
Decision-Making Needs	**Cognitive Information Based on Frequency and Purpose of Decision-making**
Year-wise number of seats approved	Every year for the last five year data to analyze the growth of university course in terms of strength
Year-wise filled seats	Every year for the previous five-year data to analyze the actual number of student enrolled in the university course in course-term of existing students
Monthly attendance analysis of students	Ever month and historical moths to analyze the attendance pattern of the institution for a particular course
Daily attendance analysis of faculty members and other staff member	Every day and all historical data for each month and a whole academic year to analyze the faculty and staff attendance pattern
Middle-Level requirements for Decision-making	
Year-wise student wise subject performance	Every year the faculties are analyzing students passing ratio in the respective subject to get the teaching/learning outcome performance
Monthly attendance analysis of students	Ever month and historical moths to explain the attendance pattern of the student to attend a particular subject that faculties are taking
Daily attendance analysis of faculty members and other staff member	Every day and all historical data for each month and a whole academic year to analyze the faculty and staff attendance pattern
Lower Level requirement for Decision-making	
Year-wise/Semester wise students' performance	Every year the students are analyzing his/her performance in the semester, cumulative performance, subject-wise performance to decide the specialization and improve the overall performance
Monthly attendance analysis of students	Ever month and historical moths to analyze the attendance pattern of the student to attend the particular subject that faculties are taking
Year-wise/semester wise account status/library status etc.	Every year and historical semester wise fees paid and pending installation, library dues to make the status clear.

TABLE 3.4 PEAS Description of the Task Environment for Student Agent

Architectural Components of Agent	Description	Conceptual Type
Student	Roll no, Name, address, city, contact no	Data Object Attributes
	Add, insert, update, delete	Data Object Methods
Performance Measure	Attendance, Promoted, Grade, Placed, Library due, Accounts due, Unclear papers	Resource Object Attributes
Environment	Student Information System (database) & co-Agents	Database Platform
Actuators	Faculty, Student, Librarian, accountant, Director	Association among data objects
Percepts	Attendance Rule, Promotion Rule, Library Rule	Resource Object Class Methods
Action	Attendance Rule-based Actions Attendance 100% Send alert Send a letter to parent	Resource Object Class Methods
Sensors	Admission process, accounting process, placement process, time table, attendance entry process	Resource Object Trigger Methods

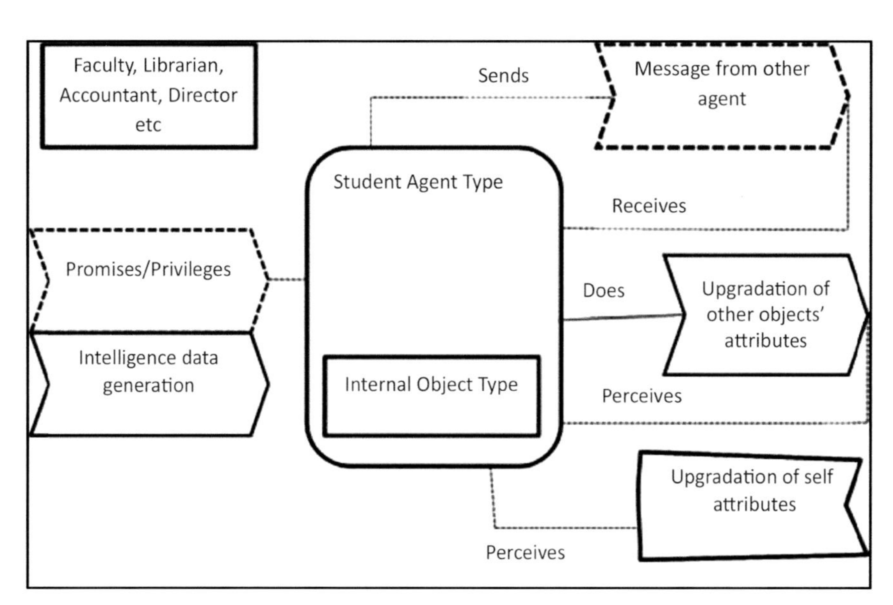

FIGURE 3.7 External AODB model for agent (STUDENT).

In addition, communication methods depict for message and data passing between two objects. Figure 3.8 illustrates how internal communication takes

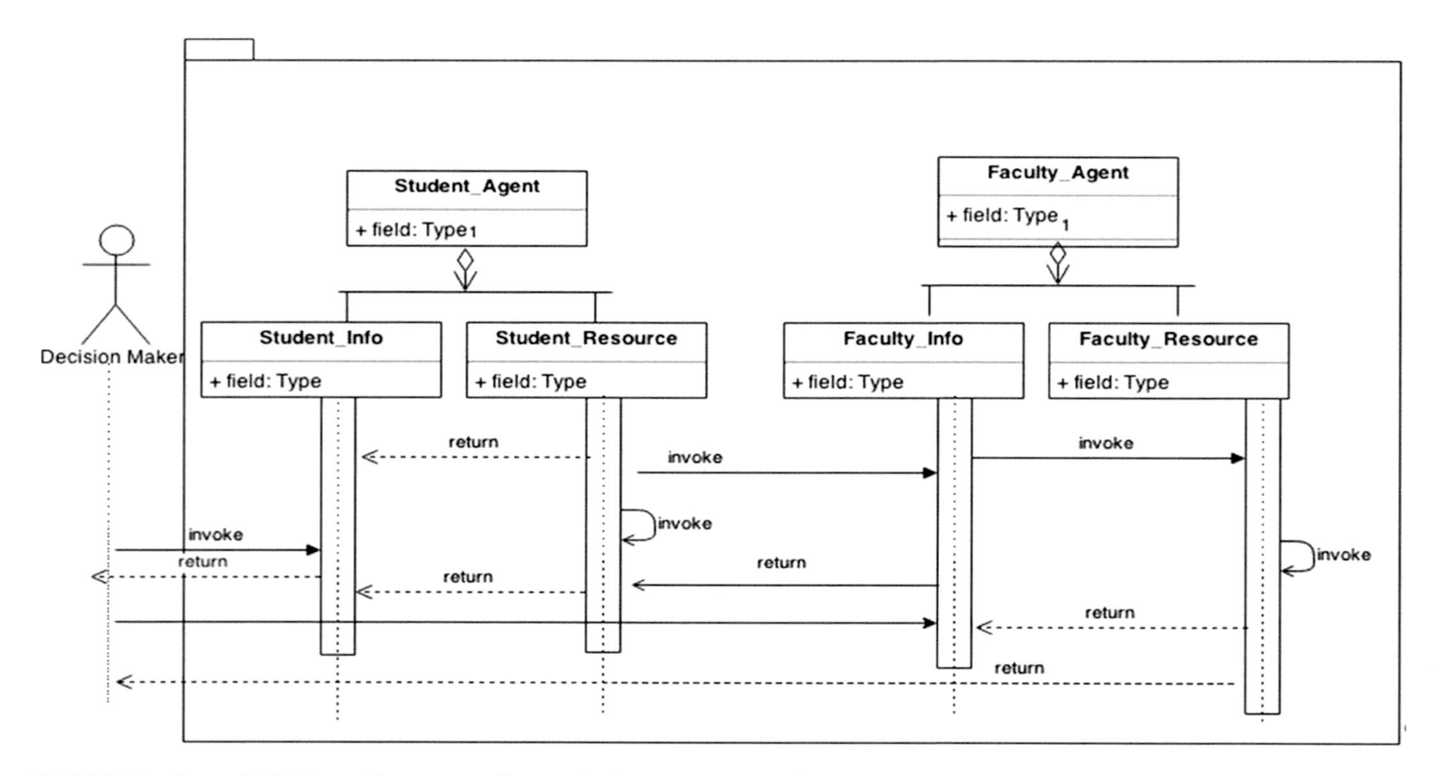

FIGURE 3.8 Internal AODB model: sequence diagram for interagent communication.

place between agents and within the agents through a sequence diagram. The class diagram is used to represent the relation between the base-data object and resource objects. To define an agent is an aggregation of both as shown in Figure 3.9.

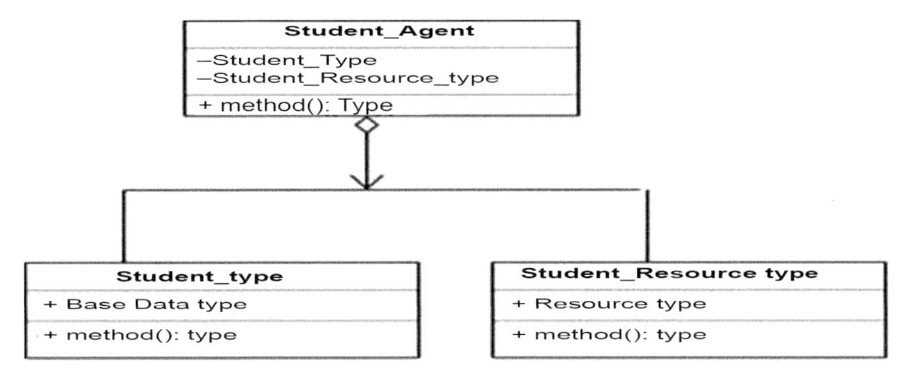

FIGURE 3.9 Internal AODB model: class diagram for inter-relation between the base-data object and resource object.

3.5.5 PHASE-III: AODB MODEL IMPLEMENTATION

In this phase, the database developer has to implement the agent as an embedded object and its programming features to automate in-data intelligence. Figure 3.10 shows algorithmic steps for iDSS Query with AODB-based IDA.

- So, which are implemented agents as an embedded data object and resource object.
- We have implemented triggers to provide communication between the different agents and base-data objects/tables.
- Implement procedures, functions, and packages to provide a data link between different agents using targeted language SQL/OQL/PL-SQL/NoSQL.

3.5.6 PHASE-IV: COGNITIVE DECISION-MAKING THROUGH IDA-BASED IDSS

In this phase, database statistics are collected based on IDA-oriented iDSS retrieval of decision-making predictive analysis. University system database's data are collected, which consist of enrollment data, courses, curriculum, timetable, attendance, and exam records of 18,000 students. Daily 75%–80% decision-making information is needed based on these records by the internal stakeholders of the university. This database is used to develop an agent-based data model for cognitive decision-making iDSS queries.

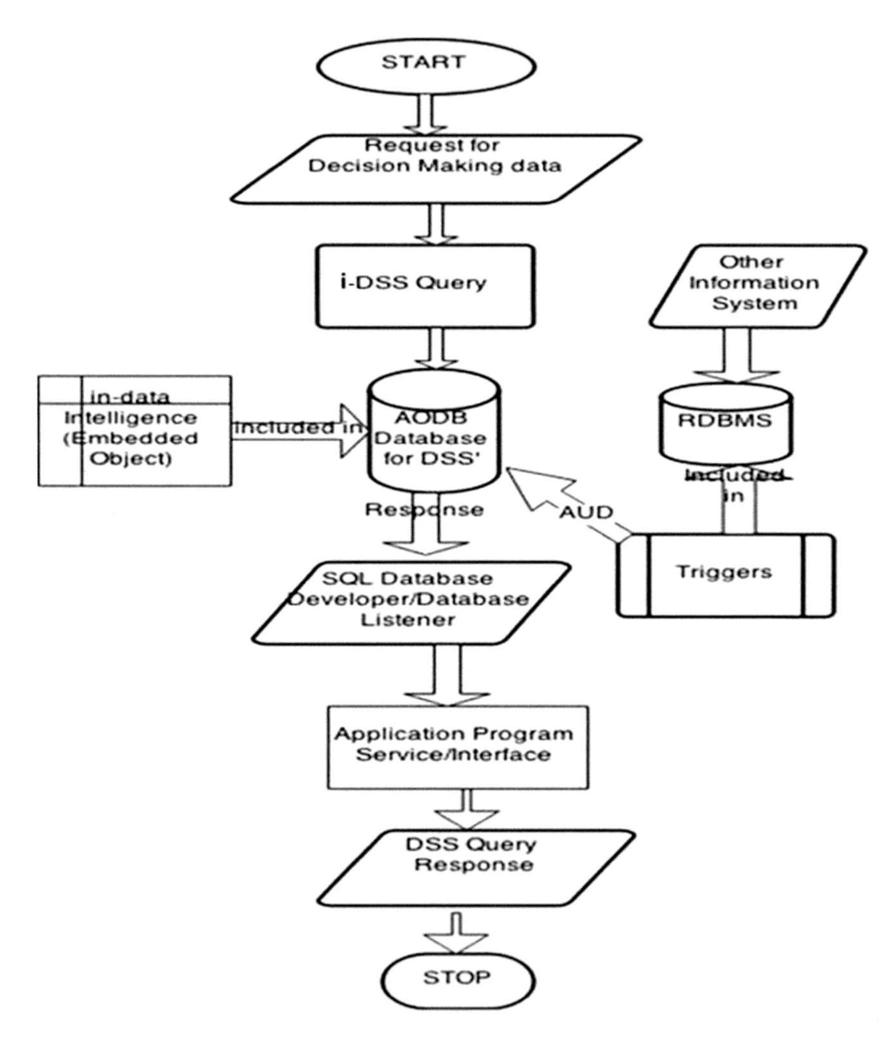

FIGURE 3.10 Algorithmic steps for iDSS Query with the AODB-based IDA.

The AODB schema implements direct access to decision-making information, through in-data intelligence. In which implementation of data as embedded objects relational database model with triggers and user-defined functions to automate the processing of data and stored the decision-making information along with data. It accelerates the data retrieval capability within the DSS. Listing 1 specifies the user-defined data type (UDTs) to define the base objects and resource objects. It contains STUDENT_TYPE as an OBJECT data type to define the attributes of students and methods to generate the data for the attributes. The STUDENT_RESOURCE_TYPE as UDT to identify RESOURCE OBJECT data type to determine the decision-making attributes/characteristics needs techniques to create that value in the base-object.

The storage of a column objects is similar to the data storage of an equivalent set of scalar columns that collectively make up the object. The difference is the additional overhead of maintaining the atomic null values of any noncollection column objects and their embedded object attributes. These values, called "null" indicators, specify every column object. This stores whether or not the column object is null and whether or not each of its embedded object attributes is null—the STUDENT_AGENT table consisting of two objects such as STUDENT_INFO and STUDENT_RESOURCE. The STUDENT_RESOURCE column object is a nested table, a collection type of column object. The storage for each object and its embedded objects attributes occupy one bit each. Thus, an object with n embedded object attributes has a storage overhead of CEIL (n/8) bytes. There is one null indicator column for each noncollection column object STUDENT_INFO and STUDENT_RESOURCE. The "null" indicator column length is one byte, as one bit represents the object itself, which translates to CEIL (1/8) or 1. Since the null indicator is one byte in size, the overhead of "null" information for each row of the STUDENT_RESOURCE table is two bytes, one for each object column.

STUDENT_INFO row objects store in object tables. An object table is a specific table that holds objects and provides a relational view of the attributes of those objects. An object table is logically and physically similar to a relational table whose column types correspond to the essential characteristics of the object type stored in the object table. The critical difference is that an object table can optionally contain an additional object identifier (OID) column and index.

In STUDENT_AGENT, "STUDENT_INFO.ENROLLMENTNO" is the primary-key-based OID. Oracle automatically created the OID for the STUDENT_RESOURCE row object. *OIDs* uniquely identify row objects in object tables. The OID cannot be accessed directly, but it can access through references (REFs). *Member methods*, DISPLAY_DETAILS,

GET_ENROLLMENTNO, GET_ACADEMIC_FACT, and GET_LIBR_FACT, provide an application with access to an object instances data. GET_ENROLLMENTNO is MAP method, called automatically to evaluate comparison between two object variables. These methods call by the triggers that are implemented to generate the STUDENT_RESOURCE type attributes value as in-data intelligence. In another way, this information is useful for the retrieval of decision-making.

3.5.7 POPULATING DATA STUDENT_AGENT

Three different sets of data volume have been created in the STUDENT_AGENT based on the described implementation strategy in Figure 3.12. The first set of data is with 400 records, the second set of information is with 50,000 records/tuples, and the third set of data with the 100,000 records/tuples. The decision support query in Figure 3.11 using object query language is executed to generate the decision maker's dashboard with visualization. The result obtained by way of implementation of earlier data models and the proposed model is filtered and tabulated in Table 3.5 for three sets of records. Table 3.5 contains the statistical data for iDSS query execution in three different data models: relational, star schema, and AODB model. Table 3.6 includes a value for execution time in seconds, memory blocks used, data volume in many records, and the level of aggregate data volume for each schema.

IDAs IDA-I–IDA-IV are developed to retrieve the data from STUDENT_AGENT. In the development of IDA-I–IDA-IV, PL/SQL features are used, such as Cursor, Stored Procedure, and Stored Procedure, using the collection to retrieve the data for query evaluation. The use of the group is required in the embedded, nested table to collect the multiple instances of the resource-data object for each base-data object instance. Table 3.6 represents the summary of query execution. It uses three different features of PL/SQL, which are a cursor, Stored Procedure, and Stored Procedure, obtained through Experiments 5–8.

The result is obtained by implementing PL/SQL features that the execution performance results are nearly similar to the SQL query performance. However, the limitation of SQL query is that only base data can retrieve. If data retrieval requires resource objects, it is necessary to implement collection in PL/SQL. These stored procedures and functions reside on the server as part of the user's schema. On "logon" to the database, these methods are called and executed to generate the decision-making data in the STUDENT_AGENT table.

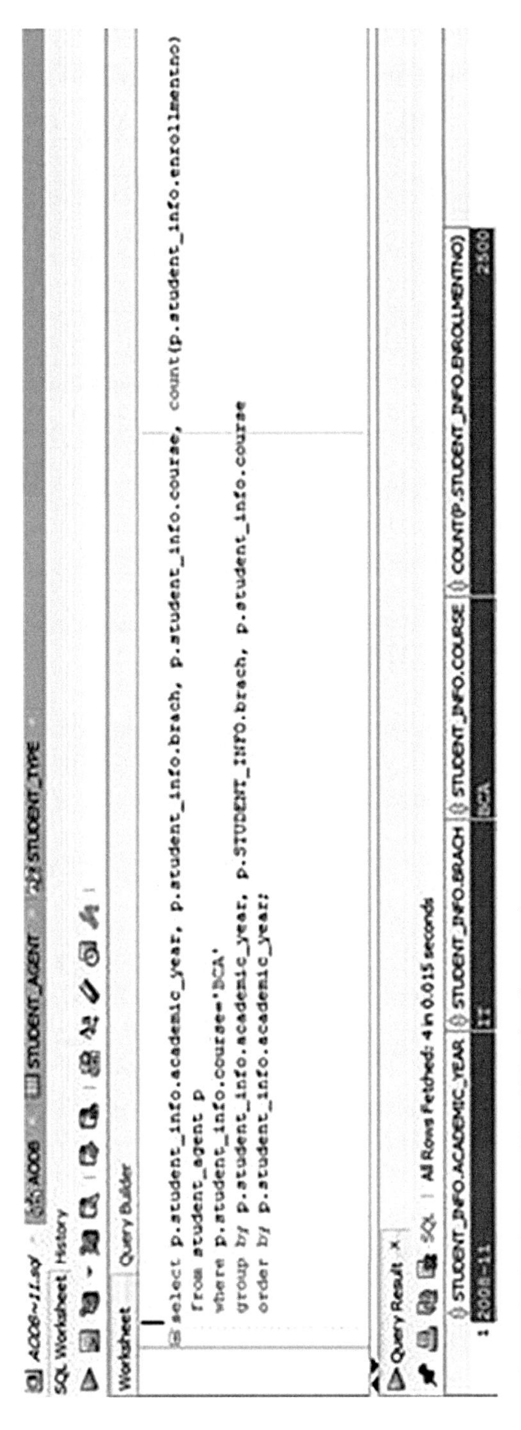

FIGURE 3.11 STUDENT_AGENT implementation.

TABLE 3.5 Tabulation of Analytical Finding from One to Three Experiments

Database Models	DV=400			DV=50,000			DV=100,000		
	ET	MB	ADV	ET	MB	ADV	ET	MB	ADV
Relational	0.0334	5	0	0.531	370	0	0.522	784	0
Star schema	0.0301	5	16	0.521	5	71	0.539	5	275
Proposed agent Oriented schema	0.012	5	0	0.014	662	0	0.016	1252	0

TABLE 3.6 Query Execution of DSS Query Using PL/SQL (Experiments 5–8)

PL/SQL Features	Is Data Retrieved from Resource Object	Execution Time in Seconds (%)
Using Cursor (IDA-I)	No	0.032
Using a Stored Procedure (IDA-II)	No	0.015
Using Collection in Stored Procedure (IDA-III & IDA-IV)	Yes	0.016

3.6 STRENGTH AND LAPS OF FRAMEWORK

The proposed model IDA based on the AODB model has an advantage over the other database models. As the iDSS query execution time reduces by 93% than the existing data models.

3.6.1 USABILITY

The IDA based on the AODB model is designed to allow the user to determine attributes. Criteria and methods are required to define based on the object for decision-making. Decision makers can submit a query or use stored PL/SQL procedure. The results are all annotated with the graphical user interface, which provides enough decision-making ready information to the decision maker to make the decision. Therefore, it satisfies the usability characteristics.

3.6.2 REUSABILITY

The object-oriented relation database is used to implement the AODB model. Reusability in object-oriented concepts includes code reuse within a single-software project and code reuse between multiple projects. The base object defined by the I-Data Engineer, if the structure of another object in different decision-making system is similar to that is applicable. Therefore, it satisfies the reusability characteristics.

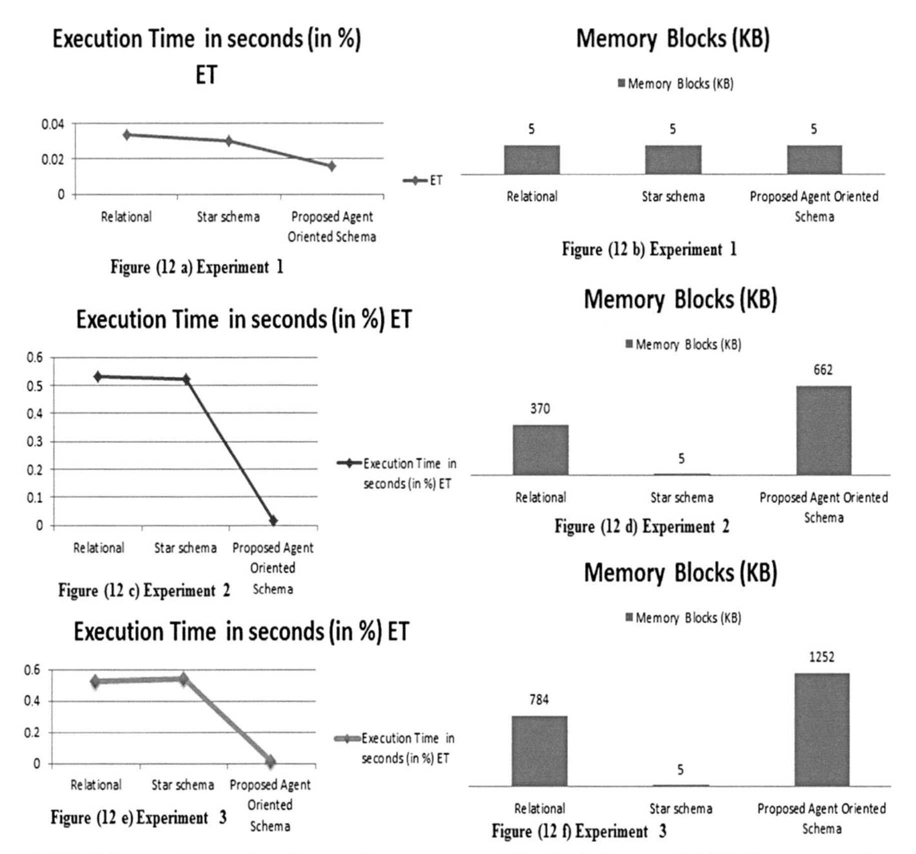

FIGURE 3.12 Execution time and memory used for IDA data model IDSS query performance in data retrieval from the AODB model using IDAs.

3.6.3 FLEXIBILITY

In this work, implemented UDT such as base-data objects and resource objects can be used to derive new "object" according to the future need using inheritance, aggregation, and alteration in the existing UDT. The newly obtained data type can add to the current agent table. Therefore, it is a flexible solution.

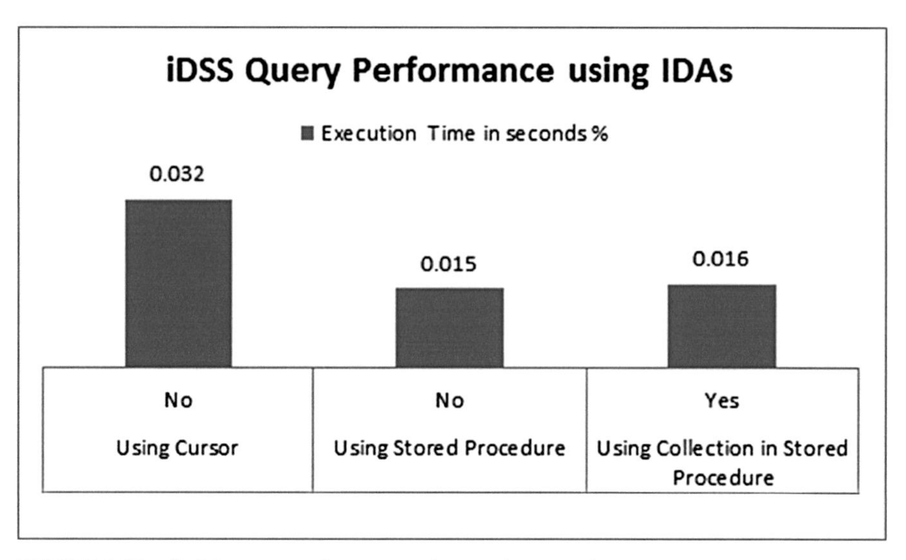

FIGURE 3.13 iDSS query performance using IDAs (Experiments 5–8).

3.6.4 MANAGEABILITY

In this work, the term manageability is considered in terms of ability to manage the decision support data model by the database programmer and I-Data Engineer, which means management capabilities of Database Engineer and i-Data Engineer that allow quickly determine, set up, and provide support for IDA. It is based on the AODB model used in cognitive decision-making. Therefore, it is an easy solution.

3.6.5 SCALABILITY AND EXPANDABLE

The consideration of scalability in the proposed model to handle the growing amount of data and the work-capable manner or its ability to be enlarged to accommodate that growth. Therefore, the proposed solution can fulfill the future needs of decision-making applications. This model extends by implementing FACULTY_AGENT to store faculty-related base-data object and resource-data object.

3.6.6 INTEROPERABILITY

In this work, interoperability is considered as the ability of implementing the proposed model to communicate and work together in a different programming environment. Performed IDAs using PL/SQL supports user-defined object types as parameters. These stored subprograms can quickly call by VB.Net, ORACLE Forms, and JADE programming interface.

3.6.7 DEGREE OF AUTOMATION

The term degree of automation is considered as the automatic generation of data retrieval and storage in Figure 3.5. Automation is implemented through triggers and calling resource object methods to generate cognitive decision-making information.

The proposed model requires more memory storage to store cognitive decision-making information. The memory utilization of blocks is found 0.39% in MDDBs, but it saves only aggregate data in the storage media, 62.61% in RDBs as compared to AODBs. The 1252-kB memory block is founded in AODB, which is the highest memory block utilization among three models. The proposed model is based on a structured data model. Currently, this solution can implement cognitive decision-making based on structured data.

The limitation of this work is that it does not support the multimedia database. For flexibility, if the AODB model is utilizing images and video information as base data, image processing and video streaming algorithms are used as resource objects. Management of a large volume of data in the distributed IDA based on the AODB model is a challenging research work for evaluating scalability. Expansion of the AODB model with more number of agents requires work to provide security for agent communication in the model.

3.7 CONCLUSION

This work highlights the result statistics, due to which the introduced model is improving information retrieval as well as storing cognitive decision-making information for an iDSS. It suggests database designers to adopt the IDA using the AODB model for developing an iDSS for improved performance in cognitive decision-making.

3.8 FUTURE SCOPE

The extension of this work may be taken to test cognitive decision-making IDA in heterogeneous distributed nonrelational database environment.

The limitation of this work and its possible extension may lead to evaluating the multimedia database. For flexibility, if the AODB model is utilizing images and video information as base data, the image processing and video streaming algorithms are used as resource objects. Furthermore, it is required to evaluate the manageability of an IDA in the development of a model-based DSS and group DSSs.

Management of a large volume of data in the distributed IDA based on the AODB model is challenging research work for evaluating scalability. Expansion of the AODB model with more number of agents requires work to provide security for agent communication in the model.

KEYWORDS

- **cognitive decision-making**
- **intelligent decision support system**
- **intelligent agent**
- **data warehousing**
- **Hadoop**
- **predictive analytics**
- **machine learning**
- **intelligent database agent (IDA)**

REFERENCES

1. A. F.-Caballero and M. V. Sokolova, "Computational agents in complex decision support systems," in *Handbook on Decision Making*. New York, NY, USA: Springer, 2009.
2. R. Agarwal, C. S. Chauhan, and R. K. Sharma, "Decision support and database management system 'AroMed' on commercially exploited medicinal and aromatic plants of India," *Int. J. Adv. Agricultural Sci. Technol.*, vol. 1, no. 1, pp. 17–22, 2012.
3. A. Armigliato, G. Pagnoni, F. Zaniboni, and S. Tinti, "Database of tsunami scenario simulations for Western Iberia: A tool for the TRIDEC project decision support system for tsunami early warning," *Geophys. Res. Abstracts*, vol. 15, 2013, Art. no. 5567.
4. L. Niu and G. Zhang, *Cognition-Driven Decision Support System Framework*. New York, NY, USA: Springer, 2007.
5. N. Lin, D. Li, and C. Bi, "Research on development of corn production decision support system," *TELKOMNIKA Indones. J. Electr. Eng.*, vol. 11, no. 7, pp. 3798–3808, 2013.

6. J. Huang, L. Antova, C. Koch, and D. Olteanu, "MayBMS: A probabilistic database management system," in *Proc. ACM SIGMOD Int. Conf. Manage. Data*, 2009, pp. 1071–1074.
7. Z. Jin and Q. Xu, "The realization of decision support system for cross-border transportation based on the multidimensional database," *J. Softw.*, vol. 7, no. 5, pp. 974–981, 2012.
8. A. K. Sharma et al., "Web-enabled decision support system on most probable producing ability and a searchable database on herd strength for livestock farm management," *Int. J. Comput. Sci. Eng.*, vol. 3, no. 11, pp. 3628–3631, 2011.
9. Fareed, "Intelligent decision-making technique for marketing using hypothetical database and fuzzy multi-criteria method," *J. Basic Appl. Sci.*, vol. 9, pp. 44–51, 2013.
10. M. Sokolova, "A review on frameworks for decision support systems," *Intell. Syst.*, vol. 30, pp. 19–45, 2012.
11. S. Alonso, E. Herrera-Viedma, F. Chiclana, and F. Herrera, "A web-based consensus support system for group decision-making problems and incomplete preferences," *Inf. Sci.*, vol. 180, no. 23, pp. 4477–4495, 2010.
12. H. Grebla, G. Moldovan, and S. Darabant, "Data allocation in distributed database systems performed by mobile intelligent agents," in *Proc. Int. Conf. Theory Appl. Math. Inform.*, 2004, pp. 164–173.
13. C. Jonquet, P. Dugenie, and S. A. Cerri, "Service-based integration of grid and multi-agent systems models," *Lecture Notes Comput. Sci.*, vol. 5006 LNCS, pp. 56–68, 2008.
14. M. Tan and J. Xu, "Study and implementation of a decision support system for urban mass transit service planning abstract," *J. Inf. Technol.*, vol. 15, pp. 14–32, 2004.
15. T. Naser and M. J. Ridley, "Two-way mapping between object-oriented databases and XML," *Int. J. Comput. Appl.*, vol. 33, pp. 297–308, 2009.
16. F. Bellifemine, G. Caire, A. Poggi, and G. Rimassa, "JADE: A software framework for developing multi-agent applications. Lessons learned," *Inf. Softw. Technol.*, vol. 50, nos. 1/2, pp. 10–21, 2008.
17. J. T., M. L. Talbert, and S. A. Deloach, "Heterogeneous database integration using agent-oriented Mcdonald information systems," *Int. J. Future Gener. Inf. Technol.*, vol. 3, pp. 1359–1365, 2000.
18. J. Sebestyénová, "Case-based reasoning in the agent-based decision support system," *Acta Polytech. Hungarica*, vol. 4, no. 1, pp. 127–138, 2007.
19. R. Vohra and N. N. Das, "Intelligent decision support systems for admission management in higher education institutes," *Int. J. Artif. Intell. Appl.*, vol. 2, no. 4, pp. 63–70, 2011.
20. M. Arora and M. S. Devi, "Role of database in multi-agent resource allocation problem," *Int. J. Comput. Sci. Inf. Technol.*, vol. 2, no. 2, pp. 668–672, 2011.
21. S. Ramanujam and M. A. M. Capretz, "ADAM: A multi-agent system for autonomous database administration and maintenance," *Int. J. Intell. Inf. Technol.*, vol. 1, no. 3, pp. 14–16, 2005.
22. M. Stonebraker *et al.*, "C-store: A column-oriented DBMS," in *Proc. 31st VLDB Conf.*, Norway, 2018, pp. 491–518.
23. J. O. Gutierrez-Garcia and K. M. Sim, "Agent-based cloud service composition," *Appl. Intell.*, vol. 38, no. 3, pp. 436–464, 2013.
24. J. Pokorny, "Databases in the 3rd millennium: Trends and research directions," *J. Syst. Integration*, vol. 1, pp. 3–15, 2010.

25. R. S. Mawale, P. A. V Deorankar, and P. V. A. Kakde, "A novel approach for converting relational database to an object-oriented database: Data migration and performance analysis," *Int. J. Res. Eng. Technol.*, vol. 3, pp. 101–104, 2013.
26. P. Trancoso, D. Othonos, and A. Artemiou, "Data parallel acceleration of decision support queries using Cell/BE and GPUs," in *Proc. 6th ACM Conf. Comput.*, 2009, pp. 117–126.
27. O. López-Ortega and M. A. Rosales, "An agent-oriented decision support system are combining fuzzy clustering and the AHP," *Expert Syst. Appl.*, vol. 38, no. 7, pp. 8275–8284, 2011.
28. S. Ceri, R. J. Cochrane, and J. Widom, "Practical applications of triggers and constraints: Successes and lingering issues," *VLDB J.*, pp. 254–262, 2000.
29. A. H. Landberg, J. W. Rahayu, and E. Pardede, "XTrigger: XML database trigger," *Comput. Sci. Res. Develop.*, vol. 29, no. 1, pp. 1–19, 2014.
30. H. Mühleisen, T. Samar, J. Lin, and A. P. De Vries, "Column stores as an IR prototyping tool," *Lecture Notes Comput. Sci.*, vol. 8416, pp. 789–792, 2014.
31. E. Serova, "The role of agent-based modelling in the design of management decision processes," *J. Inf. Syst. Eval.*, vol. 16, no. 1, pp. 71–80, 2013.
32. M. Faizal, A. P. Malai, M. F. Omar, and J. Wong, "Qut digital repository: Infrastructure project planning decision making: Challenges for decision support system applications," *J. Civil Eng.*, pp. 4–6, 2009.
33. K. Taveter and G. Wagner, "Agent-oriented business rules: Deontic assignments," in *Proc. Int. Workshop Open Enterprise Solutions: Syst., Exp., Org.*, Rome, Italy, 2001.
34. K. Taveter and G. Wagner, "Agent-oriented enterprise modeling based on business rules," in *Proc. Int. Workshop Open Enterprise Solutions: Syst., Exp., Org.*, Rome, Italy, 2001, pp. 527–528.
35. G. K. Raikundalia, Y. Zhang, and X. Yu, *ACM Tran. Comput.-Human Interact.*, vol. 15, no. 2, pp. 91–107, 2009.
36. M. Beynon, A. Cartwright, and Y. P. Yung, "Databases from an agent-oriented perspective," *Database Syst.*, pp. 1–14, 1998.
37. P. Beynon-Davies and P. Beynon-Davies, Intelligent Databases. 2015, doi: 10.1007/978-1-349-13722-0_17.
38. A. Perini, "Discussing strategies for software architecting and designing from an Agent-oriented point of view," in *Proc. Softw. Eng. Large-Scale Multi-Agent Syst. Conf.*, Portland, OR, USA, 2003.
39. V. G. Mr. R. S. Mawale, "Reengineering of relational databases to object-oriented database," *Int. J. Res. Eng. Technol.*, vol. 03, no. 01, pp. 112–115, 2015.
40. M. Malki, A. Flory, and M. Rahmouni, "Extraction of object-oriented schemas from existing relational databases: A form-driven approach," *Int. J. Inform.*, vol. 13, no. 1, pp. 47–72, 2002.
41. S. Dhawak, "eETECME constructing of object-oriented databases from relational databases: A review," *Int. J. Comput. Appl.*, pp. 31–34, 2013.
42. M. C. Nikose, S. S. Dhande, and G. R. Bamnote, "Query optimization in object-oriented databases through detecting independent subqueries," *Int. J. Adv. Res. Comput. Sci. Softw. Eng.*, vol. 2, no. 2, 2012.
43. M. Alam and S. K. Wasan, "Migration from relational database into object oriented database," *J. Comput. Sci.*, vol. 2, no. 10, pp. 781–784, 2006.

44. O. S. Benli and A. R. Botsali, "An optimization-based decision support system for a university timetabling problem: An integrated constraint and binary integer programming approach," *Math. Probl. Eng.*, no. 1, pp. 1–29, 2004.
45. M. Wang, "Using object-relational database technology," *Vis. Inform.*, vol. XI, no. 1, pp. 90–99, 2010.
46. C. Mohan, "History repeats itself: Sensible and nonsenSQL aspects of the NoSQL hoopla," in *Proc. 16th Int. Conf. Extending Database Technol.*, 2013, pp. 11–16.
47. E. E. Ogheneovo and E. Oviebor, "Open access an object-oriented approach for optimizing query processing in distributed database system," *Amer. J. Eng. Res.*," no. 12, pp. 200–206, 2016.
48. M. V. Sokolova and A. Fernández-Caballero, "Modeling and implementing an agent-based environmental health impact decision support system," *Intell. Syst. Ref. Library*, vol. 36, no. 2, PART 2, pp. 2603–2614, 2009.
49. N. M. M. Noor and R. Mohemad, "New architecture for intelligent multi-agents paradigm in decision support system," in *Decision Support Systems*, Rijeka, Croatia: InTech, 2010.
50. H. L. Zhang, C. H. C. Leung, and G. K. Raikundalia, "Topological analysis of AOCD-based agent networks and experimental results," *J. Comput. Syst. Sci.*, vol. 74, no. 2, pp. 255–278, 2008.
51. A. Tariq, "Intelligent decision support systems—A framework," *Knowl. Manage.*, vol. 2, no. 6, pp. 12–20, 2012.
52. H. Demirkan and D. Delen, "Leveraging the capabilities of service-oriented decision support systems: Putting analytics and big data in the cloud," *Decis. Support Syst.*, vol. 55, no. 1, pp. 412–421, 2013.
53. R. Alhajj and F. Polat, "Converting a legacy database to object-oriented database," in *Encyclopedia of Database Technologies and Applications*. Hershey, PA, USA: IGI Global, 2011, pp. 99–104.
54. K. Taveter and G. Wagner, "Agent-oriented enterprise modeling based on business rules," in *Proc. 20th Int. Conf. Conceptual Model.*, 2001, pp. 527–540.
55. K. Rabuzin, M. Maleković, "Active databases, business rules, and reactive agents—What is the connection," pp. 63–73, [Online.] Available: https://pdfs.semanticscholar.org/b866/ b2b80bea771e9ec2a1d28ef8b7980ec 0815 2.pdf.
56. D. Peebles, "A cognitive architecture-based model of graph comprehension," *Wiley Interdisciplinary Rev., Cogn. Sci.*, pp. 37–42, 1998.
57. Y. Wang, "The cognitive process of decision making," *Vis. Inform.*, pp. 45–52, 2014.
58. C. Lebiere and D. Morrison, "Cognitive models of prediction as decision aids," *Int. J. Virtual, Augmented Mixed Reality*, pp. 285–294, 2016.
59. S. A. Trivedi, "Implementation of agent oriented database model for accelerating DSS query," *Int. J. Sci. Res. Dev.*, vol. 2, no. 10, pp. 129–132, 2014.

CHAPTER 4

A Convergence of Mining and Machine Learning: The New Angle for Educational Data Mining

PRATIYUSH GULERIA[1*] and MANU SOOD[2]

[1]*National Institute of Electronics and Information Technology, Shimla, Himachal Pradesh 171001, India*

[2]*Department of Computer Science, Himachal Pradesh University, Shimla, Himachal Pradesh 171005, India*

[*]*Corresponding author. E-mail: pratiyushguleria@gmail.com*

ABSTRACT

In the digital era, there is an accumulation of large volume of data, and the biggest challenge being faced by humans is to derive meaningful information from these data. In such a scenario, data mining techniques become important to unearth the hitherto unknown relationship from data. There is an urgent need for scientific research in education, as it aimed at enhancing student's cognitive learning and social development. With increasing educational institutions, the large amount of data is accumulated, which is unstructured and not useful neither for students nor for teachers. Therefore, the major stress is on improving the quality of learning in schools and institutions, and there is a need to make significant progress on access to schooling and quality of learning in students. The computational intelligence is achieved through data mining techniques, which can break down information to enable it to make predictions. This chapter aims at discussing the utilization of educational data mining, learning analytics, and predictive intelligence techniques in educational field and models of machine learning, that is, unsupervised and supervised learning. There are some inbuilt libraries, such as Tensor flow, Keras, and Python libraries, discussed in this chapter, which helps in building, training, and applying it into educational data classification.

4.1 INTRODUCTION

Educational data mining is one of the most encouraging fields for predicting the educational trends with the inclusion of the latest technologies into the system. It is the field using data mining techniques in educational environments [1]. The inclusion of e-learning modes, smart learning analytics, and online learning resources in the traditional teacher-taught model results in an accumulation of large volume of data. The data collected may be unstructured, semistructured, or structured, which become a challenge to the learners as well as educational institutions in quality assurance of education and improving decision-making capabilities of management. An equally important factor that is of grave concern to the students' community at large is to choose appropriate academic courses and industrial training programs helpful in furthering their career/job prospects.

To handle the situation of big data, there is emergent need of educational data mining, which helps in converting unstructured data to structured data and disseminating meaningful information as well as knowledge.

Machine learning is a computer science field, where statistical techniques give computer system learning capability and improve the performance of a particular task continuously without being explicitly programmed. Machine learning involves cognitive and computational approaches. This field illustrates the benefits of collaboration between scientists from psychology and computer science [2]. With the help of a machine learning approach, education and personal training can be enhanced, and the following outcomes can be achieved: (a) students' dropout in distance learning can be prevented, (b) prediction of students those who can drop the course, (c) predicting students learning behavior, (d) innovative pedagogical practices, (e) clustering of students according to their learning speed, (f) predicting industry-oriented courses, and (g) real-time analysis of students feedback. There are many tools that can be used frequently for data mining and learning analytics in the educational field. The tools helped extract meaningful information from the student's dataset and find new observations related to students' learning activities in the educational system. It helps in (a) improving students' retention rate, (b) maximize educational improvement ratio, and (c) enhance students' learning results [3].

Data mining takes advantage of machine learning, statistical, and visualization techniques to find and extract knowledge. The data collected are to be analyzed using techniques such as decision trees, neural networks, naive Bayes, support vector machines (SVMs), *K*-means, etc., which help in predicting results such as students' learning styles, interest in course, learning abilities, knowledge, and interests, predict student retention, and predict course adaptability [4].

4.2 LITERATURE SURVEY

Lykourentzou et al. [5] have proposed a methodology using machine learning techniques for predicting the dropout rate of students for e-learning courses. The machine learning techniques used are feedforward neural networks and SVMs. The authors have discussed how the results of machine learning techniques are tested.

In [6], a prototype web-based support tool has been proposed, which can automatically recognize students with a high dropout probability. Zhu [7] has discussed a machine learning approach for enhancing education and personnel training. According to Liebowitz [8], adaptive/personalized learning, educational data mining, data visualization, visual analytics, knowledge management, and blended/e-learning play growing roles to better inform higher education officials and teachers. Aher and Lobo [9] have discussed the data mining techniques and machine learning algorithms for recommending courses in e-learning environment based on past data.

In [10], Beck and Woolf have developed models using machine learning for predicting student behavior and support decision making. According to Dede [11], there is a lot of development in distance education with emerging technologies and distributed learning. The author has also focused on pedagogical strategies and designs in the educational system. In [12], Demšar et al. have discussed the Orange framework for machine learning and data mining. This framework supports the following: (a) data preprocessing, (b) modeling, (c) evaluation, and (d) data mining classification and clustering algorithms. Mitchell [13] has discussed machine learning and data mining. According to him, data mining improves future decisions using historical data and discovers irregularities. The author has discussed scientific issues; basic technologies helpful in learning analytics are as follows: (a) learning from structured and unstructured data, (b) experimentations, (c) explorations, (d) optimizing decisions, and (e) inventing new features to improve accuracy. These learning approaches are helpful in applications such as healthcare, marketing, manufacturing, financial and intelligent data analysis, etc. Baker and Inventado [14] have discussed the relationship between the educational data mining and learning analytics, which are the emerging areas. In [15], Burgos et al. have used data mining techniques for modeling students' performance. They have proposed the predictive model to prevent the dropout rate in e-learning courses using knowledge discovery techniques. In educational data mining, there is a need to derive and innovate new approaches using statistical techniques, machine learning, psychometrics, and scientific computing to transform the existing system. Naren [16] has discussed data mining applications for predicting behavioral patterns of the students. In [17], the educational data mining approach is used for analyzing and predicting students learning behavior and experiences. Educational data mining helps in designing smarter and intelligent learning, which can better inform learners and educators [18].

4.3 IDENTIFICATION GAP

In the present scenario, the failure in the dissemination of quality-driven education is the biggest challenge faced by our educational institutions. The skill gap in quality parameters involves: (a) practical skilling of students, (b) placement of students, (c) adoption of latest syllabi into curricula, (d) efficient decision-making of career selection, (e) competitive environment of examinations, (f) cognitive and computational approaches, and (g) research and development methodologies in training. There are data mining techniques, but, still, their utilization in educational sector is unexplored. There is a need to implement data mining techniques in the educational sector to fill up these existing gaps and adoption of learning analytics with innovation in teaching pedagogies.

4.4 MOTIVATION

As many of the educated students involve students undergoing training, completed trainings are not getting employability, which may be due to the following: (a) fewer placements, (b) lack of knowledge, (c) skill gap, (d) lack of industry-relevant curriculum and exposure, (e) engineering concepts applied to industry are not discussed during training, and (f) lack of industry-oriented talks, workshops, and conferences. There is need of counseling to understand the factors which are as follows: (a) kinds of jobs that are available, (b) how to determine which job profiles match students' interests and skills, and (c) the skill gaps that may disqualify students and how to address those skill gaps.

As of nowadays, machine learning and data mining techniques need to be explored for answers to these questions, as this area of machine learning is least utilized in the educational sector to find the meaningful information from a huge volume of unstructured educational entities.

4.5 PROBLEM DEFINITION

With the advancement in technology, their increased utilization in the educational field has resulted in a collection of voluminous data. These data need innovative approaches to make them meaningful for students, academia, administrators, and management. The intelligent data analytics in the educational field needs to be explored with new pedagogical teaching practices and learning analytics. There is an emergent need for machine learning and data mining techniques in education because the effective utilization of data mining in the educational field is still lacking.

With the confluence of machine learning and educational data mining fields, effective decision-making can be achieved, and new academic models can be predicted. Apart from it, certain characteristics of students and academia can be predicted such as (a) learning behavior of student, (b) response time of student to answer question, (c) area of interest, (d) understanding level of student, (e) teaching aids, (f) instructor teaching strategy, (g) multimedia techniques to be included in learning or not, (h) participation of students in online activities, and (i) designing syllabi as per industry demand.

4.6 INTELLIGENT LEARNING ANALYTICS USING DATA MINING TECHNIQUES

With the advancement in the educational field, applications are being developed, where teachers, parents, and students are digitally connected in a forum. The feedback is obtained by teachers from parents and students on classroom activities and teaching aids. These inputs act as intelligent learning analytics for machines. The application helps the parents and students in the following ways: (a) classroom activities performed are available to parents on mobiles, (b) tracking of students through mobiles, (c) classroom updates are sent online, (d) parents are updated about children activity through reminders, (e) access to teachers resources to improve student outcomes, and (f) manage all the activities online on your accounts and get insights on how other teachers interact with their classrooms.

4.6.1 STUDENTS PERFORMANCE PREDICTION USING DECISION TREES

A decision tree is a mining technique, where predictions are performed after analyzing the datasets. It uses a tree-like structure for formulating predictions. The synthesized educational dataset is shown in Table 4.1. The attributes are stated as follows: (a) class performance, (b) sessional marks, (c) attendance, (d) assignment, (e) lab work, and (f) class grade. The class result of students is predicted after analyzing the attributes using the decision tree technique. The classifier model is shown in Table 4.2, which finds that the "Sessional Attribute" is having the highest Information Gain.

TABLE 4.1 Educational Data Set Description

ID	Class Performance	Sessional Marks	Attendance	Assignment	Lab Work	Class Grade
1	Good	Average	Poor	Good	Poor	C
2	Average	Good	Good	Good	Good	A
3	Good	Good	Average	Average	Average	B
4	Poor	Good	Average	Poor	Average	C
5	Good	Poor	Good	Average	Average	B
.	………	………	………	………	………	………
n	n	n	n	n	n	n

TABLE 4.2 Classifier Information

J48 pruned a tree for "Sessional" Attribute, i.e., Highest Information Gain
SESSIONAL = GOOD
| CLASS PERFORMANCE = GOOD: A (45.0)
| CLASS PERFORMANCE = AVERAGE: A (35.0)
| CLASS PERFORMANCE = POOR: C (20.0)
SESSIONAL = AVERAGE
| CLASS PERFORMANCE = GOOD: A (48.0/8.0)
| CLASS PERFORMANCE = AVERAGE: A (16.0/8.0)
| CLASS PERFORMANCE = POOR: B (16.0/8.0)
SESSIONAL = POOR
| CLASS PERFORMANCE = GOOD: C (20.0)
| CLASS PERFORMANCE = AVERAGE: C (40.0)
| CLASS PERFORMANCE = POOR: C (40.0)

4.6.2 MINING EDUCATIONAL DATA USING K-MEANS CLUSTERING

K-means clustering is the unsupervised technique for classifying the data into clusters. In *K*-means, the data are not having any specific class. The *K*-means clustering algorithm finds the *k* number of centroids and designates every input data point to the nearest cluster.

In machine learning, the *K*-means approach helps in (a) pattern matching, (b) finding clusters which are having similar characteristics, (c) advantageous for clustering of voluminous datasets, and (d) efficient way to scale down the size of data. The *K*-means clustering approach has been implemented on the dataset shown in Table 4.1, and the results are generated in the form of clusters shown in Figure 4.1. The clusters show (a) the students who are short of attendance and (b) students who have performed poorly in sessional. In the same way, the *K*-means approach can help identify the students as per their learning and understanding skills in the classroom teaching.

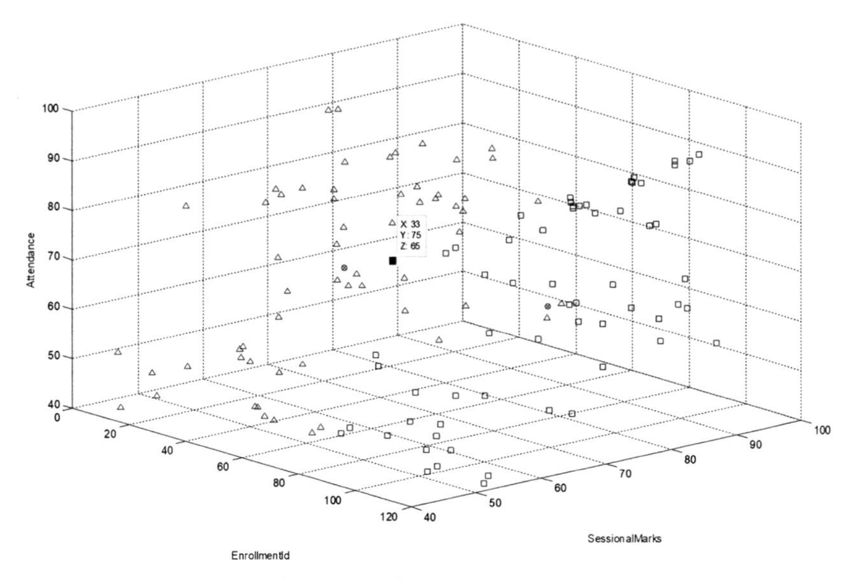

FIGURE 4.1 Visualization of attributes clusters.

4.6.3 EDUCATIONAL DATA MINING USING NEURAL NETWORKS

Neural network self-organizing maps (SOMs) clusters the student's data into classes based on their class performance, considering the same educational dataset as shown in Table 4.1. The network after getting trained shows the plots for self-organizing map neighbor distances, weight planes, and sample hits. The results displayed in Figure 4.2(a) and (b) show the cluster of students belonging to a particular class. The classes representing similar features represent the area of neurons with a large number of hits as compared to another region. The SOM distance is shown by calculating the Euclidian distance of each neurons class from its neighbors.

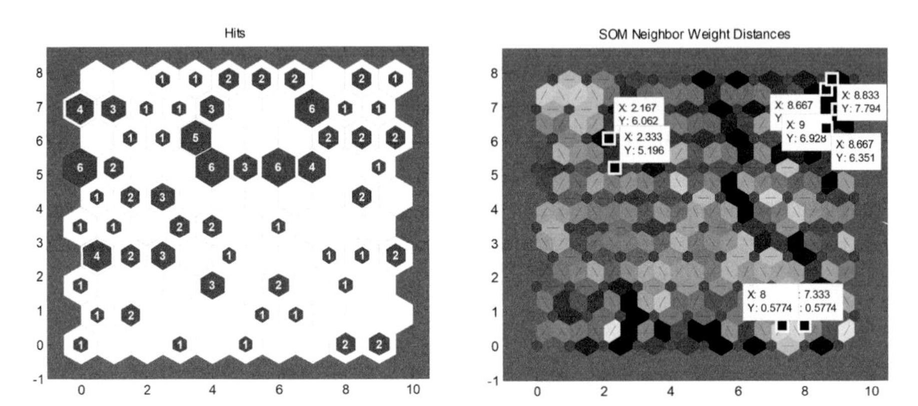

FIGURE 4.2 (a) SOM sample hits. (b) SOM neighbor weight distances.

The neural network classifies the input shown in Table 4.1 into a set of target class for pattern recognition. In neural networks, training is done multiple times to get the refined results and sampling. The desired network class is shown in Table 4.3. There are training, validation, and testing of the samples shown in the dataset.

TABLE 4.3 Desired Target Dataset

1	1	1	1	0	0	0	0	0	0	0	–	N
0	0	0	0	1	1	1	1	1	1	0	–	N
0	0	0	0	0	0	0	0	0	0	1	–	N

In the training phase, the neural network is trained, and the network is adapted conforming to its error. In the validation phase, network generalization is done, and the training of the network is stopped when generalization stops improving and is concluded.

The testing phase is an absolute part of network performance during and after training. The training performance of the network is shown in Figure 4.3(a), and the confusion matrix is shown in Figure 4.3(b). The results show that the network has learned to classify data properly with few misclassifications.

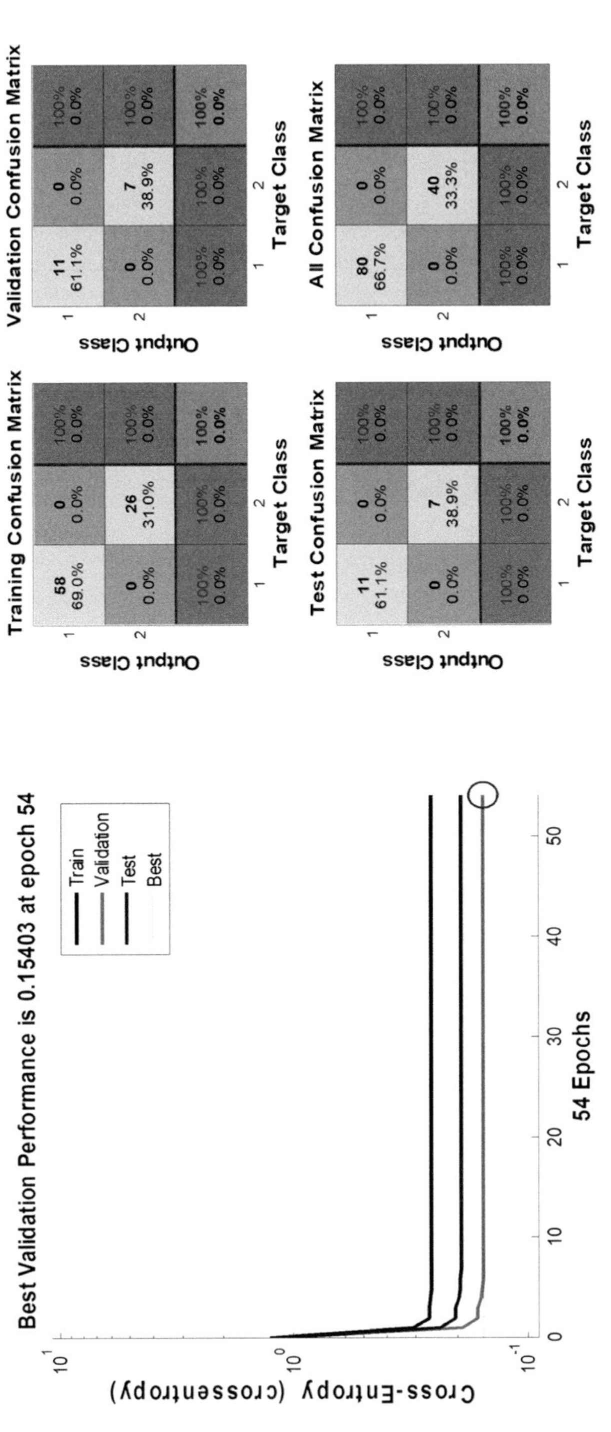

FIGURE 4.3 (a) Training performance of network. (b) Confusion matrix.

4.6.4 ACADEMIC COUNSELING OF STUDENTS USING ASSOCIATION RULE MINING (ARM)

With the help of ARM, the students' preference for the courses to undergo industrial training can be predicted. These techniques become important in academic counseling of students. The semisynthesized dataset collected from the students of engineering background is shown in Table 4.4.

TABLE 4.4 Dataset of Optional Courses

Student ID	Courses Opted(X)	Preferred Course(Y)
1	Core Java, Advance Java	Android Computing
2	Core Java, Advance Java, HTML	Android Computing
3	HTML, JavaScript	PHP
4	C, C++	DotNet
5	Advance Java, Core Java	Android Computing
6	C, C++	Core Java
7	C, C++, C#.Net	Core Java
8	C, C++	Core Java
9	Web Programming	PHP
10	HTML, Web Programming	PHP
------	------	------
n	n	n

In ARM, support and confidence are the two values obtained. The strong rules are extracted from the dataset if they satisfy minimum support and minimum confidence values. The best rules extracted using Apriori and Predictive Apriori Algorithm are shown in Tables 4.5 and 4.6.

4.6.5 PREDICTING TEACHING PEDAGOGIES USING K-NEAREST NEIGHBOR TECHNIQUE

The *K*-nearest neighbor technique is the supervised data mining technique. In this learning technique, the classification of data is done from the majority of the nearest neighbor of each point. The distance measures used for calculation are (a) Euclidean, (b) Minkowski, and (c) Chebychev distance measures. The sample dataset shown in Table 4.7 consists of a group of students categorized on the parameters as follows: (a) programming skills, (b) aptitude, (c) technical and logical skills, and (d) practical skills.

TABLE 4.5 Best Rules Found by Apriori

Rules
Courses Opted(X)=Core Java, Advance Java 1 ==> Preferred Course(Y)=Android Computing 1 conf:(1)
Courses Opted(X)=Core Java, Advance Java, HTML 1==> Preferred Course(Y)=Android Computing 1 conf:(1)
CoursesOpted(X)=HTML, JavaScript 1 ==> Preferred Course(Y)=PHP 1 conf:(1)
Preferred Course(Y) = DotNet 1 ==> Courses Opted(X)=C,C++ 1 conf:(1)
CoursesOpted(X)=Advance Java, Core Java 1 ==> Preferred Course(Y)=Android Computing 1 conf:(1)
CoursesOpted (X)=C,C++,DotNet 1 ==> Preferred Course(Y)=Core Java 1 conf:(1)
CoursesOpted(X)=Web Programming 1 ==> Preferred Course(Y)=PHP 1 conf:(1)
CoursesOpted(X)=HTML, Web Programming 1 ==> Preferred Courses(Y)=PHP 1 conf:(1)
Preferred Course(Y)=C++ 1 ==> Courses Opted(X)=C 1 conf:(1)
Courses Opted(X)=C 1 ==> Preferred Courses(Y)=C++ 1 conf:(1)

TABLE 4.6 Best Rules Discovered by Predictive Apriori

Rules
Courses Opted(X)=C,C++ 3 ==> Preferred Course(Y)=Core Java 2 acc:(0.6787)
Preferred Course(Y)=Core Java 3 ==> Courses Opted(X)=C,C++ 2 acc:(0.6787)
Courses Opted(X)=Core Java, Advance Java 3 ==> Preferred Course(Y)=Android Computing 3 acc:(0.6787)

The results are predicted for the new group of students in Table 4.9 after calculating Euclidean distance, as shown in Table 4.8. The nearest neighbors using the Euclidean distance metric are shown in Figure 4.4.

TABLE 4.7 Dataset Information

Group ID	Programming Skills(10)	Aptitude(10)	Technical and Logical Skills(10)	Practical Skills(10)	Grade
G1	8.5	9	9	8.5	A
G2	8	8.5	7.5	7	A
G3	9.5	4.5	8	8.5	A
G4	6.5	5	6	6	B
G5	6	6.5	7	5.5	B
G6	4.5	2.5	3	4.5	C
-----	-----	-----	-----	-----	-----
Gn	3.5	5.5	4.5	5	C

TABLE 4.8 Nearest Neighbor Distance Calculation

Group ID	Programming Skills	Aptitude	Technical and Logical Skills	Practical Skills	Grade	Euclidean Distance	Position
G1	8.5	9	9	8.5	A	2.783	Second
G2	8	8.5	7.5	7	A	3.5	Third
G3	9.5	4.5	8	8.5	A	2.345	First
G4	6.5	5	6	6	B	5.196	Fourth
G5	6	6.5	7	5.5	B	5.220	Fifth
G6	4.5	2.5	3	4.5	C	9.565	Seventh
------	------	------	------	------	------	------	------
Gn	3.5	5.5	4.5	5	C	8.215	Sixth

TABLE 4.9 Target Class Featured for the New Group

Group ID	Programming Skills	Aptitude	Technical and Logical Skills	Practical Skills	Grade
G 7	9	6.5	8.5	9.5	A

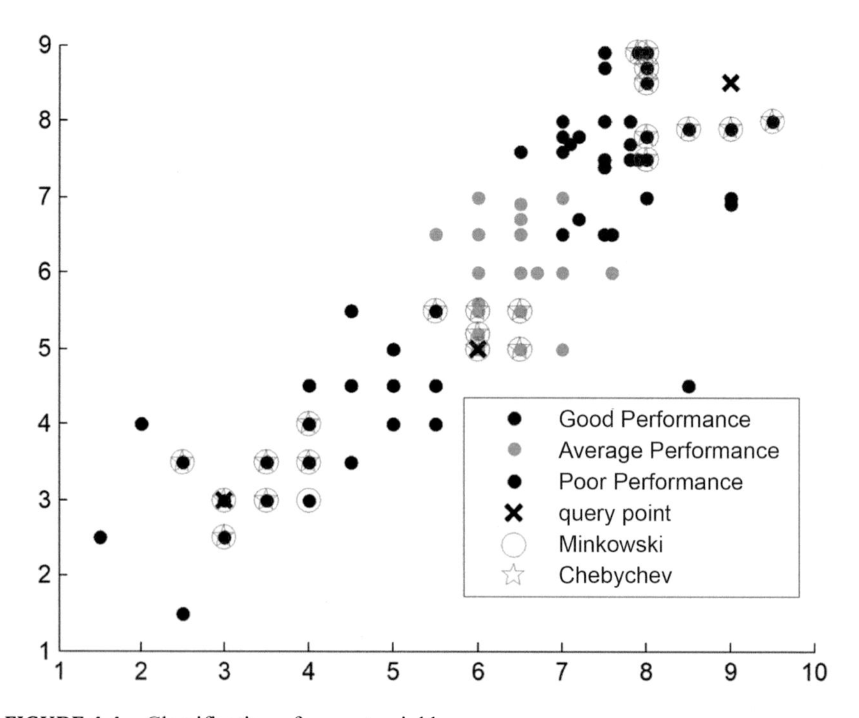

FIGURE 4.4 Classification of nearest neighbors.

4.6.6 PREDICTIONS USING SVMS AND BAYESIAN CLASSIFICATION

The techniques such as SVM and Naïve Bayesian play an important role in finding new educational traits. The SVM is a supervised data mining technique, which finds the nearest data vectors, called support vectors nearest to the hyperplane constructed by SVM, whereas Bayesian classification is also supervised learning, which finds the probability of events occurred. Both of these techniques help in improving educational trends and effective decision-making. Using both these techniques, students' placement results can be easily predicted.

4.7 EDUCATIONAL DATA MINING USING PYTHON LANGUAGE

In educational field, machine learning plays an important role in achieving the following: (a) helps faculty to improve their lecture plans by determining where cluster of students are struggling, (b) predict student's conduct, (c) helps in developing systems which can provide regular feedback to teachers, students, and parents about how the students learn, the learning pace of student, the support required to achieve learning outcome, (d) grade assessments of student's performance, (e) customize the learning schedule for each student in the classroom, (f) additional assessment to students, (g) smarter learning aid for each student, (h) organize content effectively for identifying weak students, (i) improve retention by identifying "at risk" students, (j) prepare schedules for students and teachers as per their needs, time, and availability, (k) adaptive and interactive tests that help students master each chapter, (l) personalized and experimental learning, (m) programs tailored to every student learning speed and need, (n) visualization of topics help students retain difficult concepts and terminologies, (o) interactive and engaging learning modules; the visually rich content enables conceptual clarity and lifelong term retention, and (p) strategies for seamlessly integrating technology into class.

4.7.1 MACHINE LEARNING FIELDS

Python programming language is one of the prominent fields of machine learning. It includes libraries such as PyTorch for popular deep learning frameworks. There are six classes in PyTorch that can be used for natural language processing (NLP)-related tasks using recurrent layers.

With NLP, machines are now capable of recognizing and understanding language just like humans. Examples are (a) chatting via text and (b) semantic search. Artificial intelligence and machine learning can transform education. Machine learning helps (a) teachers to focus on every student during the course teaching, (b) education for the specially abled students is possible through machine learning, and (c) grading assessment of the students. Machine learning helps us find patterns in data, and from that, predictions about new data points are made. To get those predictions right, we must construct the data correctly. There are some of the commonly used libraries in Python shown in Table 4.10 for machine learning.

TABLE 4.10 Commonly Used Libraries for Data Science

Library	Utility
Pandas	Library for data analysis
Statsmodels	It is the library for statistical modeling
scikit-learn	It is the library for classification and clustering using data mining techniques
NumPy	It is the library for Numerical routines
SciPy	These libraries are used for scientific and technical computing
matplotlib	It is the plotting library for NumPy
NLTK	It is the commonly used library for Statistical Natural Language Processing
Keras	It is Neural Network and Deep Learning Library
TensorFlow	It is the primary tool for deep learning analytics and an open-source AI Library

4.7.2 DATA DESCRIPTION

The predefined libraries of Python Machine Learning language are implemented on the dataset described in Table 4.11.

In the dataset, numerical grading is done on the internal assessment marks given by faculty members for three different subjects. The dataset is synthesized for experimentation purpose. The subjects for which numerical grading is done are (a) PC Packages, (b) Microsoft Access, and (c) VB.NET.

The input data obtained using predefined data mining libraries in Python have been shown in Table 4.12.

The classification of students' data is done on the following: (a) Marks obtained in Sessional, (b) Attendance, and (c) Assignment Submission.

TABLE 4.11 Dataset Description

Criteria	Internal Assessment			Remarks
Reg_id	651429			Random Enrollment No. of Student
Internal Marks Obtained	PC Packages (20) **17**	Microsoft Access (20) **13**	VB.NET (20) **14**	Marks obtained by students on the basis of Internal Assessment Criteria defined. Subjects: {PC Packages, Microsoft Access, VB.NET}
Sessional Marks (10)	8	4	6	Student marks in Sessional test
Attendance Marks (5)	4	4	4	Punctuality in class
Assignment Marks (5)	5	5	4	Assignment submitted by students
-----	-----	-----	-----	
N Students	N	N	N	

TABLE 4.12 Data Representation Using Python Libraries

Sessional Marks (10)			**Attendance Marks (5)**			**Assignment Marks (5)**			**Grading, i.e., Target Class (20)**		
	Predictor variables			Predictor variables			Predictor Variables		Dependent variable i.e. Target class		
PC Packages	MS Access	VB.NET	PC Packages	MS Access	VB.NET	PC Packages	MS Access	VB.NET	Target Class {Grading: "A" —Good "B" —Average "C" —Poor}		
8	6	6	5	4	4	5	3	3	A	C	C
7	8	6	3	3	4	5	5	5	B	B	B
6	6	3	4	5	3	3	3	3	C	B	C
7	8	9	4	4	4	3	2	5	B	B	A
5	9	8	2	5	5	3	4	4	C	A	A
	------			------			------			------	
N	N	N	N	N	N	N	N	N	N	N	N

4.7.3 RESULTS AND DISCUSSIONS

Python generates the following plots for three different subjects: (a) univariate plots, (b) multivariate plots, and (c) histogram plots. The input dataset in Table 4.11 is processed using python machine learning. These plots help understand the attributes of relationship and distribution. The results in the form of plots are displayed in Figures 4.5–4.7. The results have shown the following: (a) student's maximum rate of interest among three subjects, (b) understanding level of the subject, (c) assignment performance, and (d) attendance ratio of students among three subjects.

4.8 CONCLUSION AND FUTURE SCOPE

In this chapter, machine learning and its emerging roles in educational data mining are discussed. In this sequence, the predefined libraries of Python machine learning language commonly used in data science in the present scenario for effective decision-making and intelligent learning analytics are discussed. With these predefined libraries, implementation of data mining techniques and deep learning concept of neural network in Python can be used with more accuracy and effectiveness.

Furthermore, in this direction, intelligent learning analytics over educational domain has been implemented to find meaningful information in the form of knowledge from voluminous data. The classification and clustering techniques discussed in this chapter are as follows: (a) decision trees, (c) naïve Bayesian, (d) neural networks, (e) nearest neighbor, (f) *K*-means, and (g) ARM. Using these techniques, the predicted results obtained are as follows: (a) cluster students having shortfall of attendance, (b) identifying students who have performed poorly in sessional, (c) prediction of probability of students placements, (d) guiding students to choose appropriate courses to undergo industrial training, (e) discovering new pedagogical practices of teaching, (f) improving student skills and increasing employment chances for them, (g) strengthening decision-making of institution to include industry-oriented courses into curricula, (h) guiding the students to explore other skills such as aptitude, reasoning, and communication apart from regular studies to increase placement opportunities, and (i) strengthening of in-house training activities of faculty members.

Another focus of this chapter is that using Python, multiple classifiers have been implemented on an educational dataset to obtain the results. With the implementation of predefined Python libraries for data mining, the results

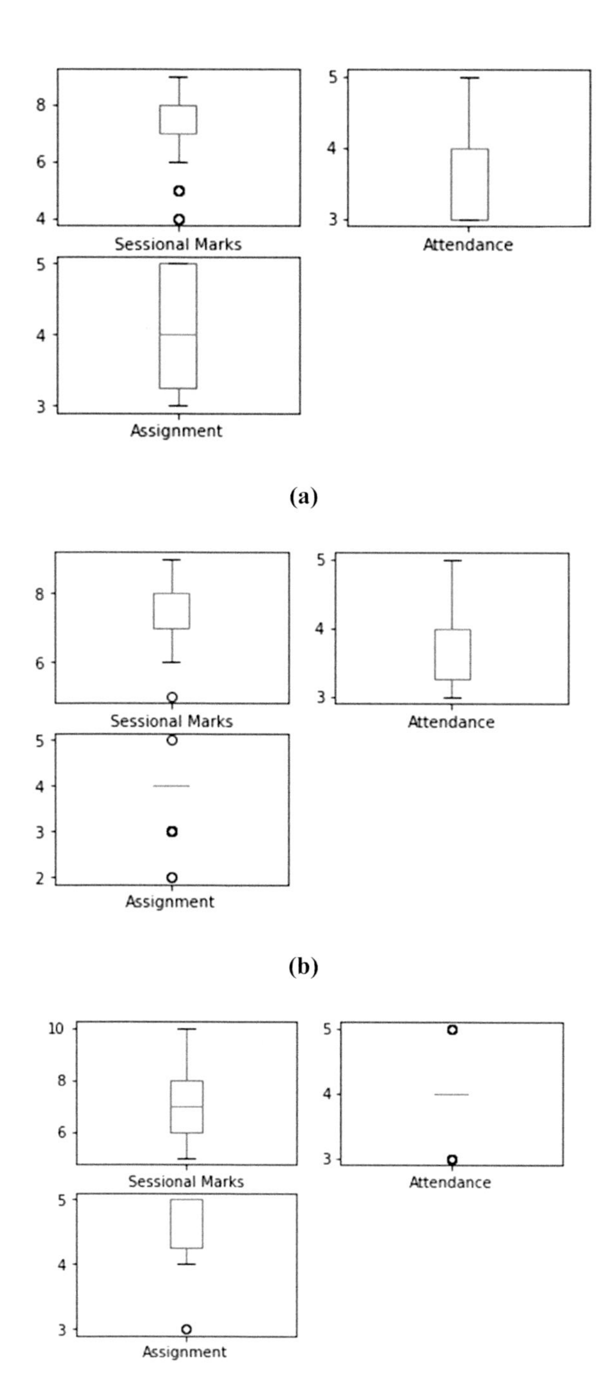

FIGURE 4.5 Univariate plots of attributes for the courses. (a) PC Packages. (b) Microsoft Access. (c) VB.Net.

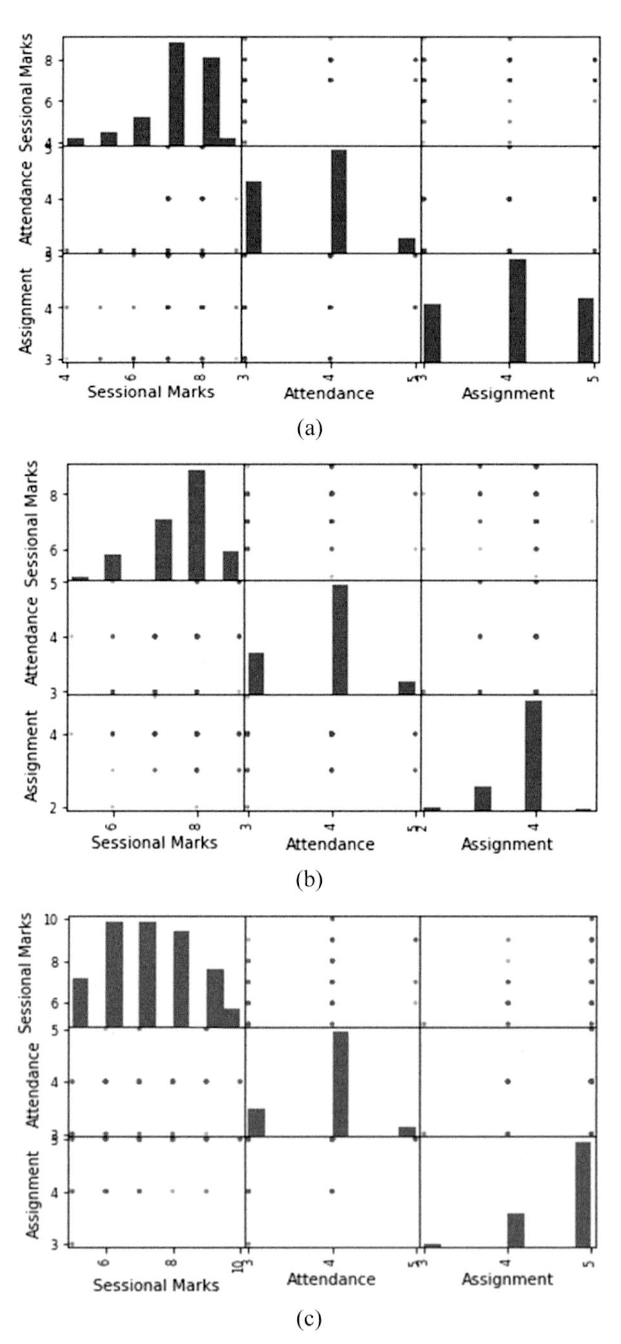

FIGURE 4.6 Multivariate plots for the courses. (a) PC Packages. (b) Microsoft Access. (c) VB.Net.

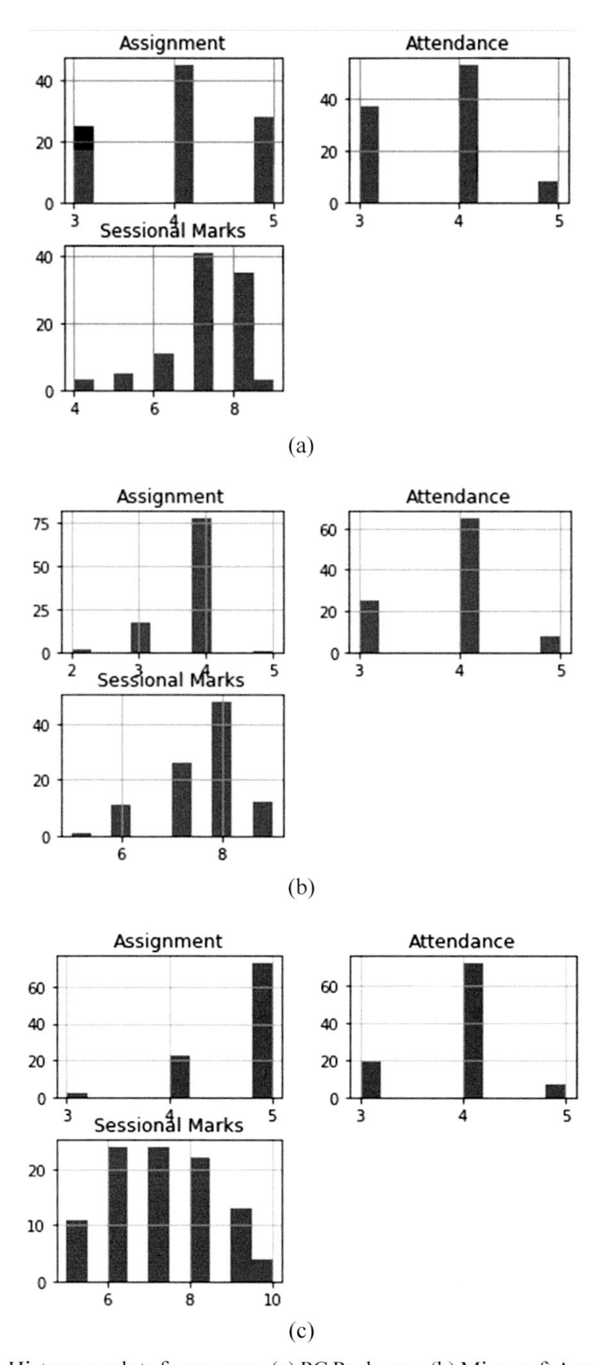

FIGURE 4.7 Histogram plots for courses. (a) PC Packages. (b) Microsoft Access. (c) VB.Net.

have been obtained, for example, the maximum interests of students for the course among other courses of the curriculum have been predicted.

The results obtained using different data mining techniques and machine learning in the proposed chapter help in efficient analysis and prediction of new academic trends for the betterment of students. In future, research can be used as follows: (a) admission seekers to analyze and choose the most suitable courses, (b) generate skill-oriented manpower, (c) fulfilling the skill gap of industry and academia, (d) develop educational software and visualization techniques for prediction of student's performance in examinations, (e) cluster those students who need exclusive focus in course, and (f) best utilization of unutilized and unstructured data to get the meaningful results.

KEYWORDS

- **cognitive**
- **education**
- **information**
- **intelligence**
- **supervised**
- **unsupervised**

REFERENCES

1. B. Bakhshinategh, O. R. Zaiane, S. Elatia, and D. Ipperciel, "Educational data mining applications and tasks: A survey of the last 10 years," *Educ. Inf. Technol.*, vol. 23, no. 1, pp. 537–553, 2018.
2. P. Dillenbourg, "Collaborative learning: Cognitive and computational approaches," in *Advances in Learning and Instruction Series.* Amsterdam, The Netherlands: Elsevier Science, 1999.
3. S. Slater, S. Joksimovic, V. Kovanovic, R. S. Baker, and D. Gasevic, "Tools for educational data mining: A review," *J. Educ. Behav. Statist.*, vol. 42, no. 1, pp. 85–106, 2017.
4. Y. Zhang, S. Oussena, T. Clark, H. Kim, "Use data mining to improve student retention in higher education—A case study," in *Proc. 12th Int. Conf. Enterprise Inf. Syst.*, 2010, vol. 1, pp. 190–197.
5. I. Lykourentzou, I. Giannoukos, V. Nikolopoulos, G. Mpardis, and V. Loumos, "Dropout prediction in e-learning courses through the combination of machine learning techniques," *Comput. Educ.*, vol. 53, no. 3, 950–965, 2009.

6. S. B. Kotsiantis, C. J. Pierrakeas, and P. E. Pintelas, "Preventing student dropout in distance learning using machine learning techniques," in *Proc. Int. Conf. Knowl.-Based Intell. Inf. Eng. Syst.*, 2003, pp. 267–274.
7. X. Zhu, "Machine teaching: An inverse problem to machine learning and an approach towards optimal education," in *Proc. 29th AAAI Conf. Artif. Intell.*, 2015.
8. J. Liebowitz, "Thoughts on recent trends and future research perspectives in big data and analytics in higher education," in *Big Data and Learning Analytics in Higher Education.* Cham, Switzerland: Springer, 2017, pp. 7–17.
9. S. B. Aher and L. M. R. J. Lobo, "Combination of machine learning algorithms for recommendation of courses in E-Learning System based on historical data," *Knowl.-Based Syst.*, vol. 51, pp. 1–14, 2013.
10. J. E. Beck and B. P. Woolf, "High-level student modeling with machine learning," in *Proc. Int. Conf. Intell. Tutoring Syst.* 2000, pp. 584–593.
11. C. Dede, "The evolution of distance education: Emerging technologies and distributed learning," *Amer. J. Distance Educ.*, vol. 10, no. 2, pp. 4–36, 1996.
12. J. Demšar, B. Zupan, G. Leban, and T. Curk, "Orange: From experimental machine learning to interactive data mining," in *Proc. Eur. Conf. Principles Data Mining Knowl. Discov.*, 2004, pp. 537–539.
13. T. M. Mitchell, "Machine learning and data mining," *Commun. ACM*, vol. 42, no. 11, pp. 30–36, 1999.
14. R. S. Baker and P. S. Inventado, "Educational data mining and learning analytics," in *Learning Analytics.* New York, NY, USA: Springer, 2014, pp. 61–75.
15. C. Burgos, M. L. Campanario, D. de la Pena, J. A. Lara, D. Lizcano, M. A. Martínez, "Data mining for modeling students' performance: A tutoring action plan to prevent academic dropout," *Comput. Elect. Eng.*, vol. 66, 541–556, 2018.
16. J. Naren, "Application of data mining in educational database for predicting behavioural patterns of the students," *Int. J. Comput. Sci. Inf. Technol.*, vol. 5, no. 3, pp. 4649–4652, 2014.
17. M. Abdous, H. Wu, and C.-J. Yen, "Using data mining for predicting relationships between online question theme and final grade," *J. Educ. Technol. Soc.*, vol. 15, no. 3, pp. 77–88, 2012.
18. R. S. Baker, "Educational data mining: An advance for intelligent systems in education," *IEEE Intell. Syst.*, vol. 29, no. 3, pp. 78–82, 2014.

APPENDIX

Apriori: It is an association rule mining algorithm to find the frequent item sets.

ARM: ARM means association rule mining. It is the data mining technique that shows the probability of relationships in the form of if–then statements among data items in a large dataset.

Bayesian classification: It is based on Bayes theorem and represents a supervised learning approach for classification of data. This technique follows the probabilistic approach.

Confusion matrix: In machine learning, a confusion matrix displays the classifier performance in a tabular format.

Decision tree: It is a data mining technique and displays the tree-like structure of decisions. This technique also follows the probabilistic approach.

J48: J48 is an algorithm for decision tree analysis and performs predictions.

***K*-means:** It is a clustering technique of data mining. *K*-means follows unsupervised learning approach, and it is used to find groups in the data.

Neural network: It is a computer system based on nervous system just like in human brains. It is the learning algorithm and can be supervised or unsupervised.

NLP: NLP means natural language processing. It is a field of artificial intelligence, where a large amount of natural language data is processed and analyzed.

Supervised learning: It is the learning approach where learning is performed using given training data, and there is response, that is, decision variable. The classification is an example of it.

Support vector machines: It is a supervised learning algorithm of data mining. In SVMs, data analysis is performed for classifications and regression analysis. In SVM, there is a separating hyperplane, which separates the hyperplane into two parts to lay class on either side of the hyperplane.

Unsupervised learning: It is the learning approach where learning is performed without past training data and no decision variable. Clustering is an example of unsupervised learning. In this learning approach, the data given are not labeled, that is, only input variables are given with no output variables.

CHAPTER 5

ApnaDermato: Human Skin Disease Finder Using Machine Learning

TARUN METHWANI*, GAUTAM MENGHANI, NITIN TEJUJA, HEMDEV KARAN, and RAGHANI MADHU

Department of Computer Engineering, Vivekanand Education Society's Institute of Technology, Chembur, Mumbai, Maharashtra 400074, India

Corresponding author. E-mail: 2016.tarun.methwani@ves.ac.in

ABSTRACT

Anyone may get skin problems at any place due to numerous reasons. Types of problems can be cracked skin, dry skin, extra growth, coloring due to infection, etc. Skin problem is faced by more than 20 million patients causing more than 10 million deaths on the globe. In this chapter, image of the affected area of skin is taken and name of the disease is predicted using a classification model which is used to detect the type of skin disease. Transfer learning model which is a research problem in machine learning is used to observe if the transfer model is useful for correct classification of skin diseases which needs urgent treatment especially for infants. In this work, MobileNet architecture is used which was trained over 1000 images. In the end, the network was tested with 500 images and achieves an accuracy of 85%.

5.1 INTRODUCTION

Image processing and machine learning have received significant implications in various fields. Nowadays, skin cancer is a significant public health problem, which approaches with the same techniques used in the dermatology field. Dermatology is an essential field of medical science, where skin diseases are considered a common illness, affecting all walks of life.

Medical imaging centers are vital, where dermatology has an endless list of conditions under treatment. The baseline method is to detect early skin disease and keep track of any change in the affected area, such as swelling, color change, or growth. Generally, doctors examine patients directly; this is still the first resource used by specialists. However, real-time skin problem detection can make life easier.

5.1.1 MOTIVATION

Many of the skin disease cases lead to death just because the patients did not get the treatment done in the early stages of the disease. Patient's ignorance, unawareness, or unable to get the appropriate diagnosis tools can be considered the reasons for not curing the disease at initial stages. With time, growth of illness conquers the body's vitals and leads to death. There is a need to develop a system that allows users to diagnose skin diseases and get it cured quickly. This chapter presents a brief review of human skin disease detection using machine learning.

5.1.2 PROBLEM DEFINITION

Earlier, people used to travel long distances to meet doctors to get a remedy for any severe ailment. Still, people prefer to meet doctors face to face. With the advancement in technology and the busy schedule of people, it becomes difficult to take out time to visit doctors before the health issue becomes severe. If anything happens to infants, their parents visit the nearest clinic. But sometimes it is not possible to visit a clinic, so parents need to reach the doctor through some other mode of communication, that is, the Internet. Reachability of patient as early as possible is also a significant issue, which leads to many deaths. Online expert advice from various doctors can play a vital role to save one's life in critical situations. Therefore, there is a need to bridge the gap between doctor and a patient by providing a single platform on which patients can send their queries to doctors online and can get a diagnosis from expert doctors, which admittedly can be a lifesaver for a patient.

In this chapter, our focus is to extract various data from the input (infected skin image), and based on which, diagnosis will be provided. ApnaDermato is the name termed for this mobile application, where "apna" means "our" and "dermato" means "skin disease helper." This application is made to enable communication between people residing in remote areas and doctors working in cities. A person suffering from skin disease can click a picture of the infected area and post it on the app for diagnosis. On the doctor's side, the machine learning model will detect the disease and suggest it to the doctor. The doctor can overwrite the automated diagnosis in case the model's prediction is inaccurate.

The baseline steps in image processing are the following.

- *Acquisition:* This is the first step in digital image processing. It consists of preprocessing such as scaling down resolution so that processing gets faster.
- *Image enhancement:* This step is used to get details that are obscured or to highlight certain features in an image. Example: changing brightness and contrast, etc.
- *Image restoration:* This step is used to improve the appearance of the image. Mathematical or probabilistic models drive the process of restoration.
- *Color image processing:* This step includes color modeling and image processing of digital images.
- *Wavelets and multiresolution processing:* Wavelets are useful for representing images in various resolutions.
- *Image compression:* This step compresses the image size for processing.
- *Morphological processing:* This step is used to extract shape-related information from the image.
- *Segmentation procedure:* This step divides the input image into subparts which the algorithm computes separately.
- *Representation and description:* Transforming raw data into a form suitable for subsequent computer processing and choosing the illustration is a part of the solution.
- *Object detection:* This step assigns labels to input images that help to predict the output (Figure 5.1).

The MobileNet model is based on the "Inception V3 classifier," which consists of two parts: one part is for data extraction, and another part is for classification of the object. Convolutional neural network (CNN) is the base of the model that we have used. The pretrained model that we have used is based on the above phases. The CNN algorithm belongs to the category of deep learning. It takes in an image as input, assigns weights and biases to the objects in the image, and can differentiate among the detected objects.

In our system, we have used a technique called transfer learning. Transfer learning is a technique in which we can improve learning in a new task through the transfer of knowledge from a related task that has already been learned. The weights and biases of the existing system are reused to train the new set of systems to give the desired results. The model that we have used is the pretrained Inception v3 classifier from Google. This model consists of two parts:

- Feature extraction part with a CNN.
- Classification part (with fully connected, softmax layers).

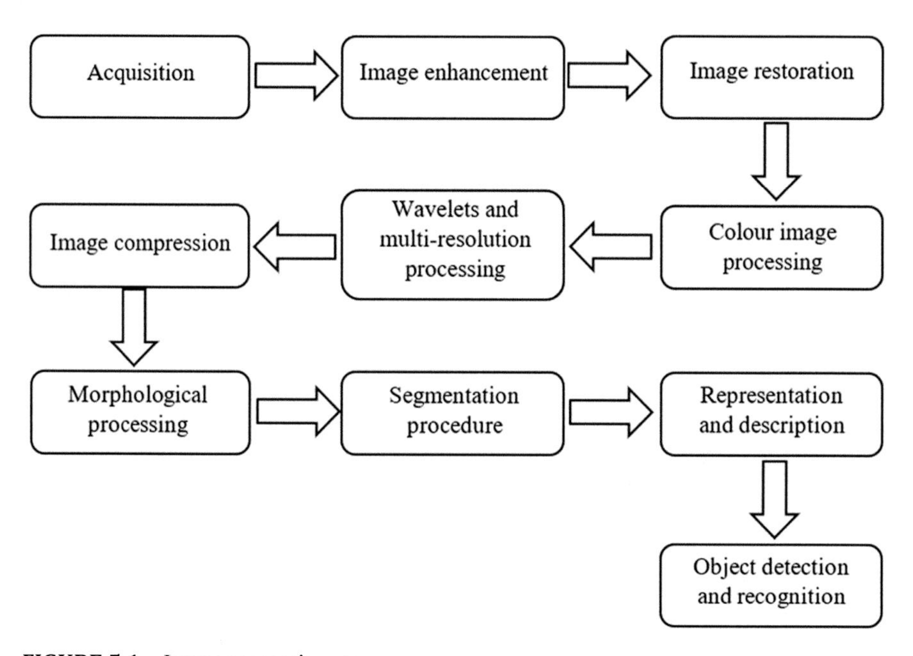

FIGURE 5.1 Image processing steps.

The pretrained Inception-V3 model achieves good accuracy while recognizing general objects with 1000 classes, such as "Zebra," "Dalmatian," "Dishwasher," etc. This model extracts general features from input images in the first phase and classifies them based on those features in the second phase. In transfer learning, when we build a new model to classify the original dataset, we reuse the feature extraction part of the model and retrain the classification part with our dataset. Since we do not have to train the feature extraction part (which happens to be the most complex part of the model), we can train the model with less computational resources and training time, which is the main advantage of using transfer learning.

During the process of transfer learning, three questions must be answered:

What to transfer?

This is the most crucial step in the whole process of transfer learning. When trying to answer this question, we try to identify which portion of knowledge has similarities with the source and target.

When to transfer?

There can be situations where transferring knowledge just for the sake of it may make matters worse than improving anything (negative transfer). We should aim at using transfer learning to improve target task performance and not degrade the performance. We need to be careful about when to transfer and when not to transfer.

How to transfer?

In this final step, we can proceed toward identifying ways of transferring knowledge across domains. This involves changes to existing algorithms and different techniques, as per requirement. In the final layer of the neural network, we can set the number of nodes accordingly based on out application domain.

5.2 LITERATURE SURVEY

Sinno and Qiang [1] have suggested the approach of transfer learning. Sometimes, we have to solve a classification problem in a particular domain, but we have an adequate amount of training data in another domain. In such cases, we can transfer the knowledge from one domain to another domain. This approach can accelerate development and reduce the effort required for creating a model from scratch. In many real-world applications, it is not feasible to collect the data needed for training. In such cases, knowledge transfer is highly advantageous. Sinno and Qiang [1] explained that they had given source domain Ds and learning task Ts, target Domain Dt and learning task Tt, transfer learning tries to improve the learning of the predictive function ft () in Dt using the feature in Ds and Ts.

The authors categorized transfer learning in three categories based on source and target domains and tasks.

- *Inductive transfer learning:* In this, the target task is different from the source task, and labeled data in the target domain are required to induce a predictive model ft ().

- *Transductive transfer learning:* In this, source and target domains are different, and feature space between the source and target tasks is the same, and the probability distribution of the feature space is different.
- *Unsupervised transfer learning:* This is used for dimensionality reduction and clustering. In this, there are no labeled data available in the source and target domains.

If there are no labeled data in the source domain, then it is self-taught learning, and if labeled data are available in the source domain, then it belongs to multitask learning where there are the source and target tasks. When a source domain data task contributes to reduced performance in the target domain, then negative transfer occurs. Clustering techniques are used to avoid the problem of negative transfer.

Guo et al. [2] proposed that the standard sigmoid needs to use some penalty factors to train a lot of close to 0 redundant data to produce sparse data; this is because the standard sigmoid output does not have sparsity. In addition, unsupervised pretraining is needed. The authors explain that the CNN includes convolutional layers, pooling layer, and fully connected layer.

The convolutional layer extracts the feature from the input given to the whole process. Input features are learned with kernel and passed into a nonlinear activation function.

The pooling layer solves the dimensionality reduction problem. Operations may be average and maximum depending on the feature space. Since the recognition rate is not high, it is a simple structure and involves less computational cost. The shallow network gives the best performance as compared to other networks. System tested on the Modified National Institute of Standards and Technology dataset, which consists of 6000 gray-scale images of handwritten digits from 0 to 9 for classification.

Nisha et al. [3] have proposed the development of an expert system for diagnosing children's skin diseases. This expert system is rule-based, and it uses a forward chaining technique to retrieve inferences from the information store. The rules apply IF–THEN structure, where the block contained in IF is related to information in the THEN block. The system also asks a set of questions regarding the disease symptoms and generates diagnosis accordingly. The algorithm converts each RGB image into color space. The *K*-means clustering algorithm receives as input the features extracted from the image. Finally, for recognizing the disease, neural networks are used.

Vinay et al. [4] explained the system into two stages. Stage 1 consists of computer vision in which the system uses computer vision for the identification of skin diseases based on the features extracted from the images using various image processing techniques. This stage comprises two substages. The first substage filters the input data and extracts the features required, and the second stage identifies the disease by using the maximum entropy model and artificial neural networks (ANNs).

For the identification of the disease, the system creates a model. They used the maximum entropy model in which features extracted were supplied with a maximum entropy model, which gave better results as compared to other models because of the number of feature functions and its pertinence to this problem. As per the principle of maximum entropy stated by Jaynes, in making inferences based on incomplete information, we must use that probability distribution, which maximizes entropy subject to whatever is known

$$H(P) = -\Sigma_a (p(x) \log (p(x)) \tag{5.1}$$

where $x = (a,b)\ a \in A,\ b \in B$ and $E = AXB$

$$v, f(\text{x,y}) = V_k\, f_k \tag{5.2}$$

where v is the model parameter. The conditional log-likelihood is used to train the model parameters as described. Theta, as described by Kevin and Noha [6], is the model parameter (vector)

$$\min_\theta \sum_{i=1}^{n} -\theta^T f\left(x^{(i)}, y^{(i)}\right) + \log \sum_y \exp\left(\theta^T f(x,y)\right). \tag{5.3}$$

The following equation computes the output by selecting the disease using the highest probability:

$$p(y = \text{Disease} \mid x = \text{Test Example} = \frac{\exp(vf(x,y))}{\sum_{y'} \exp(vf(x,y'))}. \tag{5.4}$$

For the classification part, the algorithms used were *K*-nearest neighbors and decision trees. The system extract features from the image; first feature extraction is an infected area or region of the skin disease. The system uses the Sobel operator for detecting the shape of the infected region using image segmentation algorithms, and another is the absence of bumps around hair follicles using an algorithm.

Features are considered and given a probability distribution based on max entropy.

The ANN contains two hidden layers and one input layer. The ANN uses the input layer for feature extraction, the first hidden layer comprises of sigmoid neurons, the second layer is tan function, and the output layer gives the probability distribution among diseases.

The system tested for six dermatological diseases. Due to problems of underfitting, the authors discussed that ANNs are not proper for that dataset. The use of ANN requires substantial dataset training and high computation cost. The system predicts the disease from extracting features of the image and uses various if conditions for classification. The system first converts the images for classification into gray-scale images, and some methods are applied to get the features of the infected region.

Pravin and Shirsat [5] have proposed a system for early detection of skin cancer, psoriasis, and dermatophytosis. The system uses a support vector machine with radial basis function kernel for image classification. The wavelet tool was used to remove the noise from the dataset. The system used a two-level classifier for classification. The first stage classifies the input as normal or abnormal. It also eliminates the noise. The second stage classifies the category of disease. For classification of skin disease, the system uses statistical analysis. Parameters of each image are calculated and diagnosed, therefore, to be classified. The system uses two-level classifiers; the first level classifies normal or abnormal, and the second level classifies the disease. At the first level, the system extracts the features after detecting and excluding the noise. The system can increase the number of statistical parameters for better performance. The database contains a total of 130 images of each disease. The system requires gray images; otherwise, it converts into a gray-scale image; then, the noise is removed using a median filter. Smoothening of an image is done using a median filter. The system calculates mean, standard deviation, and entropy for each image; features are extracted from images and fed into the classifier. The system uses various algorithms for improving the quality of the image. The system uses the AdaBoost classifier to correlate the mean, standard deviation, and entropy. After performing the above operations, the system will be able to classify some diseases with reasonable accuracy.

Kevin and Noha [6] used methods of task functions for training the model. The system uses a Softmax margin for the training data. The Softmax margin focuses on high-cost outputs functions. The Softmax machine offers a probabilistic interpretation of the training data. The Softmax margin is easy to implement and gives the best results compared to other models. Hyper-parameters are applied to data and used to label the data. Jensen risk bound performs similar to risk but takes less time to train. Jensen risk bound uses stochastic gradient ascent with a fixed step size.

Howard [7] has used the MobileNets model for mobile applications. The MobileNet structure contains depthwise separable filters. The system uses MobileNet models in the inception model for performing computation in layers. We can apply different techniques on the network structure, such as hashing, quantization, factoring, or compressing the networks. Depthwise separable convolution splits into two layers. The system uses the first layer for filtering. The purpose of other layer is to combine input into a new set of outputs. The system uses factored convolutions for reducing the computational cost. Depthwise separable layers handle downsampling. Before the last layer, pooling reduces the resolution to one. The system performs all computations into 1 1 convolutions called a pointwise convolution. For training the MobileNet model, the authors used Tensorflow. In addition, the system uses inception training for limiting the size of inputs, which improves performance. Since the application requires fast performance, the model should be properly compressed and perform computation faster. The system applied hyperparameters to the model for compressing and quantizing the model so that the authors can easily integrate it into mobile applications. A batch norm and rectifier Rectified Linear Unit (ReLU) precedes all layers in the MobileNet model. ReLU is an activation function used in deep learning models. The models that use ReLU are easier to train and often achieve better performance. The rectified linear activation function is a piecewise linear function that will output the input directly if it is positive, otherwise, it will output zero. For converting the model into a compressed model, a width multiplier is used. The width multiplier's role is to compress each layer uniformly. For reducing the overall computation cost of neural networks, the system applies a resolution multiplier to every input image and each layer. The MobileNet model gives the best performance as compared to several models. The MobileNet model tested on recognition of Stanford dog's dataset.

Sourav et al. [8] used pretrained image recognizers for the identification of disease. They used the transfer learning concept; features and classification parts are reused and retrained, respectively, with the dataset. In transfer learning, the last layer is retrained for the dataset so that we can use it in our application. The system uses the Inception V3 and Inception Resnet V2 networks for feature extraction. In addition, learning algorithms for the training data are used. The MobileNet model is lightweight and performs computation faster. Hence, it can efficiently work with mobile applications. Depthwise convolution is the base for MobileNet architecture. Features are extracted using CNNs, and the classification part is done using a fully connected layer. Pretrained model Inception V3 gives good accuracy while being able to recognize around 1000 classes. It extracts the feature from the image and then classified based on features. The learning algorithms can predict some diseases with good accuracy. Inception V3 gives better results as compared to Inception Resnet V2 and MobileNet.

Rahat et al. [9] used computer vision techniques for the identification of dermatological skin disease. They use a feedforward, backpropagation ANN for training and various image processing algorithms for feature extraction. Two types of features extracted are given as follows:

- Feature extracted from the image (color, area, and shape).
- Feature extracted from the user (elevation, feelings, gender, age, and liquid type).

The system uses an algorithm for finding out the color code of the infected area. In addition, the system uses a Sobel operator for detecting the edge of the infected area. The neural network consists of one input layer, one hidden layer, and an output layer. The input layer receives as input the features extracted from the image. These features are validated and tested using a 10-fold cross-validation process. The output layer gives the predicted disease. The system examines the human infected skin and detects the disease with reasonable accuracy. The system works on nine diseases.

Christian et al. [10] have different ways to use the convolutional networks for large-scale use with the goal of faster computation and factorized convolutions and aggressive regularization. The VGGNet model has the additional feature of architectural simplicity, but it comes at a relatively high cost; evaluating the network is computationally expensive. The computational cost of the inception classifier is much lower than VGGNet or other similar models. This lowered cost has made it possible to utilize the inception model in big-data scenarios, where the data are enormous. It performs computation faster and gives better efficiency in mobile devices. However, still, the inception architecture is complex, which makes it relatively difficult to make changes to the network.

General design principles are as follows.

- Voiding bottlenecks used for representation. Feedforward networks are represented by a graph, in which connections are not forming cycle. This principle defines a clear flow of information in the system. The amount of data passing through the partition between inputs and outputs can be accessed. One should avoid the bottlenecks with extreme compression for smooth functioning.
- It is easier to process higher dimensional representation locally. We can improve the performance by increasing the number of activations per tile in a convolutional network. The resulting networks will train faster than it did before.

- Spatial aggregation is feasible without loss in the representational power. For example, we can reduce the input dimension before the spatial aggregation without adverse effects. The authors have hypothesized that the reason for this is the strong correlation between adjacent unit results in less loss of information during dimensionality reduction.
- Balancing the parameters of the network. By balancing the number of filters in every stage and the depth of the network, we can optimize the performance. We can achieve a higher quality network by increasing both the width and the depth of the network. If both the width and the depth have increased in parallel, only then the optimal improvement for a constant amount of computation can be reached. Hence, the computational budget should, therefore, be equally distributed between both the depth and the width of the network.

The authors have proposed a technique to regularize the model's classifier layer by estimating the effect of label dropout during the training.

Let x be a sample training example. The following gives the probability of each label k:

$$k \in \{1, ..., k\} : p(k \mid x) = \frac{\exp(z_k)}{\sum_{t=1}^{k} \exp(Z_t)} \tag{5.5}$$

where Z_t are logits.

Cross-entropy is the loss for the example

$$l = -\sum_{k=1}^{k} \log \log\left(p(k)\right) q(k). \tag{5.6}$$

5.2.1 BUILDING BLOCKS IN THE CNN

Let $H \times W$ denote the spatial size of the output feature map, N the number of input channels, $K \times K$ the size of the convolutional kernel, and M the number of output channels; the computational cost of a standard convolution evaluates to $HWNK^2M$. The computational cost of the standard convolution depends on the following:

- the spatial size of the output feature map $H \times W$;
- size of convolution kernel K^2;
- numbers of input and output channels $N \times M$.

The system is required to calculate the computational cost mentioned above when it performs the convolution on both spatial and channel domains.

5.2.2 MOBILENET ARCHITECTURE

Let F be the input feature map, G be the output feature map, and K be the convolutional kernel. G is given by

$$G_{k,l,n} = \Sigma_{i,j,m} K_{i,j,m,n} \cdot F_{k+i-1,l+j-1,m}. \tag{5.7}$$

G is the output feature map for standard convolution while assuming stride one and padding.

Depthwise separable convolutions comprise two layers. Depthwise convolutions are used to apply a single filter per each input channel.

We can define depthwise convolution with one filter per input channel as

$$\hat{G}_{k,l,n} = \sum_{i,j,m} \hat{K}_{i,\ j,m,n} \cdot F_{k+i-1,l+j-1,m} \tag{5.8}$$

where F is the input feature map, $\hat{G}$ is the output feature map, and $\hat{K}$ is the convolutional kernel of size $D_k \times D_K \times M$ where the mth filter in convolutional kernel $\hat{K}$ is applied to the mth channel in input feature map F to produce the mth channel of the filtered output feature map $\hat{G}$.

The computational cost of a depthwise separable convolution is

$$D_K D_K \alpha M D_F D_F + \alpha M \alpha N D_F D_F \tag{5.9}$$

where α is the width multiplier and $\alpha \in [0,1]$. The width multiplier reduces computational cost and the number of parameters quadratically by around α^2 $D_K \times D_K$ is the kernel size, and $D_F \times D_F$ is the feature map size.

5.3 PROBLEM IDENTIFIED IN SKIN DISEASE DETECTION

This chapter understands problem definition, observing the possible solutions to the problem and formulating the solution to it. A research problem is a statement that specifies the area of concerns, a difficulty to be eliminated, or trouble with existing theory, or concept, a condition to be improved upon to a need for necessary investigation.

5.3.1 UNDERSTANDING THE PROBLEM

Infants are prone to skin diseases more often than adults because their skin is delicate, and their tissues can be easily ruptured. If one encounters with a skin disorder, immediate treatment is essential with proper diagnosis and medication, without which infant can die at early stages of skin disease due to a lower immune system. Remote areas where the treatment centers are not nearby to household regions hence become challenging to diagnose cases of skin disease and may lead to death due to lack of adequate treatment and faster transport. Even in the rural areas, although health centers are nearby, due to unawareness of the skin diseases, it is quite challenging to provide medication for it.

5.3.2 OBJECTIVES OF THE PROPOSED SYSTEM

- To develop a system that provides immediate or faster than existing medication.
- To build a platform that serves the patient and the doctor in their terms.
- To deploy a system with a user-friendly interface for ease of usage.
- To analyze the effect of a disease over a region of the map and visualize it for better actions to be taken.
- To provide the location of nearby hospitals, pharmacies for medicines.
- To establish a connection between a doctor and a patient for useful, healthy conversation without any leak of doctor's contact information.
- To make aware of the common solutions or home-made remedies for some trivial diseases at initial stages.
- To improve the experience level of beginner-level medical graduates.

5.3.3 FORMULATING THE SOLUTION

The solution for the problem mentioned is to build a platform for both doctors and patients, where patients can register themselves for getting a diagnosis from a doctor. It solves the problem of transportation and makes the immediate connection between the doctor and the patient. With new patients, the app is tracking diseases for further analysis by location. Later on, it can be visualized to cure the diseases of patients within the same local geographical regions. Vaccines can be provided based on geographic areas and can prevent a significant loss of humanity.

With the access of patient's location, we can provide them with their nearby location of hospitals and pharmaceutical stores, according to disease identified from image. In case the patient has additional queries, the system should provide the chat facility but ensuring the privacy of both a patient and a doctor. The system has an encrypted way to store details of patients and doctors. The patients do not need to discuss a small health issue with a doctor every time, and hence, the system itself can provide the solution in such cases. The solution is provided based on the classification of disease by the machine learning model.

This system benefits the patients to cure their diseases and also provide platform to medical graduates to provide medication to patients. Not only freshers but also higher level medical students can do practice in the system. It provides a summarized description of the spreading of certain diseases and their root cause of it from analysis of patient's locations. In addition, the system can recommend effective drugs immediately, which cures most of the patients. ApnaDermato consists of a centralized database for doctors and patients, which stores the entire information about patients and doctors, along with their queries. Few intelligent questions are asked to the patients to know how the condition of the patient is and accordingly gets treated by doctors. If there is no critical case of skin disorder, then home remedies are suggested to heal naturally. Doctors can register this system manually, which prevents the legitimate issues of medical science. With this feature, a patient can save a lot of time waiting to get the doctor's appointment and hence can save millions of lives.

5.4 COMPARATIVE STUDY WITH EXISTING SYSTEMS

This chapter gives a brief comparison between the proposed method and the existing systems. Existing systems along with their pros and cons are listed in the following.

5.4.1 FIRST DERM: ONLINE DERMATOLOGY

This system bridges the gap between an Internet search and an in-person dermatologist consultant.

Advantages:

- User-friendly graphical user interface.
- Able to get nearby clinics and pharmacies.

Disadvantages:

- Paid and nonrefundable.
- Provides only suggestions about the disease.
- Does not cure the diseases at first place.

5.4.2 PRACTO: DOCTORS, CONSULT ONLINE, ORDER MEDICINES

Practo is a fully featured app from doctors and consultants to deliver medicines; all functions are integrated into this system.

Advantages:

- Provides plenty of services.
- Provides chat with the doctor.

Disadvantages:

- Response time is huge.
- Rude doctors on chat.
- The support team is not professional.

5.4.3 mfine: CONSULT TOP DOCTORS FROM BEST HOSPITALS

The system provides seamless interaction between patients and doctors on chat and video call

Advantages:

- First-time free consultancy.

Disadvantages:

- Provides unmanaged reports.
- Multiple reschedule the appointments.
- Poor customer service.

5.4.4 DOCTOR ON DEMAND

The system provides services in 50 states and the District of Columbia, and it ensures 24 h of on-time service.

Advantages:

- User-friendly interface.

Disadvantages:

- Cannot detect a disease from an image.
- Nonprofessional advice from doctors on this platform.
- Poor response time (Table 5.1).

TABLE 5.1 Comparison with Existing Systems [3,4]

	Proposed System (ApnaDermato)	**Existing System(s)**
Emergency Cases	The app takes as input various medical conditions which patient is suffering through and based on those conditions as well as based on prediction our application had made it decides the case as an emergency case or a normal case	Existing applications on the play store does not support the consideration of emergency cases [1.4.1, 1.4.2, 1.4.3][a]
Morals And Ethics	On the start of the app, it shows that these are the morals and ethics which should be kept in mind before doing any steps which are supposed to be forbidden	Instructions for using the app are not mentioned in the existing systems [1.4.1, 1.4.3]
Personal Doctor Interaction	When identified as an emergency case, that case is forwarded to the doctor. Not only that when once a case is forwarded to the doctor, the doctor can accept that query and reply to it, else if the doctor rejects then that case is forwarded to another doctor, and even if the doctor accepts a case and then do not reply within a period of time, then that case is automatically forwarded to another doctor	Existing applications support the appointments fixing of the patients with the doctors, but it nowhere supports the online interaction with doctor personally
Nearby Hospitals, Pharmacies, Burn centers	When a medication is provided to the patient either by the doctor or some normal medicines by the application itself, our application suggests the location of nearby hospitals, pharmacies, burn centers which is very helpful for people living in villages or places (remote areas) where we cannot find the hospitals, pharmacies so easily	Existing applications show their own respective nearby clinic centers but do not show nearby hospitals, pharmacies, or burn centers [1.4.4]
Statistics	The application shows various statistics such as no of patients cured by that doctor or no of patients in a particular area cured by a particular medicine by doing analysis	No such statistics or analysis is provided in the existing applications [1.4.1,1.4.2, 1.4.3]

[a]Square brackets indicate figure numbers of the application user interface.

5.5 PROPOSED SOLUTION

5.5.1 METHODOLOGY

For the implementation of a pretrained model, that is, MobileNet, this chapter uses various TensorFlow libraries. This application will help the user to identify skin disease using affected skin images along with home remedies and instantly connect with teledermatology. In addition, it provides emergency and nearby pharmacy and hospitals. It involves two patient users and doctors. In skin disease classification, the prior data set is required to make the system learn about its features and characteristics. When the data are available, the system has to preprocess it and apply filters such as color transformation, rotation, etc. Figure 5.2 shows the complete flowchart of ApnaDermato.

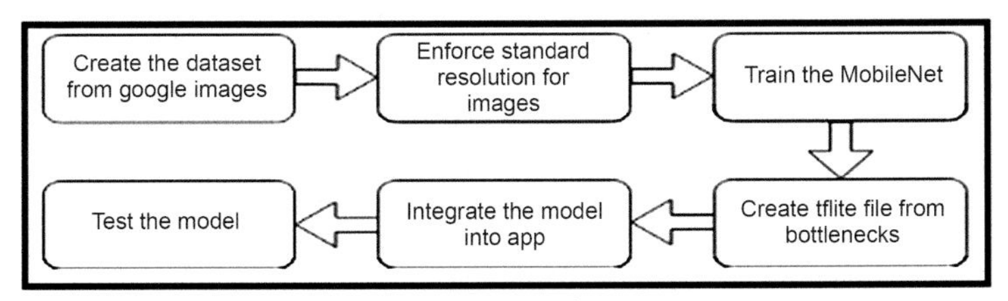

FIGURE 5.2 The complete flowchart of ApnaDermato.

5.5.1.1 CREATING THE DATASET

For this application, we downloaded images for each skin disease from Google images for training purposes. We also created a separate folder for each skin disease for classification. For each disease, we have maintained a dataset of 1000 training images.

5.5.1.2 ENFORCE STANDARD RESOLUTION

For ensuring consistency in the training process, we took all images in JPEG format and converted all images in 320 × 240 resolution.

5.5.1.3 TRAIN THE MODEL (MOBILENET)

The model used here is MobileNet. MobileNet is a small, efficient CNN, and it performs the same calculations at each location in the image. MobileNet takes input image in 128-, 160-, 192-, or 224-pixel resolution. We supply the infected skin images as input to the MobileNet model, and we will run the training using script retrain and pass parameters such as the number of training steps, output_graph, output_labels, and image_dir.

5.5.1.4 CREATING BOTTLENECKS

When retraining begins, bottlenecks are created similar to the screenshot given below. The penultimate layer of the inception model classifies the images supplied during training. This penultimate layer is called bottleneck. After the bottleneck training is over, the final layer generated gives the validation accuracy. In the end, we find two files, retrained_graph.pb and retrained_labels.txt, in the working directory. The file "retrained_graph.pb" is our final retrained model for our app that classifies diseases based on the image provided, and retrained_labels.txt is a text file that contains the labels, that is, the disease names from the dataset. For testing any input image, we use script label_image and pass the location of that image.

5.5.1.5 CONVERSION OF THE MODEL TO TFLITE FORMAT

TensorFlow Lite is TensorFlow lightweight solution for mobile and embedded devices. TensorFlow Lite comprises a runtime on which you can run pre-existing models and a suite of tools that you can use to prepare your models for use on mobile and embedded devices. Using the TFLite flat buffer script, we convert our model to the TFLite file for use on mobile.

In Figure 5.3, skin image will be provided to the MobileNet model as input, and output will be in the form of disease label with the corresponding probability.

5.5.1.6 TESTING THE MODEL

For testing the model, we created a separate test set that contained images for each skin disease. The model produced good accuracy, over 80% on every image. Figure 5.3 contains a sample test case wherein the input image was ringworm. This particular test case produced 98.8% accuracy. We integrated this model into an android app since it is convenient for various users. The model ranks the result set as per the accuracy calculated. It returns a list of top three predictions, as shown in the figure. In this mobile application, we capture the best result, query the name in the database, and display the appropriate information.

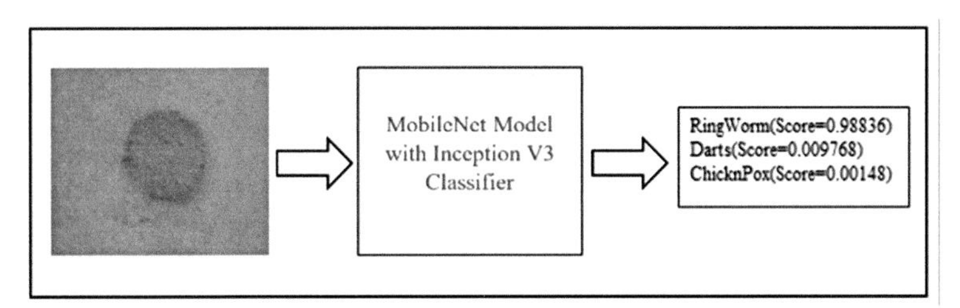

FIGURE 5.3 Diagram of MobileNet model.

5.5.2 APNADERMATO

ApnaDermato bridges the communication gap between the doctor and the patient. Earlier, patients used to wait in long queues or carry their child for skin treatment; now, using ApnaDermato, patients can get the diagnosis for skin diseases from specialized doctors across the world.

For using this android mobile application, first, the patient has to agree to the terms and conditions; as it is a health-related application, the patient has to accept the ethics and responsibilities of the app. Once the user accepted all the terms and conditions, the user has three options: patient login, doctor login, and patient register. In Figure 5.4, disclaimer is shown because this is health-centric application.

In Figure 5.5, registration form is shown for patient to fill. The patient has to register itself before using the application. Once a user registered with our application, the patient can log in into our app. After login, the patient will see a dashboard where the patient can either fill a diagnosis form or can find a nearby pharmacy store and hospital based on its current location and many other features. Apart from nearby pharmacies, patients can also search for nearby hospitals and burn hospitals.

Figure 5.6 shows login screen for once registered users and Figure 5.7 displays dashboard for patient. A doctor can only log in to the application. Doctors are registered manually, and after verification of the required documents, the doctor is allowed to log in into the system. After login, the doctor can view patient queries on which he can provide the diagnosis and send it back to the patient. Figure 5.8 shows dashboard for doctors.

If a parent or a guardian wants to get a diagnosis of his or her child's skin diseases, he or she has to fill the diagnosis form on the dashboard. The diagnosis form consists of some standard information such as an image of the

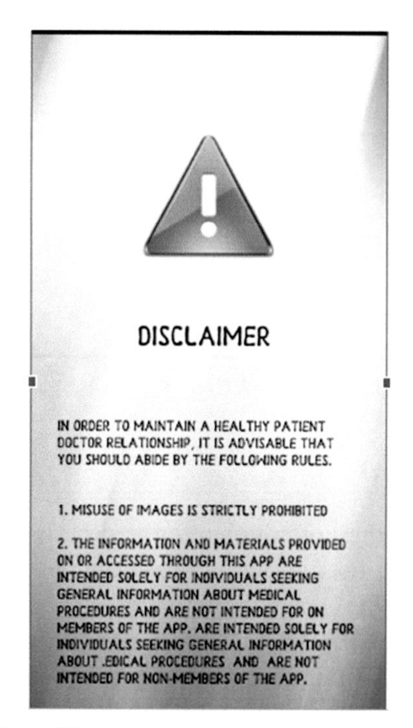

FIGURE 5.4 Terms and conditions.

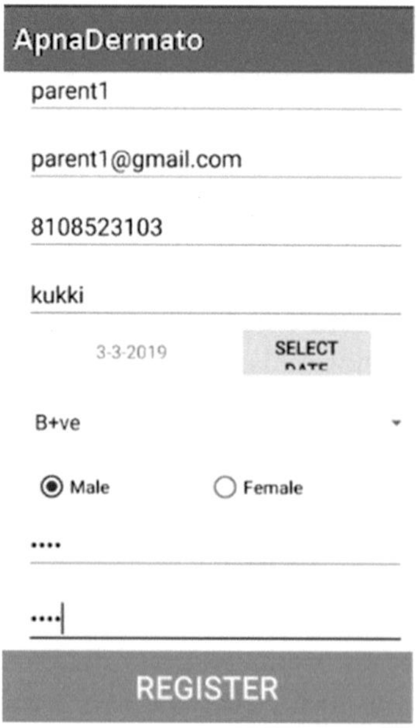

ApnaDermato
parent1
parent1@gmail.com
8108523103
kukki
3-3-2019 SELECT DATE
B+ve
Male Female
••••
••••
REGISTER

FIGURE 5.5 Registration form.

FIGURE 5.6 Home screen.

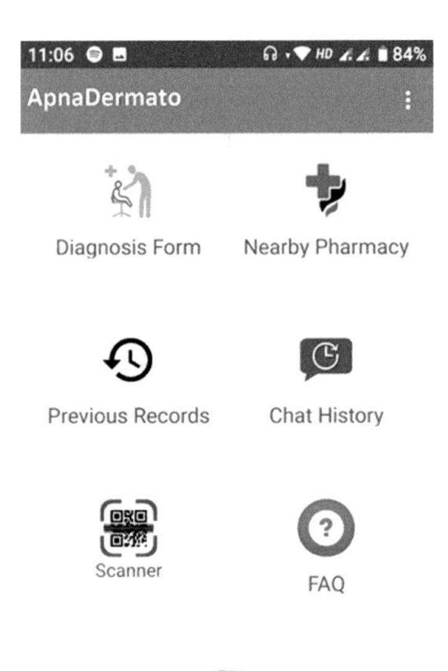

FIGURE 5.7 Patient login.

affected area, the name of a child, age, and a description of the disease. After filling the form, parent has to answer specific intelligent questions such as: Is the affected area bleeding? Is child crying? etc. Figure 5.9 reveals patient submitting query to doctor and Figure 5.10 illustrates intelligent questioning with patient to know more about their condition.

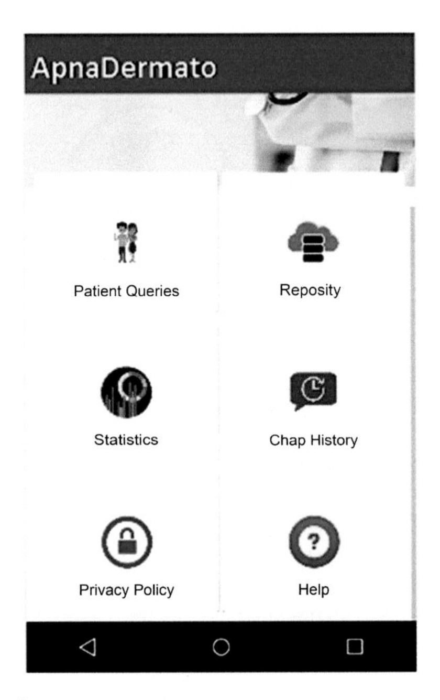

FIGURE 5.8 Doctor login.

Based on the diagnosis form and intelligent questions, this application can classify the condition of the child as urgent or not urgent. Every time, we cannot disturb a doctor for some common problems that can be cured by some home remedies. Our application ApnaDermato can identify five skin diseases trained on 1000 images and can give an accuracy of more than 85%.

The primary purpose of intelligent questions is to improve the accuracy of the system and adequately classify the condition. If the system identifies the disease as not urgent, then specific home remedies will be suggested to the patient, and after some time, a notification will be forwarded to the patient asking the condition of the child. If a parent says a child is fine, then the case will be stored in a database for future reference, but if the child's condition is serious, then the case is directly forwarded to the doctor. Figure 5.11 shows some remedies and FAQs and Figure 5.12 reveals reminder notification for patients.

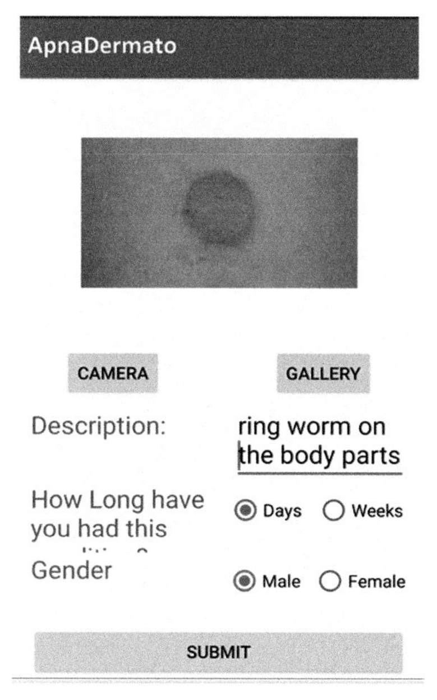

FIGURE 5.9 Patient query.

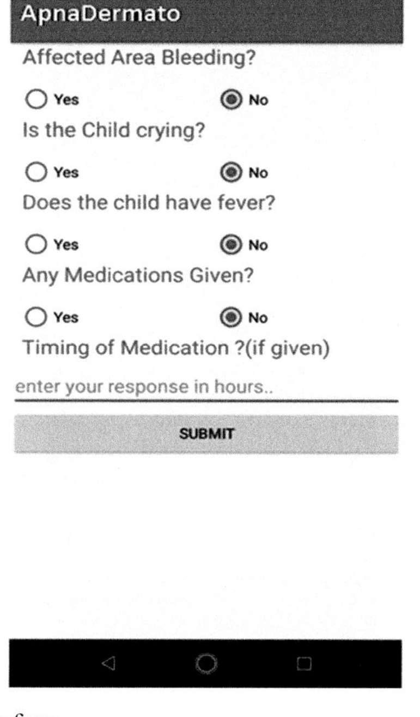

FIGURE 5.10 Diagnosis form.

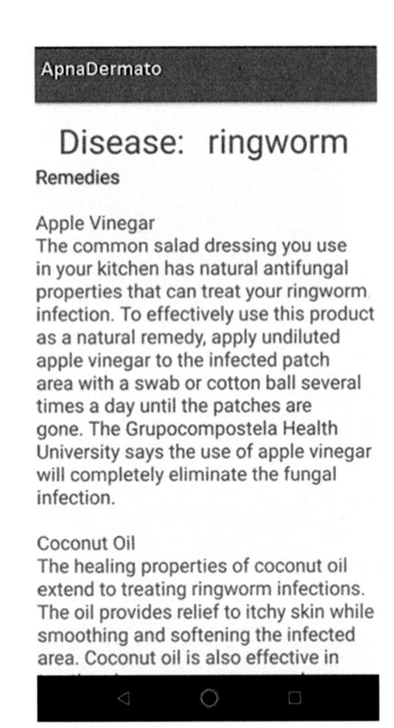

FIGURE 5.11 Remedies for nonurgent conditions.

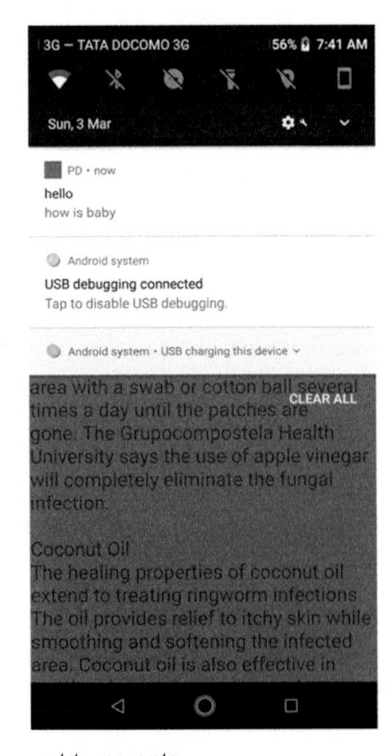

FIGURE 5.12 Notification sent to parents

If the system identifies the condition as urgent, then it will directly be forwarded to the doctor for diagnosis, and the patient will not be suggested home remedies. Figure 5.13 shows identifying emergency cases.

FIGURE 5.13 Urgent case condition.

A doctor can see all the patient queries and can accept or reject the case based on his or her availability. If the doctor accepted the requested, then all the information of the patient will be shown along with the predicted skin disease. If the doctor thinks the disease is the same as predicted, he or she can provide the diagnosis on it, else if the doctor thinks the predicted disease is not the same, then he or she can uncheck the check and provide diagnosis along with disease name. Figure 5.14 shows doctor receiving patient request and Figure 5.15 reveals doctor writing prescription to patients.

A patient can see the diagnosis in the previous record section on the dashboard where all the previous requested cases are stored. The patient can also search the nearby pharmacy and hospital based on the current location. A patient can also contact the development team for any technical help by clicking on the help tab on the dashboard. Figure 5.16 shows patients' records stored in application, Figure 5.17 showing nearby pharmacies to get medications, and Figure 5.18 reveals providing medications for common problems. Figure 5.19 displays giving patients an option to chat with doctor if they accepted their request.

FIGURE 5.14 Accept or reject.

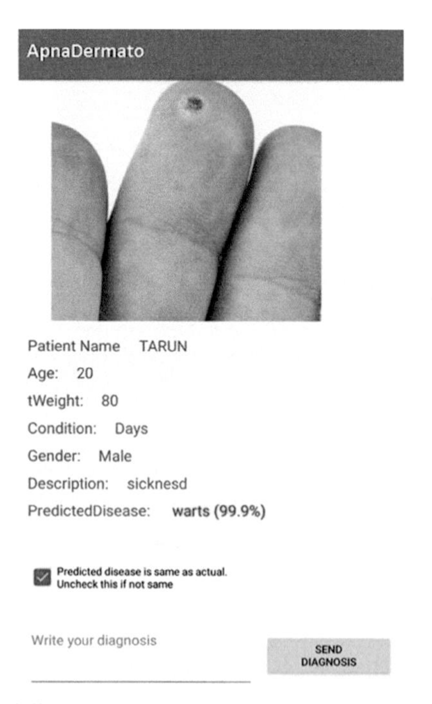

FIGURE 5.15 Predicted disease.

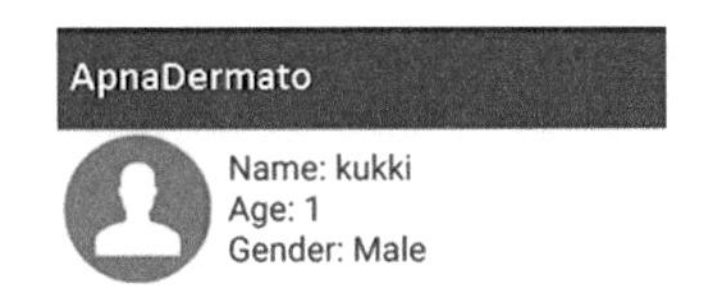

FIGURE 5.16 Nearby pharmacies.

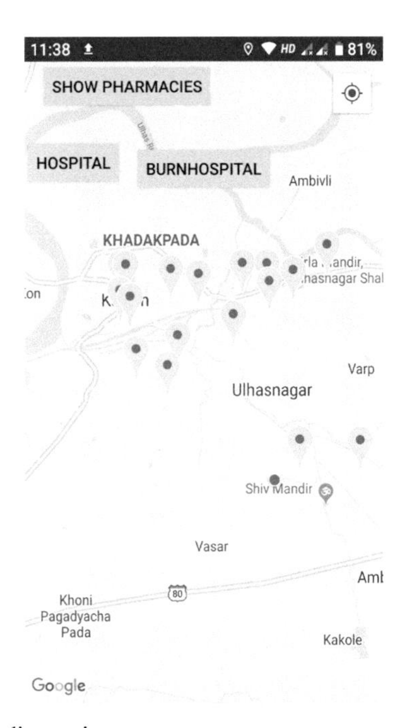

FIGURE 5.17 Previous diagnosis.

FIGURE 5.18 Common problems.

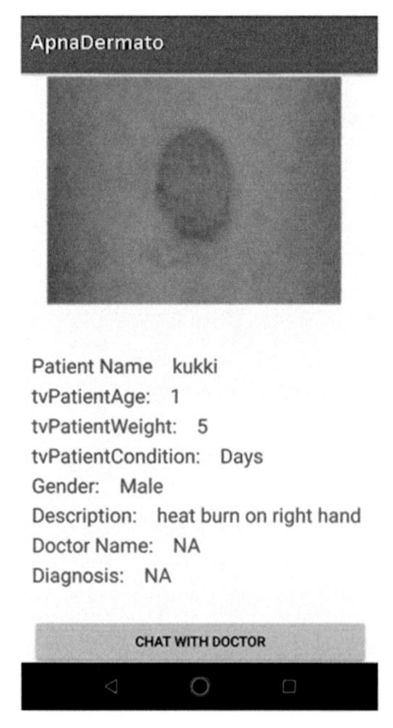

FIGURE 5.19 Diagnosis receive.

A patient can also ask diagnosis from his teledermatology by scanning the quick response (QR) code of the doctor. A doctor will have a QR code generated based on the information provided; this information will be encrypted using some encryption methods. A patient has to personally visit the doctor once for scanning the QR code. Once the code is scanned, the chat interface is set up between the doctor and the patient. Doctors' QR code is shown in Figure 5.20, which can be scanned and added by the patients and after scanned by the patient, doctor information is shown in patients' app in Figure 5.21. Figure 5.22 shows patient chat with doctor and Figure 5.23 showing that doctor can consult to patients on chat.

FIGURE 5.20 Doctors QR code.

A doctor can view the statistics for a particular disease at a particular location. These statistics will be helpful to know if there has been an epidemic in an area. The doctor also has all the history of the patient treated. Figure 5.24 reveals statistics can be given to doctor based on symptoms.

A patient can also see the home remedies and the do's and don'ts for some common skin disease problems.

5.5.3 RESULT ANALYSIS

There are five types of diseases considered, ringworm, chickenpox, fifth disease, warts, and contact dermatitis, as shown in Table 5.2.

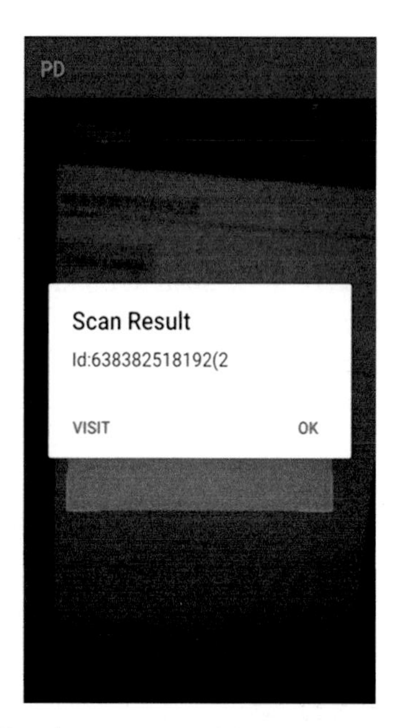

FIGURE 5.21 Scanned QR code.

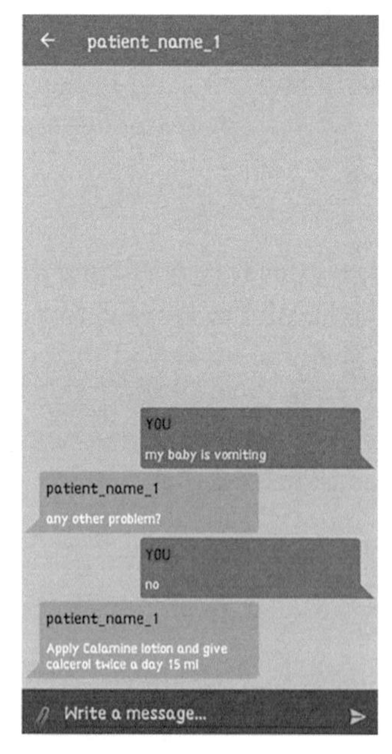

FIGURE 5.22 Chat between doctor and patient.

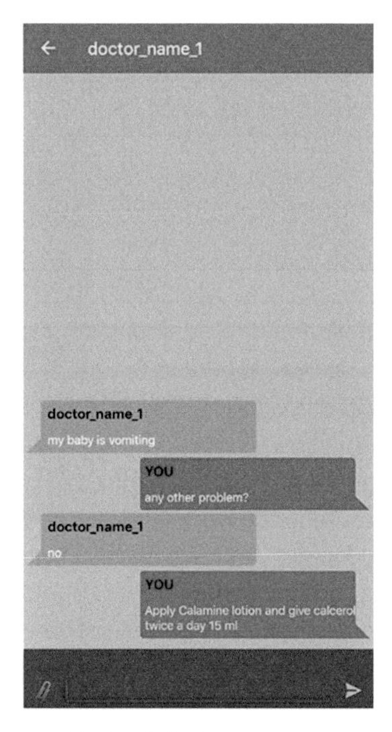

FIGURE 5.23 Chat between patient and doctor.

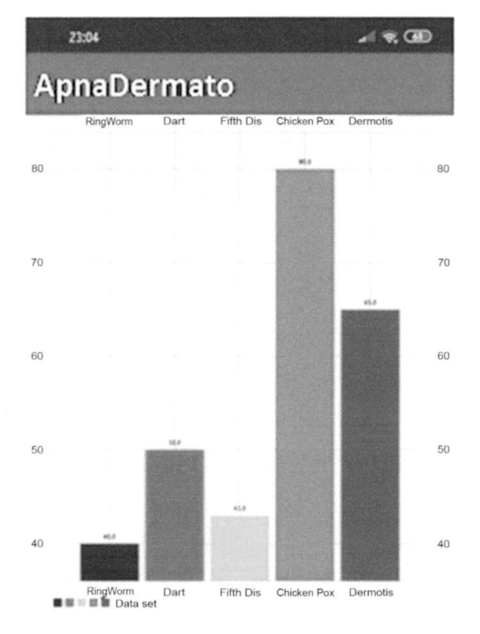

FIGURE 5.24 Diseases statistics.

TABLE 5.2 Training Accuracy

	Disease Name	Number of Training Images	Accuracy (%)
1.	Ringworm	1000	95
2.	Chickenpox	1000	93
3.	Fifth disease	1000	90
4.	Warts	1000	92
5.	Contact dermatitis	1000	89

Table 5.2 shows the accuracy achieved for each disease while training the dataset created from Google images. 3 × 3 depthwise separable convolutions are used by MobileNet, which uses around eight to nine times less computation than standard convolutions at only a minimal reduction inaccuracy. Considering the efficiency achieved in computation, this tradeoff with reduced accuracy is acceptable. The average accuracy for each disease is more than 85%.

Table 5.3 shows the accuracy achieved for each disease while testing the dataset created from Google images. The test set consisted of 150 images for each disease. The average accuracy obtained for each disease was more than 80%. Considering the computational limits of mobile devices, the performance of the MobileNet model is quite impressive. For mobile devices, a computationally inexpensive model is preferable. MobileNet satisfies these criteria in exchange for a small reduction of accuracy, which is tolerable.

TABLE 5.3 Testing Accuracy

Disease Name	Number of Testing Images	Number of Images Correctly Identified	Accuracy
Ringworm	150	129	86
Chickenpox	150	128	85
Fifth disease	150	126	84
Warts	150	123	82
Contact dermatitis	150	120	80

5.6 PROS AND CONS OF THE PROPOSED METHOD

Pros of the proposed method

- Increased access, particularly in areas where geographic barriers prevent or limit a person's ability to get dermatologic treatment.

- Increased convenience, particularly for patients who can avoid unnecessary visits.
- Reduced wait times for patients who need urgent consultation and for patients who need in-person visits.
- Improved scheduling, particularly in areas where dermatology services are in high demand.
- Lowered healthcare costs.
- Statistical data help to get the idea of epidemic conditions in an area.
- Recommendation of home remedies to common skin problems.
- Notification to doctors about emergency cases.
- The system can classify a case as urgent or not urgent based on a machine learning model and intelligent questions.
- Locating nearby pharmacies and hospitals.
- A personal chat with doctors.
- Statistics for research purpose.

Cons of the proposed method

- The application cannot detect the false information entered by the user.
- Image quality will be affected if the patient clicks the image with less megapixel.
- Blur image can affect the detection of disease, and the model may predict the wrong disease.
- The model cannot give 100% accuracy for any trained disease.
- The model cannot detect the disease in low-light conditions.

5.7 CONCLUSION

In this chapter, we have implemented an automatic classification method to find skin diseases and provided diagnosis. This chapter is capable of classifying five skin diseases in infants as well as adults. They are chickenpox, contact dermatitis, fifth disease, ringworm, and wart. This application gave us 85% accuracy having 5000 training images and 750 testing images. The app was able to classify 616 images out of 750 images correctly. The proposed system overcomes the drawback of existing applications, as mentioned in Chapter 4, with the remarkable idea of handling the less sensitive cases automatically without even disturbing the doctor for those. No other system can handle this scenario of less sensitive disease beforehand, but our system can. When the patient having a typical skin rash, then instead of approaching directly to the doctor, our system first provides the home remedies; then, on no sign of cure, the case will be sent to the doctor. This application can give statistics to the users about the overall diseases in a region for better visualization of the world's population. This application can help solve many daily issues of the common man. Even though the proposed method is not 100% effective and ready for public use, it is an effort to explore automation in healthcare for the benefit of society without any additional charges.

5.8 FUTURE SCOPE

This chapter has a vast scope in the future because it revolutionizes the medical field and the procedure doctors are following today to cure existing diseases.

- More than five skin diseases can be considered for training to detect more diseases.
- With due course of time, the system can be configured to learn from the diagnosis of the doctor and can have a detection rate of 100% and can be made available to the public. Such a system saves all the efforts humans can make to cure skin diseases in infants and eventually reduces the death rate of newly born children.
- Once this system reaches its saturation point, this application can extend to accommodate to detect other diseases that can be cured by visual diagnosis.
- This application can become the source of ranking for doctors who treated the patients in their past, which can help in achieving a higher level of expertise in their respective fields.

KEYWORDS

- **skin disease detection**
- **image processing**
- **transfer learning**
- **MobileNet**
- **ReLU**

REFERENCES

1. J. P. Sinno and Y. Qiang, "A survey on transfer learning," *IEEE Trans. Knowl. Data Eng.*, vol. 22, no. 10, pp. 1345–1359, 2010.
2. T. Guo, J. Dong, H. Li, and Y. Gao, "Simple convolutional neural network on image classification," in *Proc. IEEE 2nd Int. Conf. Big Data Anal.*, 2017, pp. 721–724
3. Y. Nisha, K. N. Virender, and S. Utpal, "Skin diseases detection models using image processing: A survey," *Int. J. Comput. Appl.*, vol. 137, pp. 34–39, 2016.
4. S. K.Vinay, S. K. Sujay, and S. Varun, "Dermatological disease detection using image processing and machine learning," in *Proc. 3rd Int. Conf. Artif. Intell. Pattern Recognit.*, 2016, pp. 88–93.
5. S. A. Pravin and A. S. Shirsat, "An image analysis system to detect skin diseases," *IOSR J. VLSI Signal Process.*, vol. 6, no. 5, pp. 17–25, 2016.
6. G. Kevin and S. Noha, "Softmax-margin CRFs: Training log-linear models with cost functions," in *Proc. North Am. Chapter Assoc. Comput. Linguistics Human Lang. Technol. Conf.*, 2010, pp. 733–736.
7. A. G. Howard, M. Zhu, B. Chen, D. Kalenichenko, W. Wang, T. Weyand, M. Andreetto, and H. Adam, "MobileNets: Efficient convolutional neural networks for mobile vision applications," *Comput. Vis. Pattern Recognit.*, pp. 1–9, 2017.
8. K. P. Sourav, S. S. Mansher, G. Yaagyanika, S. Bhairvi, and P. Muthu, "Automated skin disease identification using deep learning algorithm," *Biomed. Pharmacol. J.*, vol. 11, no. 3, pp. 1429–1436, 2018.
9. Y. Rahat, R. Ashiqur, and A. Nova, "Dermatological disease detection using image processing and artificial neural network," in *Proc. Int. Conf. Elect. Comput. Eng.*, 2014, pp. 687–690.
10. S. Christian, V. Vincent, I. Sergey, and S. Jonathon, "Rethinking the inception architecture for computer vision," *Comput. Vis. Pattern Recognit.*, vol. 3, pp. 1–10, 2015.

CHAPTER 6

Artificial Intelligence Technique for Predicting Type 2 Diabetes

RAMYASHREE* and P.S. VENUGOPALA

Department of Computer Science and Engineering,
N.M.A.M. Institute of Technology, Nitte, Karnataka 574110, India

**Corresponding author. E-mail: ramyashreebhat1994@gmail.com*

ABSTRACT

Diabetes is the most common disease experienced recently. Type 1 diabetes, Type 2 diabetes, and gestational diabetes are the most common types of diabetes. The aim of this chapter is to predict the Type 2 diabetes with various parameters. "Diabetes risk score or test system" is designed with the various risk factors such as age, waist circumference, physical activity, family history, and body mass index using an artificial intelligence technique. This chapter also aims to design a universally acceptable diabetes prediction system that predicts the possibility of diabetes risk. This process is carried out using various parameters of the patient's life style and without using the data from medical test results. The individuals who are interested to know about their risk score can use this diabetes risk score system.

6.1 INTRODUCTION

In the present scenario, diabetes is one of the common diseases. Type 1 diabetes occurs when pancreas does not produce insulin. This chapter gives information about Type 2 diabetes (T2D). In T2D, cells cannot utilize glucose proficiently for strength. This happens when the cells end up unfeeling to insulin and the glucose slowly gets excessively high. There are various reasons for causing T2D, which are being overweight, lack of physical movement, stress, genetics, and eating a great deal of sustenance's or beverages with sugar and straightforward starches. Risk factor may include history of family having the diabetes, being sedentary, being overweight, etc. The major symptoms of T2D are excess thirst, dark skin under armpits, chin, or groin, blurry vision, etc. [1].

The number of people having T2D is increasing. It is a vital factor for death. Several researchers have carried out the experiment on the T2D and proved that the prevention for this disease can be done by lifestyle modification [2].

The aim of this work is to develop the diabetes risk score system with the most used artificial intelligence (AI) methodologies. In the present scenario, AI can be applied in a variety of research areas because of various applications. In this chapter, we introduce the system that will predict the T2D based on different parameters, and it will also inform the patients about the most effected parameter for T2D based on the expert system [3]. The person who is interested to know about his T2D risk score can use this system.

6.2 LITERATURE SURVEY

The diabetes risk score system is developed by several researchers.

Mohan et al. [1] developed the Indian Diabetes Risk Score with the help of the Madras Diabetes Research Foundation (MDRF-IDRS) to help recognize the undiscovered T2D mellitus in that population. While developing the MDRF-IDRS, they took 26001 samples from 155 wards in Chennai Urban Rural Epidemiology Study. They built the system with the four simple parameters: age, waist circumference (WC), family history, and physical activity. They derived the score based on the logistic regression method and set the maximum score as 100. The IDRS score <30 is considered as low risk, 30–50 is considered as medium risk, and ≥ 60 is considered as high risk. To detect the optimum value (≥ 60), they used the receiver operating characteristic (ROC).

Lindström and Tuomilehto [2] developed the practical tool to predict the T2D risk in France. Here, 4500 samples were collected to develop this system considering various categorical variables such as age, gender, food, family history, waist, physical activity, blood pressure (BP), high blood glucose, and body mass index (BMI). Logistic regression was utilized to register β coefficients for known hazard factors for diabetes. β coefficients of the display were utilized to allot a score esteem for every factor, and the composite Diabetes Hazard Score was figured as the aggregate of those scores. While developing this tool at the end, the risk score ≤ 7 is considered as the low risk, 7–14 is considered as the moderate risk, 15–20 is considered as the high risk, and >20 is considered as the very high risk.

Katulanda et al. [3] developed the Diabetes Risk Score in Sri Lanka (SLDRISK). To develop the SLDRISK, 4276 samples were collected. Based on the variables such as age, family history, gender, WC, physical activity, and BP, the SLDRISK was developed. To identify the variables, univariate regression analysis is done. To derive the risk score, the β coefficient values are identified using the analysis called logistic regression. In this system for finding the optimal cutoff value, sensitivity, and specificity, ROC analysis is done. The authors also validate the SLDRISK with IDRS and Cambridge Risk Score (CRS). They concluded that, in the SLDRISK, sensitivity is 77. 9% and specificity is 65.6%, which are higher than those in the IDRS and CRS.

Al-Lawati and Tuomilehto [4] proposed the Diabetes Risk Score in Oman. They developed the diabetes risk score system for identifying the diabetes mellitus with 4881 samples. Here, the logistic regression method was used with different parameters such as age, gender, family history, WC, BP, BMI, and smoking. They have concluded that when the age, WC, and BMI increase, the probability of getting T2D is high. Age and family history are the strongest predictors, whereas BMI, BP, and WC are moderate parameters. The Oman risk score system is validated with the Nizwa survey, which contains same parameters, with 1432 samples, in which 145 had diabetes.

Griffin et al. [5] developed the diabetes risk score system for Cambridge. Collected the data of 1077 people of the range 40–64 years. Information was gone into a regression model. Here, specificity is 72%, sensitivity is 77%, and ROC is up to 80%.

Zhou et al. [6] developed the risk score system for T2D mellitus for the Chinese population. Here, they took 5453 samples. This system was developed based on the lifestyle and other factors such as gender, age, physical activity, family history, WC, history of dyslipidemia, diastolic BP, and BMI. They took the cutoff value as 17 with 67.9% sensitivity and 67.8% specificity. They validated the developed system with the American Diabetes Association Score (0.636), Inter99 Score (0.669), and Oman Score (0.675). It was concluded that using the area under the ROC curve (AUC), the Chinese Diabetes Risk Score is 0.723, which is higher compared to other systems.

Bang et al. [7] created and approved a patient self-evaluation diabetes screening score for US grown-ups. In National Health and Nutrition Examination Survey (in Atherosclerosis Risk in Communities/Cardiovascular Health Study, 30(40)% of people for diabetes screening yielded affectability of 79(72)%, specificity of 67(62)%, constructive prescient estimation of

10(10)%, and probability proportion constructive of 2.39 (1.89). Conversely, the examination scores yielded affectability of 44%–100%, particularity of 10%–73%, positive prescient estimation of 5%–8%, and probability proportion positive of 1.11–1.98. This new diabetes screening score, basic and effectively executed, appears to show enhancements upon the current strategies. Future examinations are expected to assess it in diverse populaces in certifiable settings.

Rigla et al. [8] proposed the AI technique that helps detect the diabetes. The AI technique is widely used in variety of applications. Based on the varieties of abilities such as learning and reasoning, it can be applied in predicting the diabetes risk score. Here, various AI techniques such as data mining, fuzzification, defuzzification, support vector machine (SVM), heuristic approach, hybrid systems, naive Bayes, supervised learning, and unsupervised learning are explained.

Chen et al. [9] developed the risk assessment tool for Australian called AUSDRISK. According to the survey, they told that by 2025, the people with diabetes will really reach 2 million. To prevent this, lifestyle should be improved. They took a sample of 6060 people from the five-year corresponding data. Based on this, they predicted and manipulated the risk score.

Unwin et al. [10] gave important messages about diabetes: diabetes is a colossal and developing issue, and the expenses to society are high and heightening; diabetes is a dismissed advancement issue, influencing all nations; there are financially savvy answers for switching the worldwide diabetes scourge; and diabetes is not just a medical problem, its causes are multisectoral, and it requires a multisectoral.

Observations: Diabetes risk score systems of different countries are studied and various parameters considered in these systems are observed as shown in Table 6.1.

6.3 OBJECTIVES

The motivation behind this work is to detect the T2D of individuals who are interested to know about their risk score. Therefore, the diabetes risk score system is designed without any laboratory tests. The objectives are as follows.

- Different diabetes risk score systems are studied to understand the parameters that are being used in risk estimation.
- Build a dataset using the parameters that are being used for the prediction.

TABLE 6.1 Comparison of Different Systems

Country	Sample Size	Age	Gender	Food	Family History	Waist Circumference	Physical Activity	BP	High Blood Glucose	BMI	Smoking
India	26001	✓	✓		✓	✓	✓	✓			
China	5453	✓	✓		✓	✓	✓	✓	✓	✓	
Sri Lanka	4276	✓	✓		✓	✓	✓	✓		✓	
Oman	4881	✓	✓		✓	✓		✓		✓	
Cambridge	1077	✓	✓		✓	✓		✓	✓	✓	✓
France	4500	✓	✓	✓	✓	✓	✓	✓	✓	✓	
UK	6186	✓	✓	✓	✓	✓		✓		✓	
Danish	6784	✓	✓	✓	✓	✓	✓	✓	✓	✓	✓
Australia	6060	✓	✓	✓	✓	✓	✓	✓	✓	✓	✓
Brazil	1224	✓	✓		✓	✓		✓		✓	✓
USA	5258	✓	✓		✓	✓	✓	✓		✓	

- Apply the suitable machine learning algorithm and find the score on this designed dataset (with some parameter).
- The built dataset should represent all the scoring systems existing and should be able to represent people from around the world.
- Add more features to the designed system and again calculate the score and compare it with the original system. (How the original system can be fine-tuned if we add some other feature?).
- Validate the data using the existing diabetes risk score system.

6.4 SYSTEM ANALYSIS

System analysis is a process of understanding facts and identifying the problems. The purpose of system analysis is to study system and understand its objectives. This helps improve the system and accomplish purpose of the system.

6.4.1 EXISTING SYSTEM

In the existing system, as discussed in the literature survey part, the diabetes risk score system with collected samples from certain region was developed. Based on age, family history, gender, WC, physical activity, BP, and smoking, various diabetes risk score tools were developed. To identify the variables, univariate regression analysis is done. To derive the risk score, the β coefficient values are identified using the analysis called logistic regression. For finding the cumulative regression coefficient, all β coefficients derived from the logistic regression are added. The authors also find the optimal cutoff value, sensitivity, and specificity. Based on the ROC analysis, they calculated the optimum value. They also validate the system using AUC.

6.4.2 PROPOSED METHOD

In the proposed method, we used IDRS as template to reverse calculation and create an imputed dataset for Asian and European countries. As the aim is to provide individual age-specific personalized T2D risk score, we calculated the β coefficient for each year instead of making an age group. Once the above step is completed, impute the data according to the average value between the ranges. While doing this, the value of β increases according to the individual specific age; similar calculation was done for China, Sri Lanka, Oman, Cambridge, France, UK, and Danish. A similar approach was taken to personalize BMI and WC. Several risk score tools are developed, but predicting the correct risk score without losing simplicity is really a challenging task. The proposed system is compared with the existing risk score system for the accuracy and performance. It can also be applied to different ethnic groups. Therefore, the diabetes risk score system is designed without any laboratory tests using AI techniques.

6.5 METHODOLOGY

Machine learning is the logical field, managing the manners by which machines gain the fact from expertise. Python apparatuses and modules are used. Here, in this case, matplotlib, numpy, and pyplots for plotting yield results additionally bolsters machine learning algorithms such as classification, logistic regression, decision tree (DT), random forest (RF), linear, and different algorithms were utilized. Here, accuracy, confusion matrix, sensitivity, and specificity are calculated using the machine learning algorithm. Specificity or true negative rate is defined as the level of patients who are accurately distinguished as being healthy. (1 – specificity) is the level of patients who are mistakenly recognized as being diseased. Sensitivity or true positive rate is defined as the percentage of patients who are correctly identified as being having the disease. In machine learning grouping models, one basic proportion of model exactness is AUC. By bend, ROC bend is inferred. ROC represents receiver operating trademark, which can be drawn as sensitivity versus 1 – specificity.

The motivation behind this work is to detect T2D of individuals who are interested to know about their risk score. Therefore, the diabetes risk score system is designed without any laboratory tests.

Its design steps are as follows.

- Different diabetes risk score systems are studied to understand the parameters that are being used in risk estimation.
- Built a dataset using the parameters that are being used for the prediction.
- Apply the suitable machine learning algorithm and find the score on this designed dataset (with some parameter).
- The built dataset should represent all the existing scoring systems and should be able to represent people from all around the world.

- Add more features to the designed system and again calculate the score and compare it with the original system. (How the original system can be fine-tuned if we add some other feature?).
- Validate the data using the existing diabetes risk score system.

The proposed model is shown in Figure 6.1.

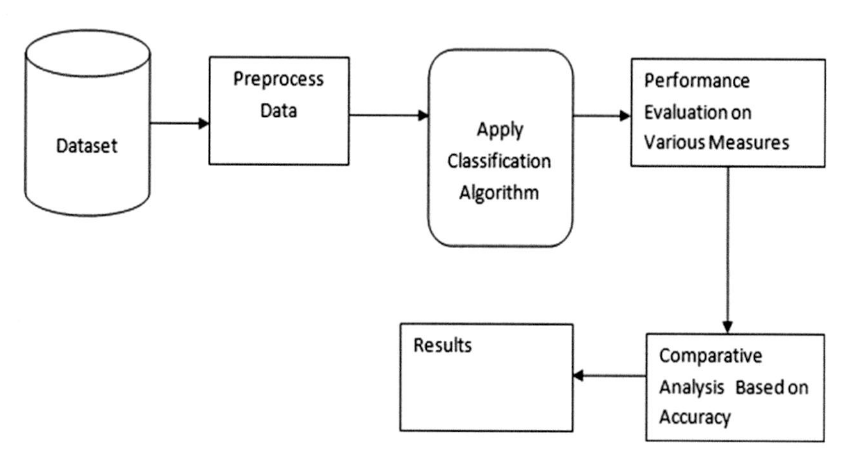

FIGURE 6.1 Proposed model diagram.

6.5.1 DATASET SELECTION

To develop a uniform T2D risk scoring system for Asians and Europeans, we used scoring values from IDRS (India), Chinese score system, SLDRISK (Sri Lanka), Omanese, Cambridge, France, UK, and Danish. The details of the scores of these four systems are given in Table 6.2.

Table 6.2 represents β coefficient values of the Asian system by considering five parameters: age, WC, physical activity, family history, and BMI. To design a uniform T2D risk scoring system for Asians among the different parameters, WC has a significant role. The protection from insulin increases as the individual progresses toward becoming overweight. The risk factor to diabetes also increases if an individual has a family ancestry, that is, if a parent or sibling of the subject has/had diabetes. As the age increases, the risk of diabetes also increases due to the lack of physical activity, yet T2D is also being observed in youths. Using all these constraints, we have identified strong parameters that affect T2D; based on these parameters, we develop the diabetes risk score system. Important parameters are age, gender, WC, physical activity, family history, and BMI.

TABLE 6.2 β Coefficient of Different Systems

Variable	India	China	Sri Lanka	Oman	Cambridge	France	UK	Danish	Australia	Brazil	US
Age											
<35	0	0	0	0		0		0	0	0	0
35–49	0.84	0.845	0.95	1.8	0	0.42	0	0.6926	0.455	0.743	0.95
>=50	1.47	1.357	1.61	2.3	0.44	0.65	0.53	1.311	0.919	1.227	1.57
60–69					0.861	0.94	0.94	1.8475	1.3		2.09
>=70					1.16		1.26		1.645		
Waist circumference											
Female < 80 Male < 90	0	0	0	0	0	0	0		0		0
Female 80–89 Male 90–99	0.44	0.952	1	0.38	0.5	1.021	0.43		0.884		0.27
Female >= 90 Male >= 100	0.81	1.493			0.64	1.424	0.56		1.411		1.12
>109		2.271			0.956		0.86				1.99
Physical Activity											
Vigorous	0	0	0			0		0	0		-0.34
Mild	1.13		0.17								
No	1.45	0.352	0.32			0.268		0.6488	0.428		0
Family History											
Two nondiabetic	0	0	0	0	0	0	0	0	0		0
Either parent	0.54	0.656	0.52	1.9	0.475	1.021	0.47	0.6835	0.624		0.67
Both parent	0.83										
BMI											
<25		0	0	0	0	0	0	0	0	0	
25–29		0.679	0.52	0.54	0.137	0.015	0.26	0.7401	0.569	0.473	
30–34		0.948	0.72	0.69	0.247	0.938	0.45	1.4672	1.224	1.802	
>=35		1.418			0.458		0.75		1.698		
		1.784									

The risk score can be easily calculated using the β coefficient value. It can be mostly used in developing countries. β coefficient calculations are explained in Section 6.2.

The risk score of probability using β coefficient value can be calculated as follows:

$$P\,(\text{diabetes}) = e^{(\beta 0 + {}_1^{\beta}{}_1^{x} + {}_2^{\beta}{}_2^{x} + \ldots)}$$

$$1 + e\,(\beta_0 + \beta_1 x_1 + \beta_2 x_2 + \ldots)$$

where $x_1, x_2, \ldots$ are independent risk factors, β_0 is the intercept, and $\beta_1, \beta_2, \ldots$ are regression coefficients.

6.5.2 DATA PREPARATION AND

The IDRS is used as template to reverse calculation and create an imputed dataset. As the aim is to provide individual age-specific personalized T2D risk score, the β coefficient is calculated for each year instead of making an age group. To achieve this, we took IDRS as reference and created the imputed dataset. In IDRS, values of β coefficient for age groups <35, 35–49, and ≥50 are 0, 0.84, and 1.47, respectively. We created a continuous dataset for individual ages from 21 to 80 using these β coefficient values. To do so, the considered lowest value is –0.4 for 21–34 years and highest value for these 21–34 years is calculated based on the next value of the category. Therefore, here, the highest β coefficient value is determined as 0.2. A similar technique is applied for the age categories 35–49 and 50–80. Therefore, values obtained for the ages 35, 49, 50, and 80 are 0.699, 1.1, 1.2, and 1.64, respectively. Once the above step is completed, impute the data according to the average value between the ranges. While doing this, the β value increases according to the individual specific age. Similarly, the calculation is done for China, Sri Lanka, and Oman. A similar approach was taken to personalize BMI and WC.

6.5.2.1 FOR PHYSICAL ACTIVITY

Physical activity is one of the important parameters for predicting the T2D. Three categories of physical activity are considered according to the IDRS: vigorous exercise with the β coefficient of 0, no exercise with the β coefficient of 1.45, and mild exercise with the β coefficient of 1.13. In view of inquiries shaped by the International Physical Activity Questionnaires, physical activity was separated as low, moderate, and high. Here, lively physical exercises are the exercises that require hard physical exertion and influence you to inhale a lot harder than typical. Such physical exercises resemble hard work, burrowing, high impact exercise, and quick bicycling. Moderate exercises are exercises that require moderate physical exertion and influence you to inhale fairly harder than ordinary.

6.5.2.2 FOR FAMILY HISTORY

Family history is another important parameter for predicting the T2D. We have considered three categories of family history according to the IDRS: two nondiabetic parents with the β coefficient of 0, either parent with the β coefficient of 0.54, and both parents with the β coefficient of 0.83. All these categories are included for dataset creation.

6.5.3 COMPUTATION FOR DATA IMPUTATION

Once β coefficients are calculated as explained in the data computation part, in the next stage, imputing the data is very much essential. Here, Python library Scikit learn is used, and also, there is a Python module dedicated to permutations and combinations called itertools. It is one of the greatest corners of the Python 3 standard library: itertools. Itertools. *Product ():* This tool computes the Cartesian product of input timetables. This module implements a number of iterator building blocks in a form suitable for Python. This is the efficient tool that can be used for a variety of combinations. Initially, we took four parameters, namely, age, waist, physical activity, and family history; later, BMI was also included in the list. Once all the values are added in the particular list, the product (*) with the itertools module was used. Therefore, it is acting like nested for loop, and we got all the combinations of four parameters so that the total number of samples obtained is 514,384. A similar approach was taken to create the dataset of India (IDRS) with BMI, China, SLDRISK, and Oman. The steps involved in creating the dataset are shown in Figure 6.2.

6.5.4 DESCRIPTION OF THE DATASET

The dataset is created for India, China, Sri Lanka and Oman based on certain important attributes and does not contain any missing values.

All variables were categorized as age (21–34 years versus 35–49 and ≥50 years), WC (men <90 cm, 90–99 cm, ≥100 versus women < 80 cm, 80–89, >90 cm), BMI (weight in kg divided by height in m2) (BMI < 25 vs. 25–29 vs. 30–34 and ≥35), family history of diabetes (two nondiabetic parents versus either parent having diabetes and both parent having diabetes), and physical activity (vigorous exercise versus mild and no exercise is considered). Based on the outcome of the diabetes (that is 0/1), training data should be classified such that 0 indicates no diabetes and 1 indicates diabetes. Table 6.3 represents the attribute of the created dataset.

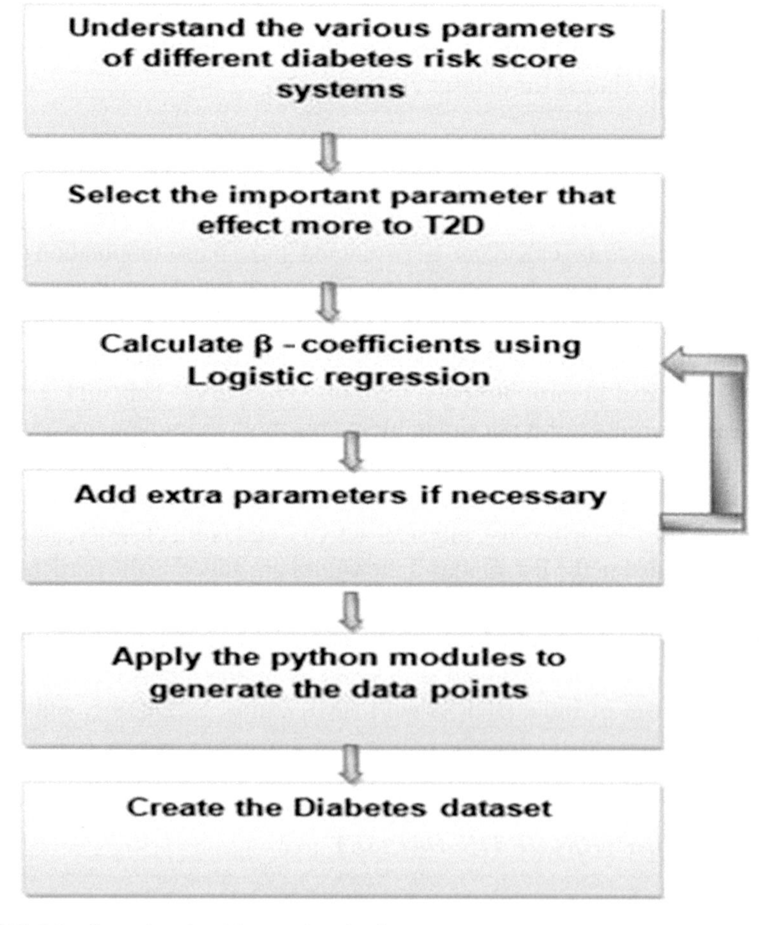

FIGURE 6.2 Steps involved in creating the dataset.

TABLE 6.3 Attributes

Attribute number	Attribute
1	Age
2	Waist
3	Physical activity
4	Family history
5	BMI
6	Outcome

6.5.5 DATA MODELING AND ALGORITHMS USED FOR PREDICTION

Different algorithms, namely, multiple logistic regression, Gaussian Bayes (GB), RF, and DT [21] are applied to the imputed Indian, Chinese, SLDRISK, and Oman diabetic datasets. The data were grouped into training (70%) and test sets (30%) comprising of 50% of T2D. Two combinations of parameters, such as (i) age, gender, physical activity, family history, and WC and (ii) age, gender, physical activity, family history, WC, and BMI, were used in predicting the efficacy (specificity and sensitivity) of each algorithm using ROC and AUC with 95% confidence interval. Furthermore, the outcomes of each algorithm were compared with each other, and the best model is selected. In another approach, we used a consensus algorithm [22] to get the average of scores from the entire algorithm. Similarly, we developed the consensus-based Asian score, as described in Figure 6.3. The essential issue to decide positioning accord is an issue to join a few rankings, which are chosen by at least two decision makers into positioning agreement. For different Asian countries, machine learning algorithms were initially applied. The average value of each method is identified; then, final prediction is done using the consensus-based average rank algorithm, as shown in Figure 6.3.

6.6 RESULTS AND DISCUSSION

6.6.1 FOR ASIAN COUNTRIES

Three combinations of parameters are used for prediction using different algorithms for different Asian countries such that accuracy, precision, sensitivity, and specificity are calculated. Sensitivity and specificity rates

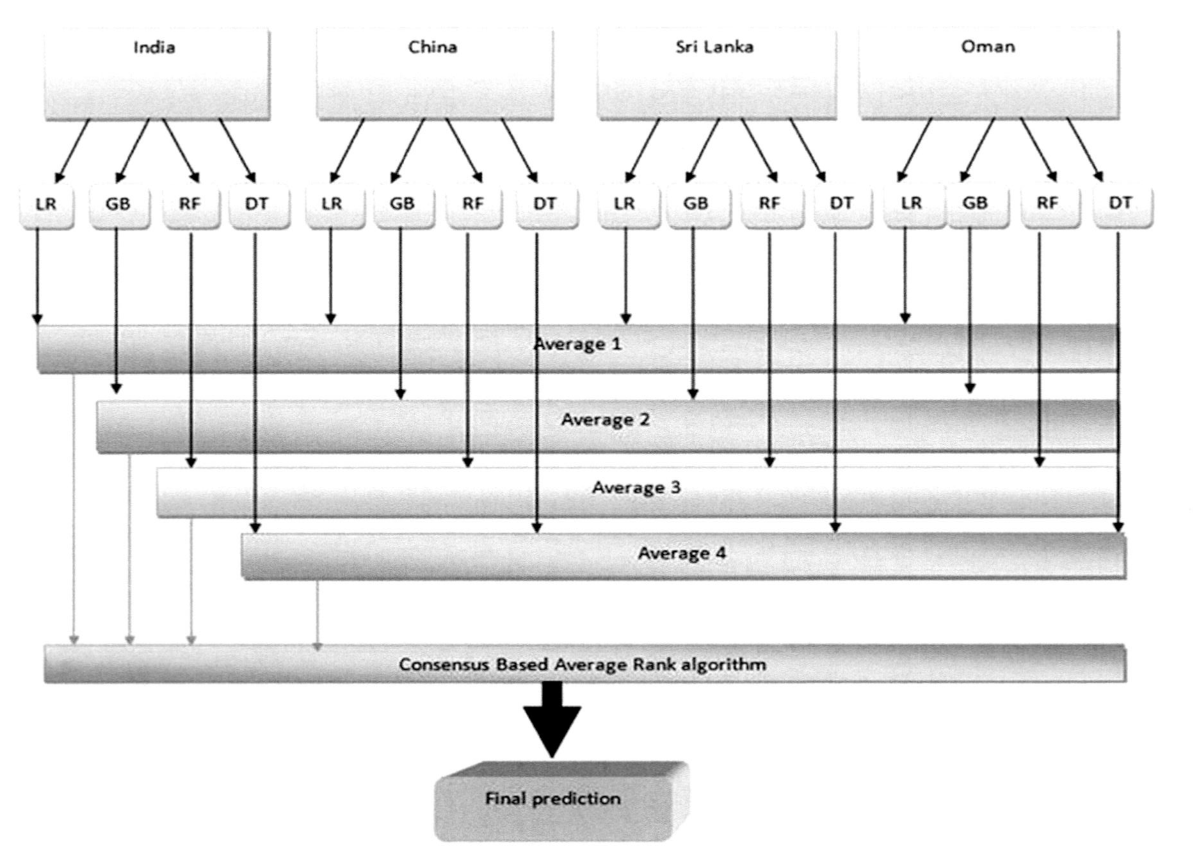

FIGURE 6.3 Workflow of the consensus-based final prediction.

are needed to draw the ROC. Three combinations of parameters were used for prediction: *Combination 1:* age, gender, physical activity, family history, and WC; *Combination 2:* age, gender, physical activity, family history, WC, and BMI; and *Combination 3:* age, gender, physical activity, family history, and BMI. While developing the model with logistic regression, initially, X and Y values are defined. X is the matrix that contains the attributes from the dataset and Y is the vector, based on which prediction can be done. X is defined as the corresponding β values of age, waist, physical activity, family history, and BMI. Y is the vector based on the outcome. Once the X and Y values are defined, split the X and Y values into corresponding training set and testing set. Here, sklearn splitting is done such that the random state value is set as zero. Once the classification of training set and testing set are over in the next stage, on the training set, we have to train a logistic regression model and fit the model on the X_train and Y_train. Once the model is fit, prediction based on the testing set, that is, X_test, should be carried out and calculate the accuracy such that in this dataset, the accuracy for the combinations 1–3 for the Asian countries is up to 0.8505, 0.9779, and 0.960, respectively. Once the model is built, now, the confusion matrix is created such that it will give the number of real and false prediction in the form of array. The confusion matrix for our dataset for combination 2 of Asian countries is shown in Figure 6.4, which indicates that the dimension is 2×2. For example, in the Indian system, real prediction values are 3153 and 122,587 (diagonal values) and inaccurate prediction values are 1846 and 1010. Similarly, the confusion matrix for different countries is calculated, which is fully used for the prediction.

Therefore, from the logistic regression model, the classification rate, precision, recall, sensitivity, and specificity are shown in Table 6.4 for combination 2. Similarly, values are identified for remaining combination. True negative rate is determined by the specificity, which defines the percentage of patients who are correctly identified as being healthy, so using a logistic regression model for combination 2, it is almost up to 63.07%, 70.94%, 69.38%, and 80.18%, respectively. True positive rate is determined by the sensitivity, which defines the percentage of patients who are correctly identified as being disease so using the logistic regression model for combination 2, it is almost up to 99.18%, 98.64%, 98.74%, and 98.54%, respectively, for the Asian countries. Therefore, similarly, these values are identified using four different algorithms for Asian countries, as shown in Tables 6.5–6.7.

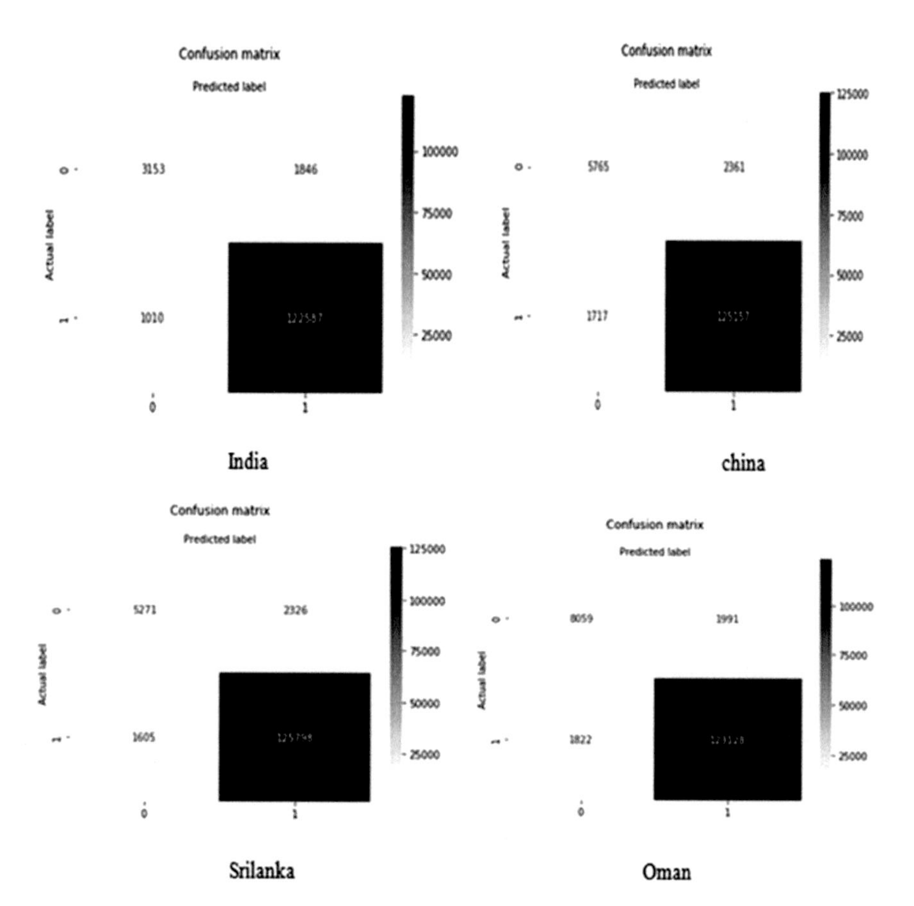

FIGURE 6.4 Visualizing the confusion matrix.

TABLE 6.4 Accuracy, Sensitivity, and Specificity Percentage Using Machine Learning Algorithms

Logistic Regression	Accuracy (%)	Precision (%)	Recall (%)	Sensitivity (%)	Specificity (%)
India	97.78	98.51	99.1	99.18	63.07
China	96.98	98.14	98.6	98.64	70.94
Sri Lanka	97.08	98.18	98.7	98.74	69.38
Oman	97.175	98.41	98.5	98.54	80.18

The optimum value (≥95%) is considered as the high risk score for the diabetes that is detected based on the ROCs for India, China, Sri Lanka, and Oman, as shown in Figure 6.5.

TABLE 6.5 Combination 1 (Without Considering BMI)

	Accuracy	Precision	Sensitivity	Specificity
India				
Logistic Regression	85.05	87.40	91.39	71.17
Gaussian Bayes Model	78.4	78	79.0	77.0
Random Forest	71.1	69.5	71.9	69.3
Decision Tree	72.7	67	67.9	67.8
China				
Logistic Regression	93.98	92.97	72.17	70.81
Gaussian Bayes Model	95.30	96	96.0	95.0
Random Forest	96	95	95.0	92.0
Decision Tree	89.2	85.5	70.2	65.8
Sri Lanka				
Logistic Regression	94.37	94.28	74.19	61.53
Gaussian Bayes Model	94.50	89	94	92
Random Forest	94.0	88.2	94.8	91.7
Decision Tree	92.0	77.7	70.3	65.6
Oman				
Logistic Regression	92.55	90.19	73.18	69.72
Gaussian Bayes Model	92.62	86.7	93.0	89.9
Random Forest	93.0	86	93.0	89.0
Decision Tree	92.0	79.0	79.0	66.1

The Indian system is cross-validated using AUC and its value is 0.9873. The China system is cross-validated using AUC and its value is 0.9870. The Sri Lanka system is cross-validated using AUC and its value is 0.9879. The Oman system is cross-validated using AUC and its value is 0.9883.

There might be unpredictable and unknown connections between the factors in the dataset. It is critical to find and evaluate how many factors in the dataset are dependent on one another. This information can enable to more readily set up the information to meet the desires for machine learning calculations. Factors inside a dataset can be connected for a number of reasons. Relation could be true, neutral, or zero depends on the movement of the two variables. Association can similarly be neural or zero, inferring that the variables are insignificant. Relation between different features for the Indian system is shown in Figure 6.6. The graph is plotted according to the pair using the correlation feature, as shown in Figure 6.7. A similar approach is applied to all the remaining T2D systems of Asian countries and then analyzed.

TABLE 6.6 Combination 2 (With Considering BMI)

	Accuracy	Precision	Sensitivity	Specificity
India				
Logistic Regression	97.78	98.51	99.18	63.07
Gaussian Bayes Model	96.06	92.00	96.12	94.3
Random Forest	96.0	92.0	96.0	94.0
Decision Tree	93.7	98.1	74.8	72.9
China				
Logistic Regression	96.98	98.14	98.64	70.94
Gaussian Bayes Model	94.7	89.23	94.80	91.0
Random Forest	94.0	89.0	94.0	91.0
Decision Tree	94.0	90.01	75.7	56.0
Sri Lanka				
Logistic Regression	97.08	98.18	98.74	69.38
Gaussian Bayes Model	97.01	97.0	97.0	97.0
Random Forest	99.9	99.8	98.7	98
Decision Tree	96.7	97.7	66.2	64.0
Oman				
Logistic Regression	97.175	98.141	98.54	80.18
Gaussian Bayes Model	95.90	96.28	96.0	95.0
Random Forest	96.0	96.0	96.0	95.0
Decision Tree	92.14	72.47	76.05	78.20

TABLE 6.7 Combination 3 (Without Considering WC)

	Accuracy	Precision	Sensitivity	Specificity
India				
Logistic Regression	96.0	95.8	96.2	94.5
Gaussian Bayes Model	96.24	96.0	96.0	94.0
Random Forest	99.0	99.0	99.2	98.0
Decision Tree	92.5	95.0	84.0	78.0
China				
Logistic Regression	96.24	97.51	98.52	60.74
Gaussian Bayes Model	95.50	95.25	96.0	94.0
Random Forest	98.0	98.0	98.0	98.0
Decision Tree	96.0	96.0	96.0	94.0
Sri Lanka				
Logistic Regression	96.52	97.78	98.46	64.07
Gaussian Bayes Model	95.47	92.5	91.0	89.5
Random Forest	98.0	97.8	96.4	92.2
Decision Tree	80.0	93.0	84.0	79.0
Oman				
Logistic Regression	85.8	91.5	85.8	81.0
Gaussian Bayes Model	94.48	94.0	94.0	93.0
Random Forest	98.0	98.0	98.0	98.0
Decision Tree	95.0	92.0	90.15	89.7

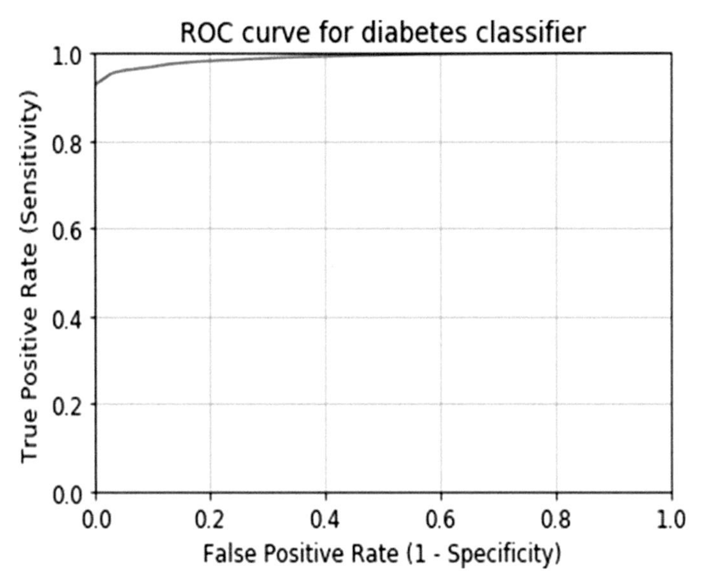

FIGURE 6.5 Optimum value using ROC.

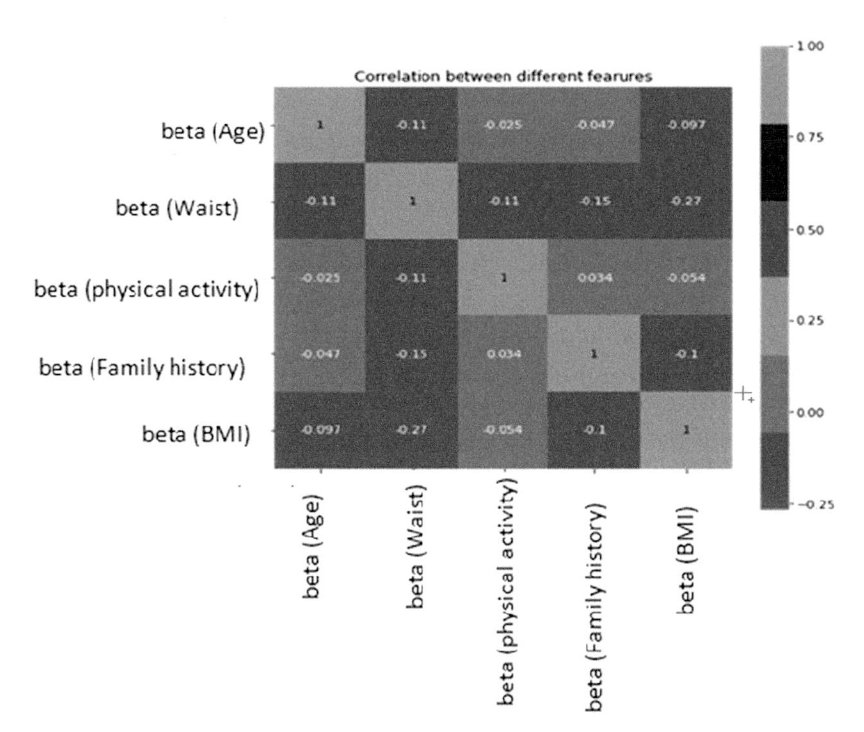

FIGURE 6.6 Correlation between the features for the Indian system.

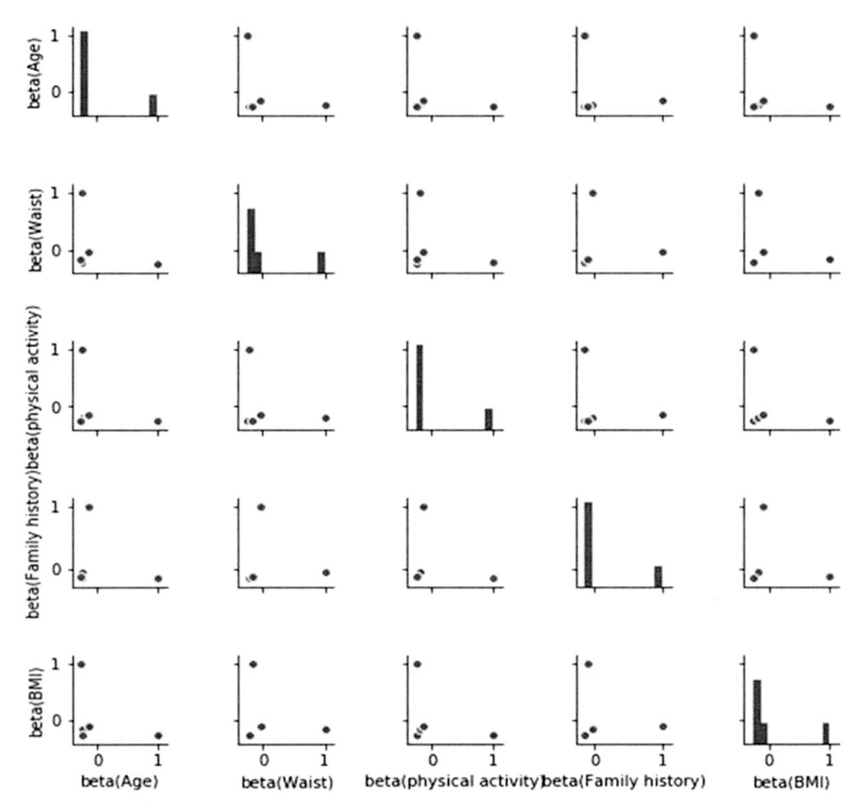

FIGURE 6.7 Pair plot according to correlation feature values.

Using the ROC curve, the cutoff value for the risk score can be identified. The ROC curves were plotted for the diabetes risk score, the sensitivity was plotted on the y-axis, and the false positive rate (1 – specificity) was plotted on the x-axis. The more precise segregating the test, the more extreme the upward part of the ROC bend and the higher the zone under the bend (AUC). The optimum value is considered as the high risk score for the diabetes that is detected based on ROCs. ROC curves using different combinations for the Indian system are demonstrated in Figure 6.8. Therefore, it is observed that logistic regression performs better compared to all other algorithms. Since the AUC is constantly used to identify how well the test is performed between the two gatherings like if the value of AUC increases, which indicates, the better is the test. Therefore, the Indian system is validated using the AUC, and its value is 0.98. Similarly, China, Sri Lanka, and Oman systems are also validated, and the obtained results are 0.98, 0.97, and 0.94, respectively. A similar approach is applied for remaining countries.

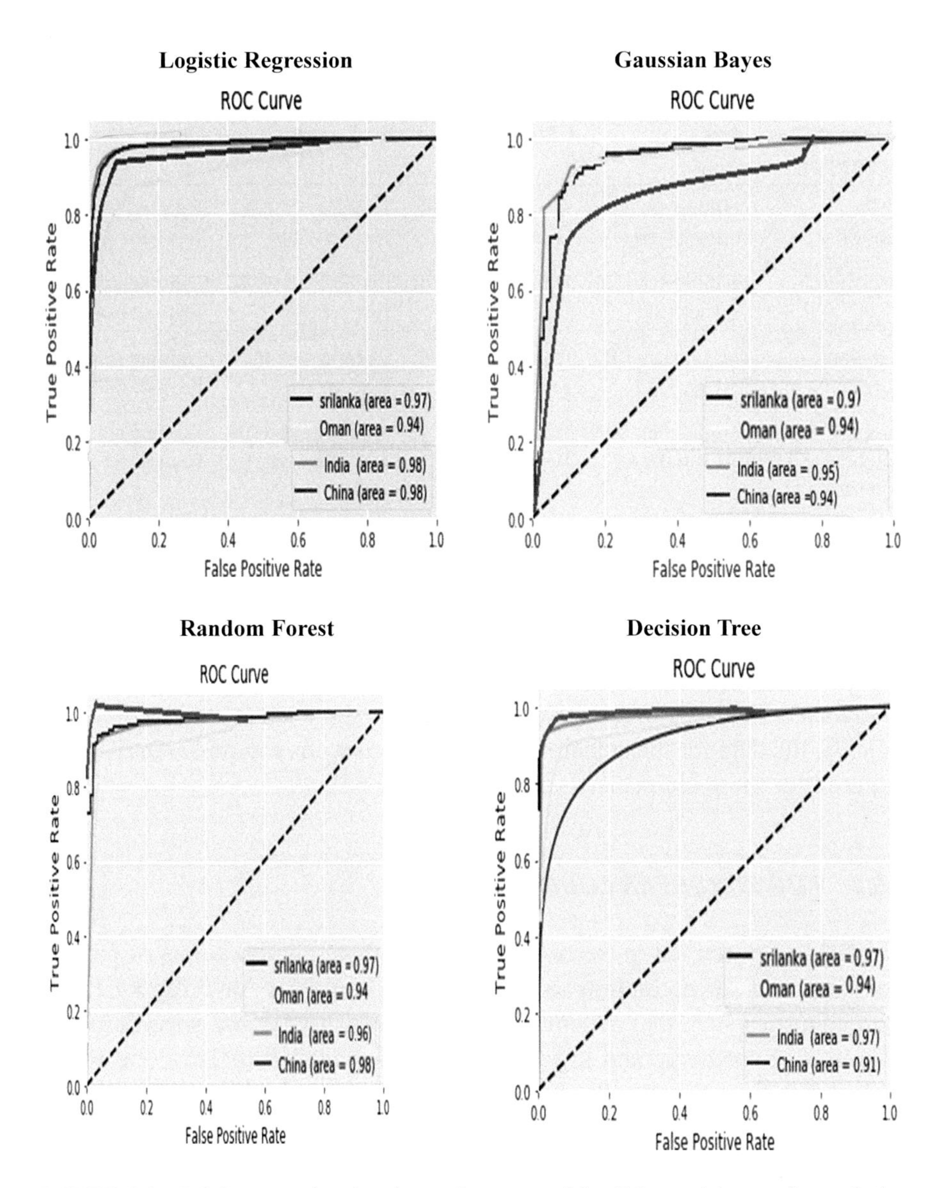

FIGURE 6.8 ROC curves showing the performance of the diabetes risk score in predicting diabetes.

Similarly, about 514,384 samples are used for analysis. Optimization of diabetes data using *k*-nearest neighbors (KNN) classifier, DT, RF, SVM, and GB model is compared as shown in Table 6.8.

TABLE 6.8 Analysis of Optimization of Different Machine Learning Algorithms

Machine Learning Algorithm	Accuracy—India	Accuracy—China	Accuracy—Sri Lanka	Accuracy—Oman
KNN Classifier	Training set: 1.00 Test set:1.00	Training set:1.00 Test set:1.00	Training set:1.00 Test set:1.00	Training set:1.00 Test set:1.00
Decision Tree	Training set:1.00 Test set:1.00	Training set:1.00 Test set:1.00	Training set:1.00 Test set:1.00	Training set:1.00 Test set:1.00
Random Forest	Training set:1.00 Test set:1.00	Training set:1.00 Test set:1.00	Training set: 0.94 Test set: 0.94	Training set: 0.93 Test set: 0.93
Support Vector Machine	Training set: 0.99 Test set: 0.99	Training set: 0.95 Test set: 0.95	Training set: 0.965 Test set: 0.965	Training set: 0.95 Test set: 0.95
Gaussian Bayes Model	Training set: 0.9611 Test set: 0.9608	Training set: 0.9404 Test set: 0.9414	Training set: 0.9443 Test set: 0.9450	Training set: 0.9261 Test set: 0.9262

By considering the simple five parameters, namely, age, family history of diabetes, WC, physical activity, and BMI, the system was developed. Quickly, the data for these factors were obtained by five inquiries and scores acquired for these elements, as shown in Table 6.9.

6.6.2 FOR EUROPEAN COUNTRIES

Similarly to explained in Section 6.4.1, here, three combinations of parameters are used for prediction using different algorithms for different European countries such that accuracy, precision, sensitivity, and specificity are calculated. Sensitivity and specificity rates are needed to draw the ROC. The confusion matrix for the European dataset for combination 2 is shown in Figure 6.9.

Therefore, from the logistic regression model, the classification rate, precision, recall, sensitivity, and specificity are shown in Table 6.10 for combination 2. Similarly, the values are identified for remaining combinations. True negative rate is determined by the specificity; therefore, using the logistic regression model for combination 2, it is almost up to 55.70%, 82.55%, 56.62%, and 55.83%, respectively. True positive rate is determined by the sensitivity, which defines the percentage of patients who are correctly identified as being disease; therefore, using the logistic regression model for combination 2, it is almost up to 97.70%, 90.35%, 93.07%, and 91.98%, respectively, for European countries. Therefore, similarly, these values are identified using four different algorithms for European countries, as shown in Tables 6.11–6.13.

TABLE 6.9 Diabetes Risk Score for Asian Countries

Parameters	Risk Score
Age	
<35	0
35–49	22
≥50	34
Obesity	
Waist Circumference Female <80 cm, Male <90 cm	0
Female 80–89 cm, Male 90–99 cm	11
Female ≥90 cm, Male ≥100 cm	20
Physical Activity	
Vigorous Exercise	0
Mild Exercise	13
No Exercise	18
Family History	
Two Nondiabetic Parents	0
Either Parent Having Diabetes	18
Both Parent Having Diabetes	29
BMI	
<25	0
25–29	11
30–34	16
≥35	29
Maximum Score	130

Score ≥95:Very High Risk, 70–95:High Risk, 35–69: Medium Risk, <35: Low Risk

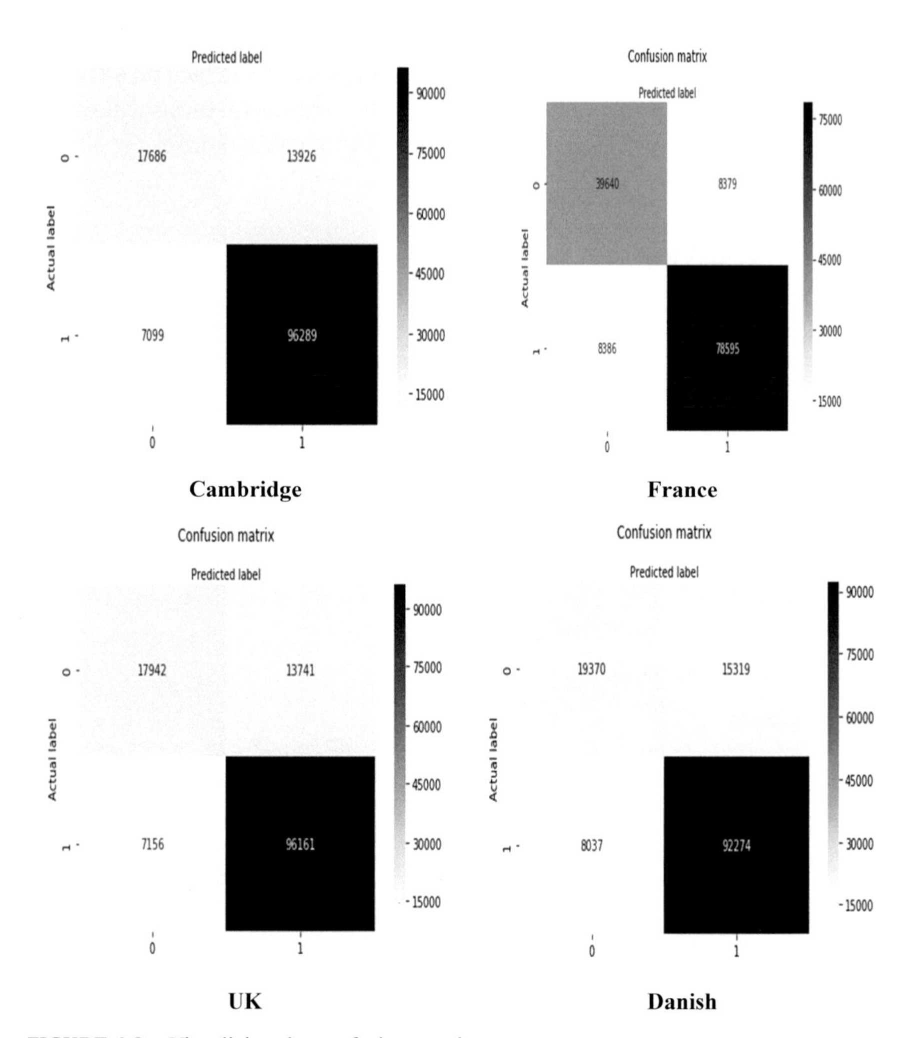

FIGURE 6.9 Visualizing the confusion matrix.

TABLE 6.10 Accuracy, Sensitivity, and Specificity Percentage Using Machine Learning Algorithms

Logistic regression	Accuracy (%)	Precision (%)	Recall (%)	Sensitivity (%)	Specificity (%)
Cambridge	84.42	87.36	93.13	97.70	55.70
France	87.58	90.36	90.35	90.35	82.55
UK	84.52	87.49	93.07	93.07	56.62
Danish	82.69	85.76	91.98	91.98	55.83

The optimum value (80%) is considered as the high risk score for the diabetes that is detected based on the ROCs for Cambridge, France, UK, and Danish, as shown in Figure 6.10.

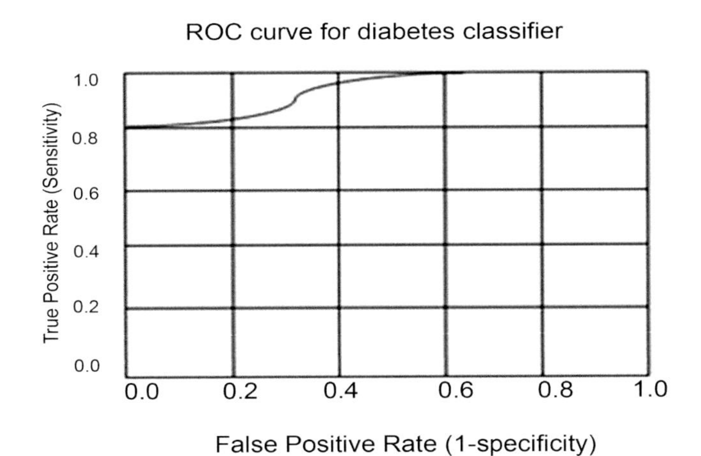

FIGURE 6.10 Optimum value using ROC.

TABLE 6.11 Combination 1 (Without Considering BMI)

	Accuracy (%)	Precision (%)	Sensitivity (%)	Specificity (%)
Cambridge				
Logistic Regression	78.87	82.06	97.89	92.66
Gaussian Bayes Model	83.54	83	84	81
Random Forest	78	75	78	75
Decision Tree	72	80	72	74
France				
Logistic Regression	73.94	78.76	81.55	60.17
Gaussian Bayes Model	71.3	73	71	72
Random Forest	77	74	76	73
Decision Tree	70	75	70	71
UK				
Logistic Regression	79.05	82.15	97.75	92.77
Gaussian Bayes Model	84	84	84	82
Random Forest	78	75	78	75
Decision Tree	80	73	75	73
Danish				
Logistic Regression	76.53	80.38	97.13	90.48
Gaussian Bayes Model	80.81	81	81	78
Random Forest	75	72	75	72
Decision Tree	76	70	72	70

TABLE 6.12 Combination 2 (With Considering BMI)

	Accuracy (%)	Precision (%)	Sensitivity (%)	Specificity (%)
Cambridge				
Logistic Regression	84.42	87.36	97.70	55.70
Gaussian Bayes Model	77.4	74	77	70
Random Forest	86.5	90	86	87
Decision Tree	65.5	74	64	66
France				
Logistic Regression	87.58	90.36	90.35	82.55
Gaussian Bayes Model	67.50	66	67	61
Random Forest	80	87	80	80
Decision Tree	64	65.8	65.3	62
UK	84.52	87.49	93.07	56.62
Logistic Regression	84.52	87.49	93.07	56.62
Gaussian Bayes Model	77.45	75	77	70
Random Forest	86.5	90	86	87
Decision Tree	76	67	76	68
Danish				
Logistic Regression	82.69	85.76	91.98	55.83
Gaussian Bayes Model	75.3	73	75	68
Random Forest	84.5	89	84	85
Decision Tree	74	65	74	64

TABLE 6.13 Combination 3 (Without Considering WC)

	Accuracy (%)	Precision (%)	Sensitivity (%)	Specificity (%)
Cambridge				
Logistic Regression	82.09	84.84	97.55	93.27
Gaussian Bayes Model	83.16	84	83	80
Random Forest	87	86	86	86
Decision Tree	77	72	77	71
France				
Logistic Regression	87.58	90.36	90.35	82.55
Gaussian Bayes Model	71.40	73	71	72
Random Forest	69	70	69	69
Decision Tree	70	75	70	71
UK				
Logistic Regression	82.01	85.00	92.88	86.32
Gaussian Bayes Model	81.90	83	82	78
Random Forest	88	86	86	86
Decision Tree	76	71	76	71
Danish				
Logistic Regression	81.46	84.53	91.86	86.54
Gaussian Bayes Model	80.05	80	80	76
Random Forest	90	89	89	89
Decision Tree	75	71	75	71

Relation between different features for the Cambridge system is shown in Figure 6.11. The graph is plotted according to the pair using the correlation feature, as shown in Figure 6.12. A similar approach is applied to all the remaining T2D systems of European countries and then analyzed.

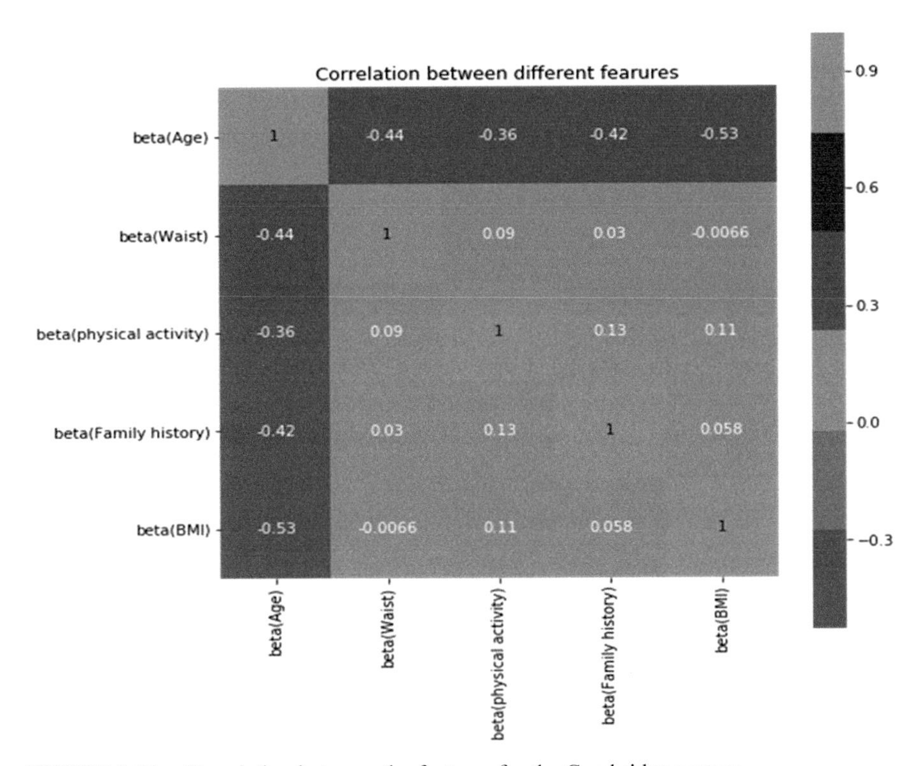

FIGURE 6.11 Correlation between the features for the Cambridge system.

The ROC curve is drawn for European countries with different algorithms, which is fully used to identify the risk score, as demonstrated in Figure 6.13.

The Cambridge system is cross-validated using AUC and its value is 0.8829. The France system is cross-validated using AUC and its value is 0.9408. The UK system is cross-validated using AUC and its value is 0.8962. The Danish system is cross-validated using AUC and its value is 0.8848.

Similarly, about 514,384 samples are used for analysis. Optimization of diabetes data using KNN classifier, DT, RF, SVM, and GB model is compared, as shown in Table 6.14.

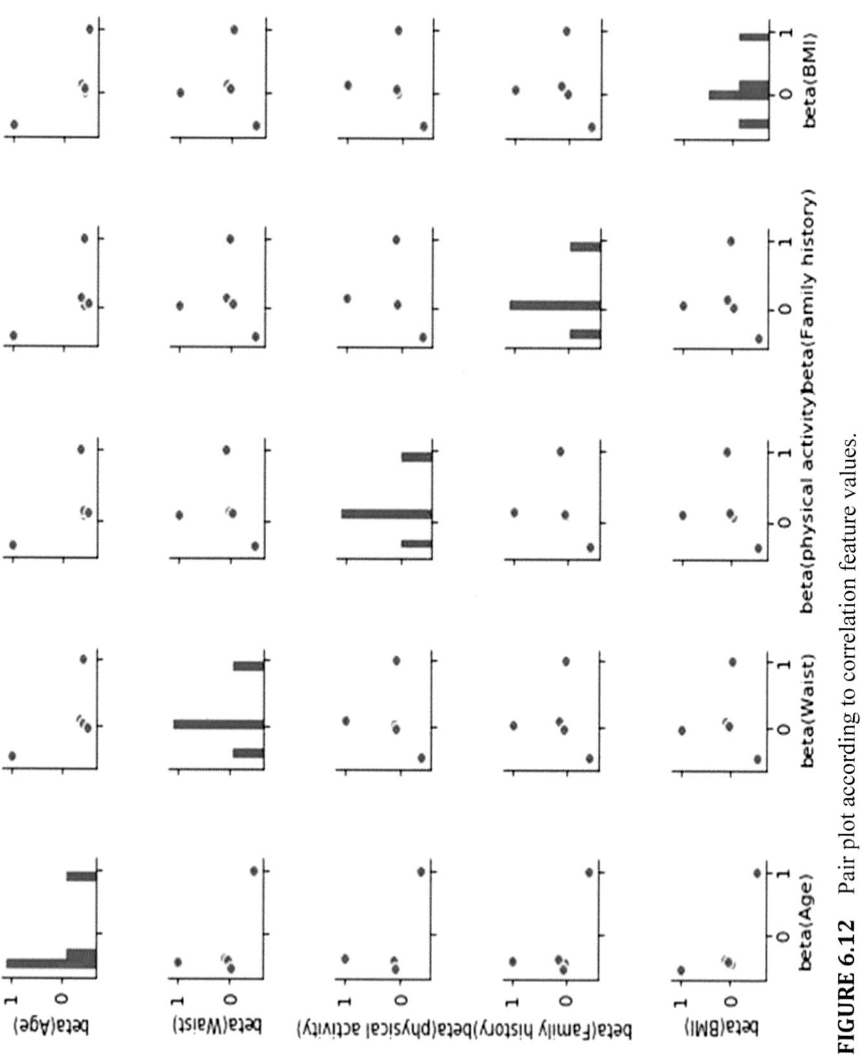

FIGURE 6.12 Pair plot according to correlation feature values.

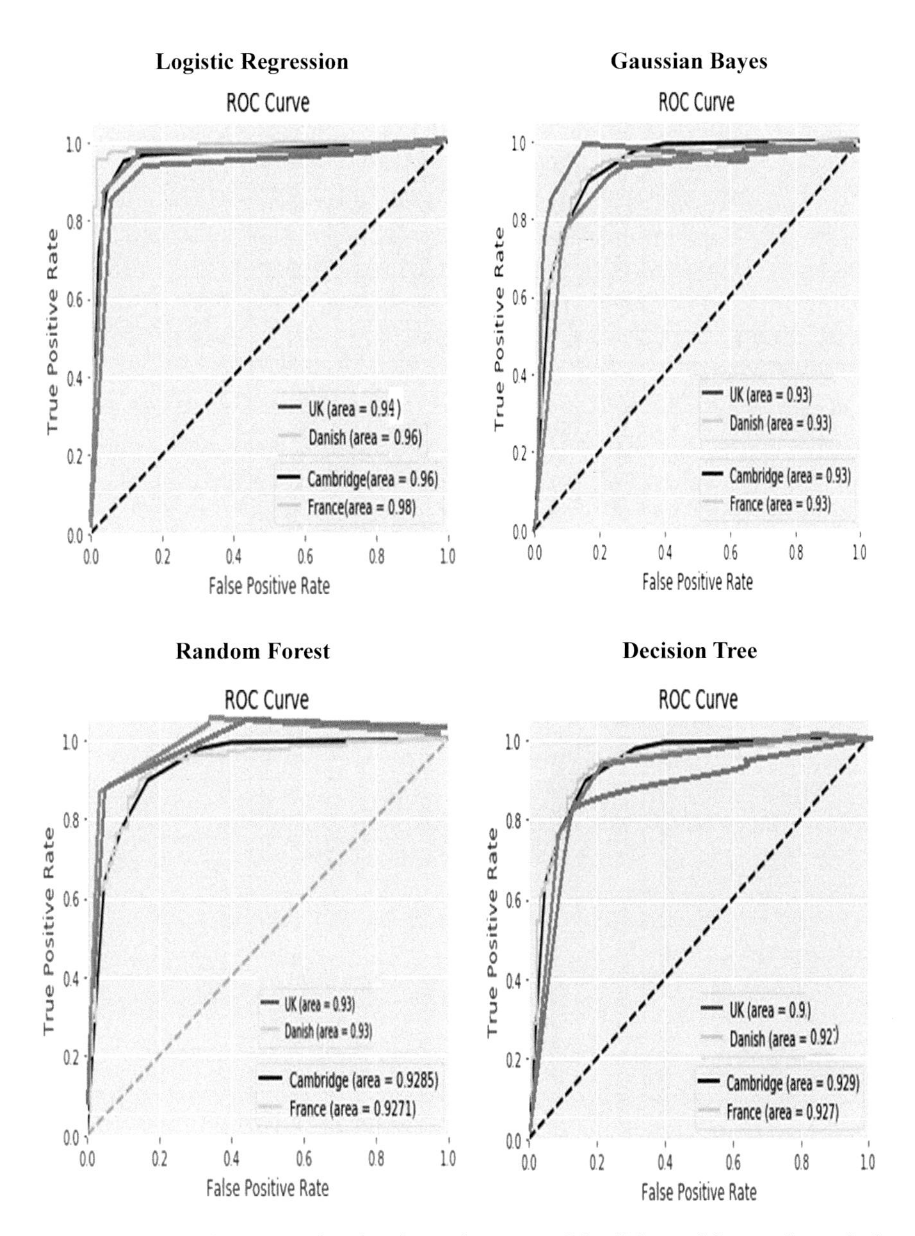

FIGURE 6.13 ROC curves showing the performance of the diabetes risk score in predicting diabetes.

TABLE 6.14 Analysis of Optimization of Different Machine Learning Algorithms

Machine Learning Algorithm	Accuracy (Cambridge)	Accuracy (France)	Accuracy (UK)	Accuracy (Danish)
KNN classifier	Training set: 0.99 Test set: 0.98	Training set: 1.00 Test set:1.00	Training set: 0.99 Test set: 0.99	Training set:1.00 Test set: 0.99
Decision tree	Training set: 1.00 Test set: 1.00	Training set: 1.00 Test set: 1.00	Training set:1.00 Test set:1.00	Training set: 1.00 Test set: 1.00
Random forest	Training set: 1.00 Test set: 1.00	Training set: 1.00 Test set:1.00	Training set: 0.87 Test set: 0.86	Training set: 0.827 Test set: 0.827
Support vector machine	Training set: 0.87 Test set: 0.86	Training set:0.875 Test set: 0.876	Training set: 0.965 Test set: 0.965	Training set: 0.95 Test set: 0.95
Gaussian Bayes model	Training set: 0.7744 Test set: 0.7738	Training set: 0.6760 Test set: 0.6745	Training set: 0.7745 Test set: 0.7748	Training set: 0.7526 Test set: 0.7534

By considering the simple five parameters, namely, age, family history of diabetes, WC, physical activity, and BMI, the system was developed. Quickly, the data for these factors were obtained by five inquiries and scores acquired for these elements, as shown in Table 6.15.

TABLE 6.15 Diabetes Risk Score for European Countries

Parameters	Risk Score
Age	
<35	0
35–49	15
≥50	23
Obesity	
Waist Circumference Female <80 cm, Male <90	0
cm	17
Female 80–89 cm, Male 90–99 cm	20
Female ≥90 cm, Male ≥100 cm	
Physical Activity	
Vigorous Exercise	0
Mild Exercise	8
No Exercise	12
Family History	
Two Nondiabetic Parents	0
Either Parent Having Diabetes	19
Both Parents Having Diabetes	26
BMI	
<25	0
25–29	11
30–34	15
≥35	19
Maximum Score	100

Score ≥70:Very High Risk, 51–69: High Risk, 30–50: Medium Risk, <30: Low Risk

From this work, it is observed that performance of the system is better using logistic regression compared to other machine learning algorithms for both Asian and European countries. The corresponding bar chart of accuracy calculation using logistic regression for Asian and European countries is shown in Figures 6.14 and 6.15. For India, China, Sri Lanka, and Oman, the accuracy is 97.78%, 96.98%, 97.08%, and 97.175%, respectively. Similarly, for Cambridge, France, UK, and Danish, the accuracy is 84.42%, 87.58%, 84.52%, and 82.69%, respectively.

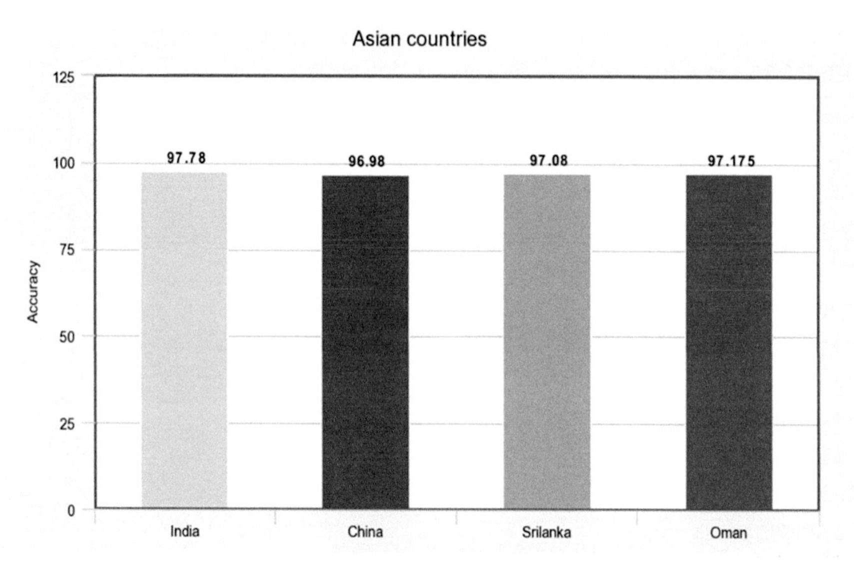

FIGURE 6.14 Accuracy using logistic regression for Asian countries.

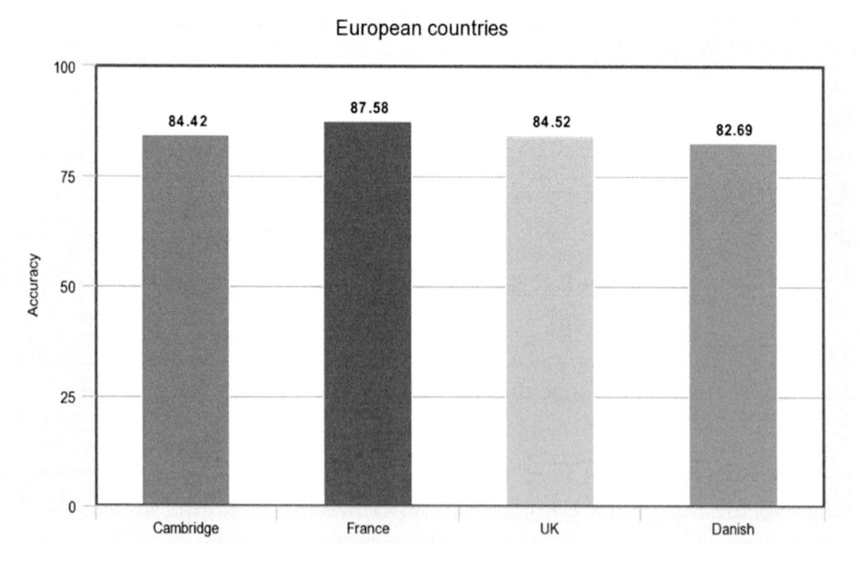

FIGURE 6.15 Accuracy using logistic regression for European countries.

6.7 CONCLUSION AND SCOPE OF FUTURE WORK

A simple diabetes risk assessment tool is developed and validated. A diabetes risk score system is developed for Asian and European countries using five different parameters, namely, age, WC, family history, physical activity, and BMI.

Several risk score tools are developed, but predicting the correct risk score without losing simplicity is really a challenging task. The proposed system is compared with the existing risk score system for the accuracy and performance. It can also be applied to the different ethnic groups. From this work, it is observed that performance of the system is better using logistic regression compared to other machine learning algorithms. In conclusion, the diabetes risk score system is developed that can be used in a stepwise screening strategy for T2D to provide individual age-specific personalized T2D risk score, using β coefficient for each year instead of making an age group. From Tables 6.9 and 6.15, it is concluded that score with <35 for Asian countries and <30 for European countries is considered as low risk, 35–69 for Asian countries and 30–50 for European countries is considered as medium risk, 70–95 for Asian countries and 51–69 for European countries is considered as high risk, and finally ≥ 95 for Asian countries and ≥ 70 for European countries is considered as very high risk. Optimum values ≥ 95 for Asian countries and ≥ 80 for European countries are identified using the ROC, and these values are validated using AUCs. This tool also provides the information about which factor affects T2D more. In future, the precautionary measure for this T2D will be provided by the expert panel.

KEYWORDS

- **type 2 diabetes**
- **risk score**
- **artificial intelligence**

REFERENCES

1. V. Mohan, R. Deepa, M. Deepa, S. Somannavar, and M. Datta, "A simplified Indian Diabetes Risk Score for screening for undiagnosed diabetic subjects," *J. Assoc. Phys. India*, vol. 53, pp. 759–763, Sep. 2005.

2. J. Lindström and J. Tuomilehto, "The diabetes risk score: A practical tool to predict type 2 diabetes risk," *Diabetes Care*, vol. 26, no. 3, pp. 725–731, Mar. 2003.
3. P. Katulanda, N. R. Hill, I. Stratton, R. Sheriff, S. D. N. De Silva, and D. R. Matthews, "Development and validation of a Diabetes Risk Score for screening undiagnosed diabetes in Sri Lanka (SLDRISK)," *BMC Endocr. Disord.*,vol. 16, no. 1, Jul. 25, 2016, Art. no. 42.
4. J. Al-Lawati and J. Tuomilehto, "Diabetes risk score in Oman: A tool to identify prevalent type 2 diabetes among Arabs of the Middle East," *Diabetes Res. Clin. Pract.*, vol. 77, no. 3, pp. 438–444, 2007.
5. S. J. Griffin, P. S. Little, C. N. Hales, A. L. Kinmonth, and N. J. Wareham, "Diabetes risk score: Towards earlier detection of type 2 diabetes in general practice," *Diabetes Metab. Res. Rev.*, vol. 16, no. 3, pp. 164–171, 2000.
6. H. Zhou, Y. Li, X. Liu, F. Xu, L. Li, K. Yang, X. Qian, R. Liu, R. Bie, and C. Wang, "Development and evaluation of a risk score for type 2 diabetes mellitus among middle-aged Chinese rural population based on the RuralDiab Study," *Sci. Rep.*, vol. 7, Feb. 17, 2017, Art. no. 42685.
7. H. Bang, A. M. Edwards, A. S. Bomback, C. M. Ballantyne, D. Brillon, M. A. Callahan, S. M. Teutsch, A. I. Mushlin, and L. M. Kern, "A patient self- assessment diabetes screening score: Development, validation, and comparison to other diabetes risk assessment scores," *Ann. Internal Med.*, vol. 151, no. 11, pp. 775–783, 2009.
8. M. Rigla, G. García-Sáez, B. Pons, and M. E. Hernando, "Artificial intelligence methodologies and their application to diabetes," *J. Diabetes Sci. Technol.*, vol. 12, no. 2, pp. 303–310, 2018.
9. L. Chen, D. J. Magliano, B. Balkau, S. Colagiuri, P. Z. Zimmet, A. M. Tonkin, P. Mitchell, P. J. Phillips, and J. E. Shaw, "AUSDRISK: An Australian Type 2 diabetes risk assessment tool based on demographic, lifestyle and simple anthropometric measures," *Med. J. Aust.*, vol. 192, no. 4, pp. 197–202, 2010.
10. N. Unwin, D. Whiting, L. Guariguata, G. Ghyoot, and D. Gan, *Diabetes Atlas*, 5th ed. Brussels, Belgium: International Diabetes Federation, 2011.
11. L. J. Gray, N. A. Taub, K. Khunti, E. Gardiner, S. Hiles, D. R. Webb, B. T. Srinivasan, and M. J. Davies, "The Leicester risk assessment score for detecting undiagnosed Type 2 diabetes and impaired glucose regulation for use in a multiethnic UK setting," *Diabet Med.*, vol. 27, no. 8, pp. 887–895, 2010.
12. C. Glümer, B. Carstensen, A. Sandbaek, T. Lauritzen, T. Jørgensen, and K. Borch-Johnsen, "A Danish diabetes risk score for targeted screening: The Inter99 study," *Diabetes Care,* vol. 27, no. 3, pp. 727–733, 2004.
13. [Online.] Available: https://www. medicinenet. com/type_2_diabetes/article. htm#what_ medications_treat_type_2_ diabetes
14. A. G. P. de Sousa, A. C. Pereira, G. F. Marquezine, R. M. do Nascimento-Neto, S. N. Freitas, R. L. de C. Nicolato, G. L. L. Machado-Coelho, S. L. Rodrigues, J. G. Mill, and J. E. Krieger," Derivation and external validation of a simple prediction model for the diagnosis of type 2 diabetes mellitus in the Brazilian urban population," *Eur. J. Epidemiol.*, vol. 24, no. 2, pp. 101–109, 2009.
15. P. Adhikari, R. Pathak, and S. Kotian, "Validation of the MDRF—Indian Diabetes Risk Score (IDRS) in another south Indian population through the Boloor Diabetes Study (BDS)," *J. Assoc. Physicians India*, vol. 58, pp. 434–436, 2010.

16. V. V. Shanbhogue, S. Vidyasagar, M. Madken, M. Varma, C. K. Prashant, P. Seth, and K. S. Natraj, "Indian Diabetic Risk Score and its utility in steroid induced diabetes," *J. Assoc. Physicians India*, vol. 58, 2010, Art. no. 202.
17. K. M. Sharma, H. Ranjani, H. Nguyen, S. Shetty, M. Datta, and K. M. Narayan, "Indian Diabetes Risk Score helps to distinguish type 2 from non-type 2 Diabetes Mellitus (GDRC-3)," *J. Diabetes Sci. Technol.*, vol. 5, pp. 419–425, 2011.
18. M. J. Kim, N. K. Lim, S. J. Choi, and H. Y. Park, "Hypertension is an independent risk factor for type 2 diabetes: The Korean genome and epidemiology study," *Hypertens Res.*, vol. 38, pp. 783–789, 2015.
19. A. S. Shera, F. Jawad, and A. Maqsood, "Prevalence of diabetes in Pakistan," *Diabetes Res. Clin. Pract.*, vol. 76, no. 2, pp. 219–222, 2007.
20. M. I. Schmidt, B. B. Duncan, H. Bang, et al., "Identifying individuals at high risk for diabetes: The atherosclerosis risk in communities study," *Diabetes Care*, vol. 28, pp. 2013–2018, 2005.
21. I. Contreras and J. Vehi, "Artificial intelligence for diabetes management and decision support," *J. Med. Internet Res.*, vol. 20, no. 5, Art. no. e10775, 2018.
22. J. J. Nalluri, D. Barh, V. Azevedo, and P. Ghosh, "*miRsig*: A consensus- based network inference methodology to identify pan-cancer miRNA-miRNA interaction signatures," *Sci. Rep.*, vol. 7, 2017, Art. no. 39684.
23. N. Razavian, S. Blecker, A. M. Schmidt, A. Smith-McLallen, S. Nigam, and D. Sontag, "Population-Level prediction of type 2 diabetes from claims data and analysis of risk factors," *Big Data*, vol. 3, no. 4, 2015.

PART II

Designing IoT-Based Smart Solutions for Monitoring and Surveillance

CHAPTER 7

Smart Interfaces for Development of Comprehensive Health Monitoring Systems

DINESH BHATIA

Department of Biomedical Engineering, North Eastern Hill University, Shillong, Meghalaya 793022, India, E-mail: bhatiadinesh@rediffmail.com

ABSTRACT

With the rapid increase in the older population coupled with enhancement in their life span, the number of patients who require continuous monitoring rises tremendously. It would lead to higher costs of hospitalization and patient care globally. Hence, the requirement of smart interfaces and systems for observing health may be employed in lessening the hospitalization stay, weight on clinical staff, counseling time, holding up records, and, in general, social insurance costs. Different smart interfaces frameworks are characterized into three subcategories: remote health monitoring system (RHMS), mobile health monitoring system (MHMS), and wearable health monitoring system (WHMS). The RHMS alludes those with remote access or frameworks to communicate back and forth information or multiple patient parameters from a remote location or region. MHMSs refer to mobile phones, personal digital assistants, and pocket-personal-computer-based monitoring systems, which are utilized as the principle preparing station or at times as the primary working modules. The RHMS and MHMS are considered to be more advantageous and practical than the traditional institutional care mechanism. They empower patients to stay at their respective locations while getting access to proficient healthcare. WHMSs refer to wearable gadgets or biosensors that can be worn by patients comprising of WHMS, RHMS, and MHMS. Shrewd health monitoring systems are referred to as trendsetting innovations with regard to patient's continuous health monitoring (R. Roine, A. Ohinmaa, and D. Hailey, "Assessing telemedicine: A systematic review of the literature,"

Can. Med. Assoc. J., vol. 165, no. 6, pp. 765–771, 2001). They comprise smart gadgets that could be employed to address several health-related issues. The devices measure heart rate, blood pressure, electrocardiogram, oxygen saturation levels, body temperature, and respiratory rate. This chapter would explore the development of smart interfaces for available comprehensive health monitoring systems built on artificial intelligence tools, cognitive computing systems, and machine learning algorithms.

7.1 INTRODUCTION

In today's advanced technological era, continuous health monitoring plays a vital role due to the rapid rise in the elderly population that enjoys a long life span. It is more prominent due to the increase in a large number of nuclear families, wherein the elderly adults are living separately from their children. Continuous health monitoring of such a population would lead to higher costs of hospitalization and patient care globally. In developed countries such as the USA, the death rate for the older population is more than 770,000 persons every year. Mistaken diagnosis, measurement errors, and delay in intercessions lead to increased hospitalization. The treatments costs are ranging between $1.5 billion and $5 billion every year [2]. Hence, development of smart healthcare interfaces and systems for continuously observing the health of such population can play an essential part in reducing the duration of hospitalization, number of caretakers, counseling time, and maintaining records with reduced social insurance costs. Smart healthcare frameworks could be classified as remote health monitoring system (RHMS), mobile health monitoring system (MHMS), and wearable health monitoring system (WHMS). The RHMS alludes those with remote access or frameworks to communicate to and fro information from a remote location or region. This framework can help in monitoring single or multiple parameters covering different patient symptoms employed at the user homes and a doctor's clinical facilities. MHMSs could be referred to mobile phones, personal digital assistants, and pocket-personal-computer-based monitoring systems, which are utilized as the principle preparing station or at times as the primary working modules [17]. RHMSs and MHMSs are thought to be more advantageous and practical than traditional institutional care systems. They empower patients to stay at their respective locations while availing better healthcare facilities [17,22]. WHMSs may refer to as wearable gadgets or biosensors that can be worn by patients comprising of WHMS, RHMS, and MHMS. Shrewd health monitoring systems (HMSs) consist of smart gadgets to address multiple health-related issues and referred to as trendsetting innovations in continuous health monitoring [34,35]. General HMSs refer to systems that monitor different parameters and general symptoms. The devices measure heart rate (HR), blood pressure (BP), electrocardiogram (ECG), oxygen saturation (SpO_2) levels, body temperature, and respiratory rate (RR) [39]. This chapter discusses the development of smart interfaces for comprehensive HMSs built on artificial intelligence tools, cognitive computing systems, and machine learning algorithms.

7.2 LITERATURE SURVEY ON THE CURRENT STATE-OF-THE-ART MONITORING SYSTEMS

Rapid development in the cutting-edge health monitoring procedures and techniques in the past decade has enabled healthcare experts to precisely monitor grown-ups or adults in connection to age-related diseases such as dementia, Alzheimer's, and Parkinson's [7,25,30]. Since there are no confinements to HMS applications, they can be utilized in the clinic [10], home [14], and outdoor settings using either global positioning system [39] or radio frequency identification technology [9]. In spite of slow advancement of innovation, there are worries with regard to the nature of medical information, the security of patient data, reliability of sophisticated monitoring systems, ease of usage, adequacy by the therapeutic staff and patients, and the recurrence of false alerts [10]. However, various studies and research investigations over the past two decades have primarily diminished such worries. For example, Imhoff and Kuhls [16] have figured out that up to 90% of all alarms in critical care monitoring are false positives. Different scientists proposed measures to lessen these false positives by incorporating adjusting the scope of parameters, decreasing limit esteems, or joining a period delay in producing such alerts [40].

7.2.1 WEARABLE HEALTH MONITORING SYSTEMS

A smart vest [6] is an example of a WHMS, which is a wearable physiological monitoring framework, fused in a jacket. The device is an assortment of different biological sensors coordinated into the piece of clothing's texture that gathers biosignals in a noninvasive and simple way without causing subject discomfort or pain. The parameters estimated by the vest comprise of ECG, photoplethysmography (PPG), HR, BP, body temperature, and galvanic skin reaction. The patient's ECG can be recorded without using electrode jelly and is free from benchmark confusion and development collectibles due to execution of high pass, low pass, and notch filters in the device. BP is measured from acquired ECG noninvasively by employing either the auscultatory or oscillometric methods. Results from approval preliminaries affirm the precision of estimated physiological parameters. LOBIN [26] presented an e-textile wearable wireless healthcare monitoring system comprising of sensors to record ECG, HR, and body temperature. Similarly, Blue Box [4] developed a novel hand-held device capable of collecting and wirelessly transmitting vital cardiac parameters such as ECG, PPG, and bioimpedance. It could assist in measuring patient's RR intervals and QRS duration, HR, and systolic time intervals, as well as assessing their values in correlation with cardiac output measured by an echo-Doppler. An in-shoe device was developed by Saito et al. [20] to monitor plantar pressures under real-life conditions. A pressure-sensitive conductive elastic sensor measures plantar weight and approval performed by the F-scan framework. SMARTDIAB [29] device intends to help the monitoring, administration, and treatment of patients with type 1 diabetes mellitus by combining with a patient unit and patient administration unit. The pilot form of the SMARTDIAB was actualized and assessed in a clinical setting. TELEMON [11] is an electronic informatics-telecom and versatile framework permitted automatic and ongoing telemonitoring by mobile correspondences for monitoring the essential indications of constantly sick elderly patients.

7.3 DESIGN METRICS AND DESIGN FLOW OF HEALTH MONITORING SYSTEM DESIGN

The design of the above-discussed systems is governed by a specific set of rules called design metrics [37] that were defined for the ease of design engineers and helped in improving the overall design of an HMS. The rules are applicable for any healthcare device as well as for any other embedded system. However, some additional constraints that are specific to the HMS design are discussed with design goals. The design metrics are explained in detail in the following sections.

7.3.1 DESIGN METRICS

The design of a system is influenced by parameters [19], for example, consider the design of house floor if we want to improve the quality of the flooring, by improving the quality of the tiles. It will increase the cost of the building. Additionally, consider the computational system in which decreasing the price of a processor may lower the processing capability. Similarly, for designing health or any other embedded systems, parameters are interdependent and affect the performance of the design. Increasing one parameter may decrease or compromise on the other aspect [36]. These parameters are referred to as design metrics and have shown conflicting requirements that need to be addressed before any system design. During the designing stage, a system design engineer tries to find out an optimal set of solution for these metrics. Earlier, the design metrics solution was evaluated after the prototype development of the device, which was a time-consuming, stressful, costly, and tedious process. At that stage, if constraints were not as per our expectations and required modification, the design engineer was bound to change the developed prototype or model. However, nowadays, availability of advanced system simulation software allows us to evaluate the design metrics of the concerned system by creating a virtual model [24]. The following points and Figure 7.1 shows design metrics for a modern HMS, which needs to be carefully optimized during any design development process.

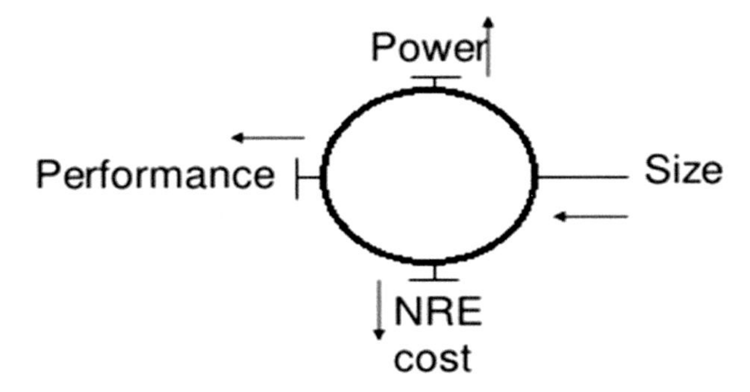

FIGURE 7.1 Design metrics constraints involved in the design of an HMS.

7.3.2 POWER CONSUMPTION

Nowadays, in the era of wearable technologies, the amount of battery and power embedded within the device is limited [31]. It necessitates engineers to develop devices that are capable of operating with ultralow-power consumption. The amount of power dissipated by the device requires careful evaluation before marketing. Consider a commonly employed method in cardiac cases, namely, a pacemaker. If the pacemaker under development has better accuracy, although it requires frequent charging, the demand for it may be quite low. Therefore, design engineers need to incorporate a longer power life in their design to allow the recommending doctors to suggest the device to the patients. As per the design metrics depicted in Figure 7.1, a better model of the invention is expected to have a longer life of the embedded power source.

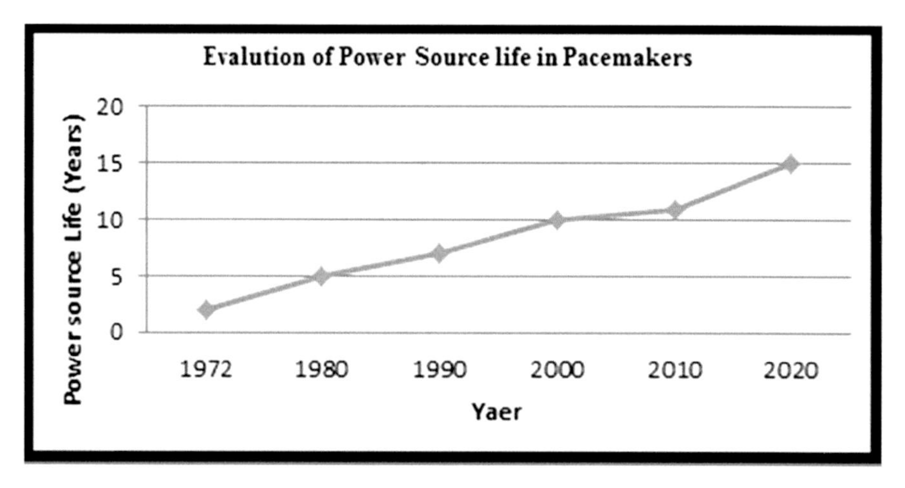

FIGURE 7.2 Evaluation of the power requirements of the healthcare devices (illustrated using pacemaker battery life).

To improve the power requirement of HMSs, before implementing them on hardware level, a design engineer typically assesses the power consumed by different modules, components, connecting wires, energy consumed in wireless transmission and reception of the data, etc., via simulating the design on simulation software [38]. This simulation software has the real-time behavior of components, modules encoded into functions, and hence, via accessing them, the power of the system can be estimated [34]. By looking at the requirement, specific to the application, we can adjust the value by replacing the components through suitable alternatives. Figure 7.2 shows how the life of the embedded power sources in implantable pacemakers has increased over the years by using different design methods and would further improve by the year 2020. Several other ways to increase the battery life of the healthcare devices and monitoring systems are to reduce the number of sensors, to use appropriate fabrication technology, to decrease the size of transmission and reception data bits, etc. [3].

7.3.3 SIZE

Another constraint that affects the suitability and popularity of an optimized HMS is the size of the system [19]. For a design engineer, the size of the system does not mean physical dimensions but measured in bits for any software design and the number of gates for any hardware design. An engineer always tries to reduce the size concerning these two parameters [18]. For this purpose, several optimizing techniques and software have been developed by the researchers, which evaluate their design before actual implementation. The functionality coding of the models in hardware descriptive languages such as Verilog, VHDL, etc., allows a designer to evaluate and optimize the requirement of the gate without implementing it on hardware. Other techniques employed for minimizing the size are logic optimization, technology scaling, device modeling, etc. [3]. Figure 7.3 illustrates the development using the dimensions of different electrocardiograph machines, from the first Einthoven's ECG machine to modern ECG machine available today.

7.3.4 NONRECURRING COST

Besides, the power requirement and size of the nonrecurring cost (NRC), the cost involved in the designing of the system is considered to be an essential parameter. Again, this is a conflicting cost that depends upon the number of units to be manufactured. If the number of units to build is more, then we can compromise with higher NRC if it can reduce the manufacturing cost in a considerable amount [19]. Hence, the design engineer needs to put their efforts in reducing the NRC without degrading the other constraints. The alternatives to lessen the NRC are design reuse, shared design programs, design modification, etc. [3].

7.3.5 PERFORMANCE

Performance is a broader category of constraints for HMSs. Performance is the ability of the device to correctly present the parameters of interest with minimum delay [19]. To get the idea of the device performance, engineers employ several advanced design software tools such as MATLAB, LabView, etc., to evaluate the performance of the design before use [36]. It helps the designer to improve the performance of their device and other design metric constraints. However, the larger goal for the engineer is to find out an optimized metrics for their application.

(a) Einthoven ECG machine

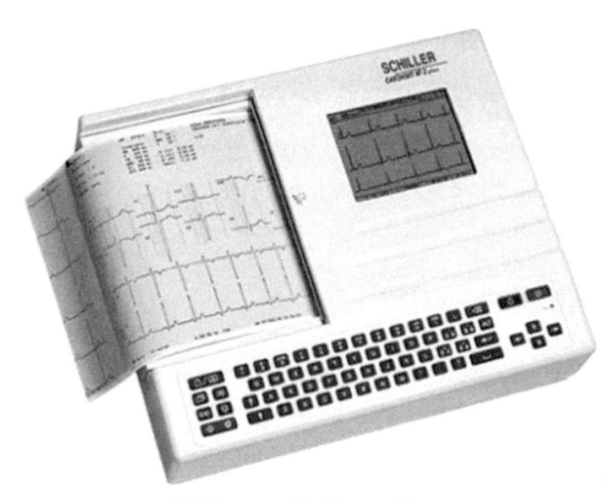

(b) Modern ECG machine

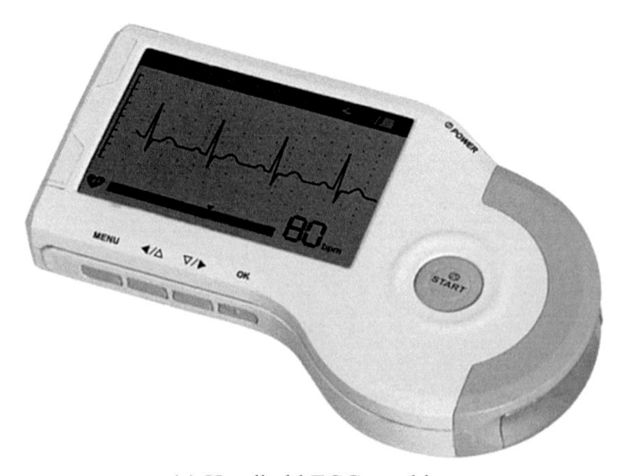

(c) Handheld ECG machine

FIGURE 7.3 Development in design methodologies has drastically reduced the size of health monitoring devices (pastmedicalhistory.co.uk, tradeindia.com, and farlamedical.co.uk).

7.4 DESIGN FLOW

The above section discussed steps in deciding on device design and various constraints involved in the design of a comprehensive HMS. It also discussed how design engineers could evaluate these constraints via virtual design techniques with the help of available sophisticated design software. The present section helps in the understanding of design methodology and the design of a comprehensive HMS explained.

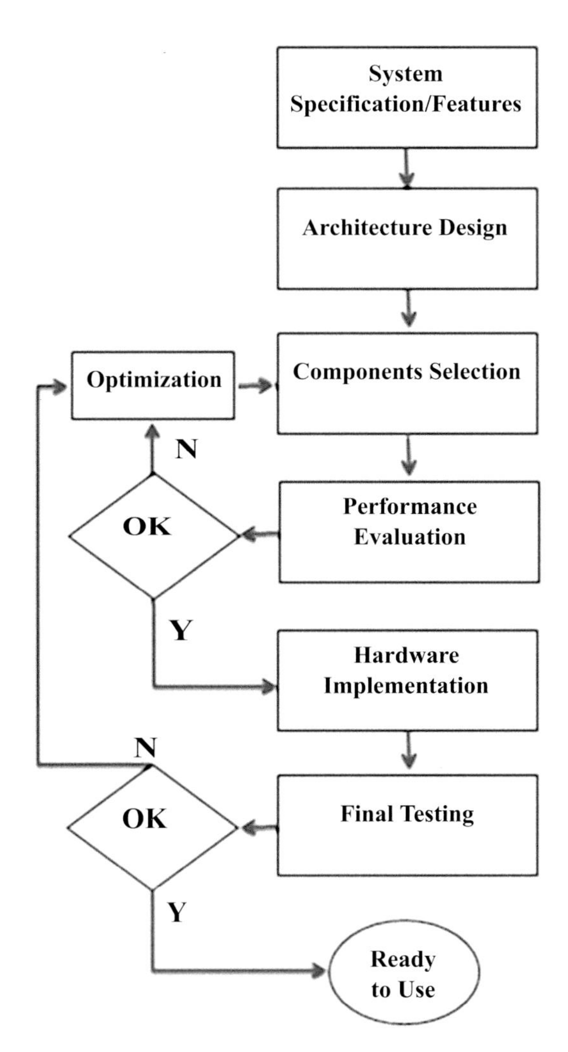

FIGURE 7.4 Design flow of a health monitoring system.

7.4.1 SYSTEM SPECIFICATION AND FEATURES

Before going into specific details, the design of a comprehensive HMS presented with the help of a simple ECG monitoring system is widely employed in different healthcare setups. Similarly, the design can be extended to any number of parameters such as body temperature, activity measure, SpO_2, etc. The design of any HMS needs to accomplish in several steps, and the complete design flow is represented in Figure 7.4. The design initiation is done by exploring the specification and features that need to incorporate into the overall system. If we consider the design of an ECG monitoring system, the specifications such as the number of leads to be monitored, the number of electrodes to be placed, sampling frequency, size of the analog-to-digital converter (ADC), and cutoff rates of filters [18] need to decide at this stage before implementation and system designing. The specifications of the single-channel ECG monitoring system are presented in Table 7.1.

TABLE 7.1 Specifications of the ECG Monitoring System Presented in the Chapter

Specification of the System (ECG Monitoring System)	
Parameter/Feature	**Specification**
Number of ECG leads to be monitored	1, Lead-2
Number of electrodes to placed	3, RA, LA, RL
Sampling frequency	128 Hz
ADC	8-bit
Filters cutoff frequencies	LPF—40 Hz, HPF—0.6 Hz, BSF—50 Hz

7.4.2 SYSTEM ARCHITECTURE

Based on the specification and features decided for the system, an architecture is developed for the desired system functionality. The purpose is to conceptualize the specs and features of the system functionality imagined at the first stage. The architecture describes the comprehensive behavior of any system. During the development of the architecture, the system designer models different blocks and units along with the connection circuitry and helps the engineer in determining the selection of components and other materials required for design [21]. The architecture of a comprehensive HMS is shown in Figure 7.5. The complete architecture can be divided into three sections, as labeled in Figure 7.6. These sections consist of a body sensor network (BSN), cloud environment consisting of Internet and cloud databases, and finally a monitoring station for accessing the parameters. In the following sections, the implementation of each component is discussed in detail. From here onward, the steps depicted in Figure 7.4 do not require a separate discussion, since component selection and performance evaluation are presented according to the concerned section.

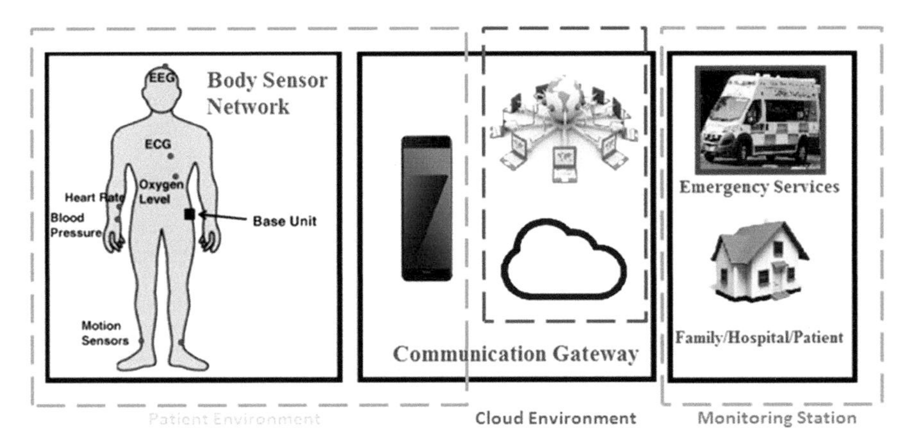

FIGURE 7.5 Architecture of a comprehensive health monitoring system.

7.4.3 PATIENT ENVIRONMENT

Patient environment section is responsible for picking up various signals of interest from the human system. For physiological sensing, sensors are placed over the specific part of the body and termed as a node. By employing various sensor nodes, a network of sensors called BSN formed with the help of a microcontroller-based networking module. The networking module is then interfaced to the Internet using a smartphone or any other Wi-Fi network. By interfacing the BSN to the Internet, data can be transmitted, stored, and processed using cloud-based algorithms in real-time or offline modes. The architecture used at the patient location or environment is depicted in Figure 7.6.

7.4.4 SENSORS

The first block of the architecture shown in Figure 7.6 represents different sensors employed for recording and sensing different patient parameters for continuous monitoring of human. The front-end panel of the patient environment consists of several sensors for detecting various physiological parameters. For understanding, we are presenting a system developed for monitoring three parameters, namely, ECG, body temperature, and patient activity, by employing three individual sensors. The details of sensors used are discussed in the following sections.

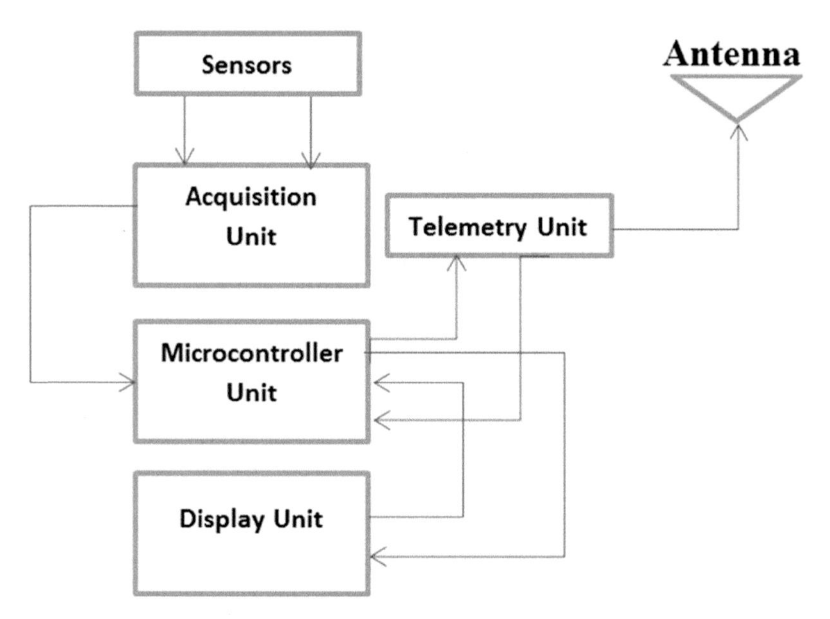

FIGURE 7.6 General architecture of the patient environment of the HMS.

7.4.4.1 AD-8232 (ECG SENSOR)

The AD8232 is a small integrated chip used to record electrical signals produced by the heart. It uses as a single-lead HR monitor and a cost alternative of conventional recording methods. AD 8232 integrated circuit produces the analog values of the electrical voltage seen at the surface of the body and can be plotted on a graph to represent the ECG signal. It has an op-amp-based electric circuitry that helps to show any noisy ECG signal as a clear waveform at the output stage [13], thus reducing processing time.

An AD8232 ECG sensor is depicted in Figure 7.7, comprising of nine pins for connection as GND, 3.3V, OUTPUT, SDN, LO+, LO-, RA, RL, and LA [13]. The relationship between the microcontroller unit (MCU) and the ECG sensor pins is detailed in Table 7.2.

FIGURE 7.7 AD8232 ECG sensor for HR monitoring (Devices, 2013).

TABLE 7.2 Connection between the ECG Sensor Unit and the MCU

ECG Sensor Pin	Connection
GND	The Ground of MCU
3.3 V	3.3 V of MCU
OUTPUT	ADC Pin of MCU
LO+	GPIO-1 of MCU
LO-	GPIO-2 of MCU
SDN	GPIO-3 of MCU
RA	Right arm of the patient
LA	Left arm of the patient
RL	Right leg of the patient

7.4.4.2 DS18B20 (BODY TEMPERATURE)

DS18B20 is a one-wire digital temperature sensor, comprising three wires each for GND, VCC, and DATA. It provides the maximum 12 bit of resolution in temperature measurement. DS18B20 has the feature of communicating over a one-wire bus that requires only one data line (and ground) for communication with a central controller unit. The DATA pin produces a digital output that is red, incorporating the predeveloped libraries by the manufacturers [33]. A DS18B20 temperature sensor used in the design of the HMS is shown in Figure 7.8.

FIGURE 7.8 DS18B20 temperature sensor for measuring the body temperature (Resolution, 2008).

7.4.4.3 ADXL335 (ACCELEROMETER)

To measure the activity of the patient to determine whether they are stationary or moving, ADXL335 is employed. It is a complete, power-efficient accelerometer sensor that can measure the dynamic acceleration in *X*, *Y*, and *Z* directions regardless of the cause of the acceleration. It might be due to motion, vibration, shock, or a static acceleration (tilt or gravity). The accuracy of the sensor can be affected by the maximum of ±3 g range with 0.3% nonlinearity characteristics [13]. An ADXL335 sensor used in the system is depicted in Figure 7.9, and the connection of the sensor to the MCU is explained in Table 7.3. It is important to note here that in order to use the ADXL335sensor, the AREFF pin of the MCU must connect to 3 V.

7.4.4.4 MICROCONTROLLER-BASED ACQUISITION UNIT

Arduino Uno is a microcontroller board that comes with the open-source facility. It has a six-channel built-in ADC of 10-bit resolution connected to pin A0-A5 [1]. Hence, these pins can be utilized to acquire analog data. AD-8232 and ADXL335 both the sensors produce analog output, and they are attached to analog pins. A0 pin of the MCU is connected to the output side of the ECG sensor, that is, AD8232, and pins A1–A3 are attached to the output pins of a three-channel accelerometer. Apart from this, it has 14 digital I/O pins; one of the digital pins in the reading mode is connected to the data pin of DS18B20, which produces a digital signal or output and performs the role of a data acquisition unit. The Arduino Uno board has a pair of RX and TX pins that can be utilized to interface the ESP-8266 module to the board and send the data over the cloud using the Internet [32] (Figure 7.10).

FIGURE 7.9 ADXL335 accelerometer for monitoring the activity of the patients (Devices, 2012).

TABLE 7.3 Connection between the Accelerometer and the MCU

ADXL335 Pin	Connection
GND	The Ground of MCU
3 V	–
5 V	5 V of MCU
Z_{out}	ADC Pin of MCU
Y_{out}	ADC Pin of MCU
X_{out}	ADC Pin of MCU
Test	–

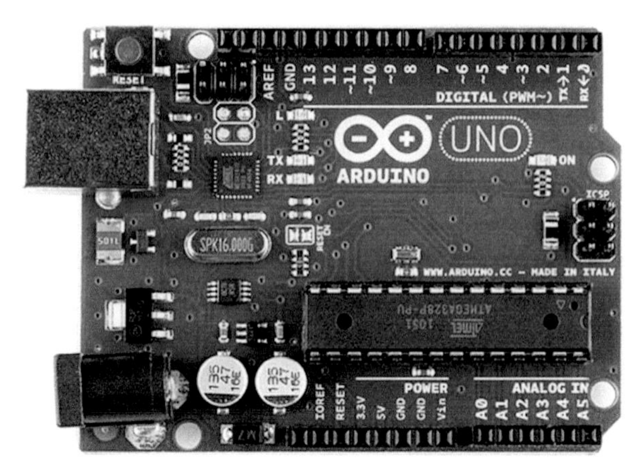

FIGURE 7.10 Arduino Uno MCU for data acquisition and controller applications (Arduino, 2015).

7.4.4.5 TELEMETRY UNIT (ESP8266)

For transmission of sensor data over a cloud server, a Wi-Fi module is interfaced through the Arduino board. The Wi-Fi module gives quick and advantageous access to the Internet to transmit the on-going ECG signal to the Internet of things (IoT) cloud server for storage and retrieval as required. Due to the utilization of the ground-breaking MCU, information is packetized and transmitted by specific interchanges conventions [27]. Esp-8266-01 module is employed as a Wi-Fi module in our design, which is shown in Figure 7.11.

FIGURE 7.11 ESP 8266 Wi-Fi module with an embedded antenna [27].

7.4.4.6 CLOUD ENVIRONMENT

To send the physiological data over the cloud, the telemetry unit, that is, ESP 8266, is interfaced to the Internet using the Arduino Uno MCU. An open-access IoT application programming interface (API) "ThingSpeak" is utilized for logging the data from physiological sensors that enable us to implement a real-time processing algorithm on the sensors data using the MATLAB script. To record the physiological data at the "ThingSpeak," which is a cloud platform over the Internet, the application of the HTTP protocol is used, and data sent via a local area network are readily available in a hospital setup or at clinician facility. After processing and necessary operations on data, the results and recordings easily visualized through a web browser, by using the concepts of "Message Queuing Telemetry Transport" (MQTT). This facility may be expanded to the mobile browsers for viewing of recorded signals on a mobile platform, thereby allowing access even when traveling provided functional network connectivity is available. The diagnostic results and recordings can be accessed by the patient, their relatives, healthcare services providers, or anyone else who has the authentication credentials to view the recorded patient data [41]. The primary steps in the process of sending the sensor data to the cloud station and visualization of data at remote monitoring station using HTTP protocol are depicted in the following steps.

Request: The user starts the communication by sending an HTTP server request to the cloud station for access to the webpage.

Response: The sensor data or HTML file is directed to the cloud station or user site in the reaction of the request. The WEB BROWSER converts the content of the HTML file into the webpage for compatibility.

Subscribe: By the application of the API over the IoT cloud, the webpage becomes capable of providing specific topics related to the ECG monitoring node.

Store: The sensor data deposited into the specific databases are created and managed by the dedicated storage server.

Publish: The ECG monitoring node distributes information to the MQTT server on specific subject information stored in the system. This information is sent to different website pages that present similar information for other authenticated users.

4.4.4.7 PROCESSING UNIT

At the IoT cloud, we have created a channel with three fields; each of them is dedicated to one individual sensor. Sensor data are coming at channels processed by the software-based unit implemented with the help of MATLAB algorithms. The architecture of the implemented processing algorithm is depicted in Figure 7.12.

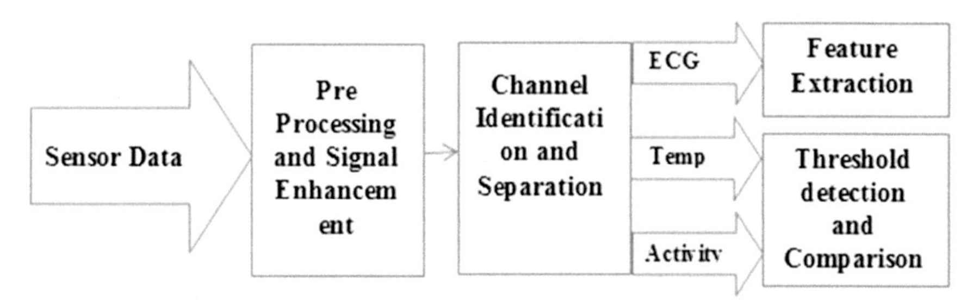

FIGURE 7.12 Architecture of the processing unit.

The processing unit processes the incoming channel data coming through different sensors and performs the necessary operations over it. The primary aim is to remove various noise and artifacts. A MATLAB-based bandpass filtering and smoothing is performed to remove the noise and objects [8]. The processed data is collected through the physiological sensors that have been classified based on the source. The electrical signal is coming from the source converted into physiological quantities such as temperature, HR, and activity via predefined algorithms. To calculate the body temperature from the electrical signal, we have employed predefined libraries of Dallas, etc. [33]. The activity of the person, whether in stationary or motion mode, is estimated by finding the absolute deflection in X, Y, and Z directions using previously developed methods [13]. The body temperature and activity (measured in absolute deviation) are compared to a predefined threshold to estimate the standard value and movement, respectively. The ECG data coming through the ECG sensor are normalized to voltage level of 0–5 mV from the level 0–1024 to convert it into voltage. Now, the feature extraction algorithms are applied to extract the six ECG features, namely, QRS duration, PR-interval, QT-interval, T-interval, P-interval, and HR. The features pass to the intelligence unit, which is a trained artificial neural network (ANN) model that estimates the current health status of the patient. Presently, the intelligence unit is only assessing and classifying the ECG signals into three different categories of patient condition; however, soon, the assessment of the body temperature and HR activity is expected to integrate into the ANN unit [15].

7.4.4.8 INTELLIGENCE UNIT

The "ThingSpeak" is an IoT platform that supports the implementation of MATLAB-based algorithms over the real-time data stream coming from the sensor nodes [42]. In this way, the designer can incorporate processing algorithms into HMSs to analyze physiological parameters measured from the subjects. Implementation of the script of the trained neural network model at the IoT platform constitutes the intelligence unit of the HMS. To implement the intelligence unit, we have trained an ANN using the pattern recognition algorithm in the MATLAB and exported the model as the MATLAB script. The ANN model inputs the six ECG features as input and performs the classification between three different ECG rhythms, namely, normal, bradycardia, and tachycardia rhythms. The ECG features used as input are features extracted by the cloud-based processing unit described in the above section. Based on the test data, the performance of ANN is accessed; the ANN has shown the overall accuracy of 99.3% in successful classification of the ECG abnormality in the patient(s) [43]. The confusion matrix for the same is depicted in Figure 7.13. The other two monitoring parameters are monitored for their reference values, and when these values are higher than the reference levels, they are categorized in the abnormal category, for example, the reference value of 98.3 F for comparing the body temperature of the human being could be considered as a reference value.

7.5 REMOTELY ACCESSIBLE MONITORING STATION

As the physiological data are stored at the cloud-based station, they can be easily accessed through remote locations. To retrieve information, Internet connectivity, and a web browser required, and by employing the access credentials, one can retrieve the desired parameters of the patients. Again, to retrieve the data, one needs to establish the communication using HTTP protocol [44]; the stepwise procedure of the data transfer and retrieval is depicted in Figure 7.14.

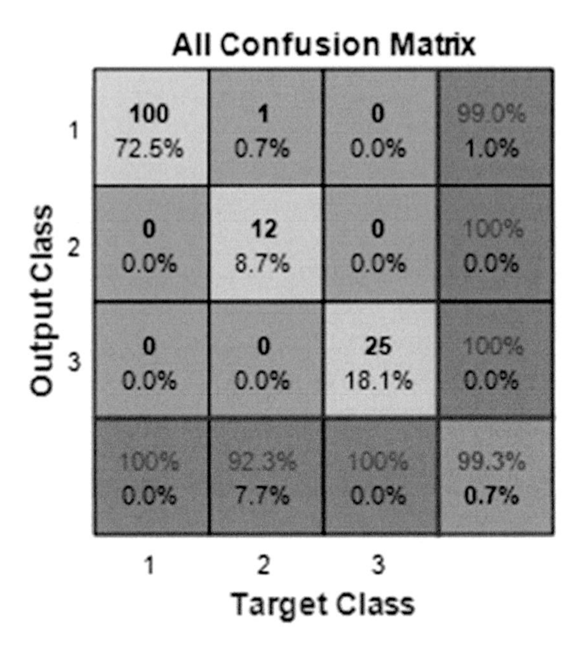

FIGURE 7.13 Confusion matrix of the ANN designed for ECG abnormity.

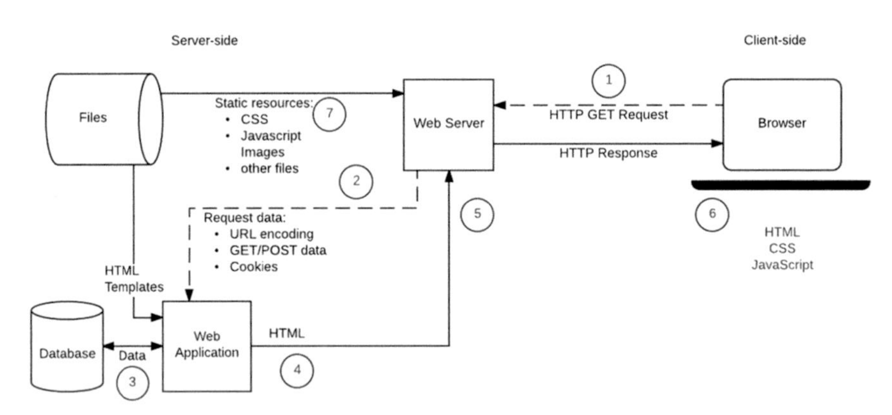

FIGURE 7.14 Procedure for accessing the patient data at the remote station using the HTTP protocol [44].

7.6 PROTOTYPE DEVELOPMENT

To develop the prototype of the above-discussed system, we have integrated the individual modules, as depicted in the sketch diagram represented in Figure 7.15. The output pin of the AD-8232 module that senses the ECG signal is attached to the A0 pin of the Arduino Uno, A0 pin is an ADC pin, which converts the analog voltage sensed by the AD-8232 into a digital value. All other pins of the AD-8232 and accelerometer connected in the same way, as depicted in the sketch diagram.

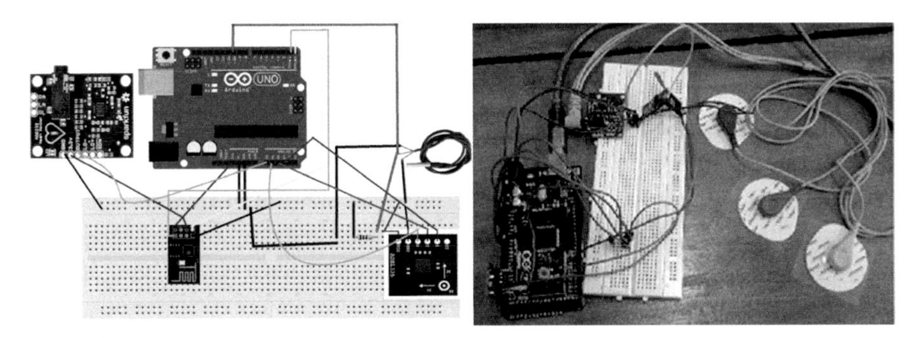

FIGURE 7.15 Sketch diagram of the HMS and its implantation using Arduino.

The Arduino function is to read and transmit the physiological sensor data over the IoT cloud that written and uploaded into the MCU unit. The DS18B20 temperature sensor is a digital sensor and hence to integrate this with MCU, a different topology used. A 4.7-kΩ value register is employed between the data and the V_{CC} pin. The data pin of AD8232 is connected to the digital pin of the MCU, and a function for reading and transmitting the numeric value to "ThingSpeak" is written separately. At last, an Arduino program to connect ESP-8266 to the Internet and to initiate the communication between cloud and hardware is developed, in which both of the above-mentioned functions are called for execution. The program successfully writes the sensor data to the "ThingSpeak" IoT software.

At the IoT platform, as per the above-discussed method, a MATLAB-based processing script is implemented over the live channel data coming through the sensor network, which is visualized to the remote clients with proper authentication.

7.7 LIMITATIONS OF THE PROPOSED APPROACH

For any healthcare systems, the essential governing features are the demonstration of its effectiveness, efficiency, and quality to users, the community, and funders. Applying remote medical diagnosis techniques and employing continuous patient monitoring system can help significantly to reduce healthcare costs and correct performance management, specifically in the management of chronic diseases. Particular challenges in implementation of these systems could be in continuous patient monitoring, threats to patient confidentiality and privacy, poor system design and its implementation, system malfunctioning leading to medical errors and misrepresentation of facts, technology acceptance and lack of system interoperability with electronic health records and other IT tools, decreased face-to-face communication between the doctor and the patient, sudden interruptions of telecommunication networks, scalability in terms of data rate, power and energy consumption, antenna design, quality of service, energy efficiency, weight of wearable devices, difficulty in data processing due to the instruments used in patient monitoring, the location of data collected, which affects accuracy and consistency of information, user training to use wearable system, market penetration with device, and sensor type for specific monitoring aspects [28]. All the above limitations need to be addressed to design a robust, comprehensive patient HMS.

7.8 CONCLUSION AND FUTURE SCOPE

This chapter introduces the design of smart embedded systems for health monitoring applications, which is an upcoming area in the field of patient healthcare diagnosis and monitoring. It encompasses the concepts starting from the basics of the system design, including the design metrics, to the hardware implementation of the HMS for a better and practical understanding of the system design approach. The first part of the chapter focuses on covering the introductory topics related to the HMSs such as background, history, current trends, and types. This chapter provides a brief overview of the smart HMSs and their involved design processes. Further sections are dedicated to concepts that are considered during the design of any system, such as size, cost, speed, etc., and the discussions concerning the improvement of the constraints are provided in this chapter.

Furthermore, discussion is extended to practical aspects of the system design and implementation of a smart system for monitoring ECG, body temperature, and patient activity presented using analysis of the data collected from physiological sensors. This chapter introduces the procedure of developing smart interfaces for health monitoring using concepts of IoT, ANN, and embedded system design, which is an upcoming area presently. In the future, the method is expected to improve for more compact, feasible, and robust systems. In the near future, the intelligent advanced algorithm systems may automatically direct the patient for a particular health issue.

This would be a tremendous medical boon for the patient(s) as well as their clinical care provider. The remote access of the physiological parameters is expected to enhance the diagnostics performance and can improve the life of the patient and their attendants.

ACKNOWLEDGMENT

The authors would like to graciously acknowledge the financial assistance provided by the Department of Biotechnology, Government of India, under Grant BT/PR15673/NER/95/22/2015 dated 09.12.2016.

KEYWORDS

- **smart interfaces**
- **comprehensive health monitoring**
- **artificial intelligence**
- **machine learning**
- **personal digital assistant**

REFERENCES

1. S. Arduino, Arduino. *Arduino LLC*, 2015.
2. M. M. Baig and H. Gholamhosseini, "Smart health monitoring systems: An overview of design and modeling," *J. Med. Syst.*, vol. 37, no. 2, pp. 1–14, 2013.
3. S. P. Balasundaram, "Increasing the battery life of a mobile computing system in a reduced power state through memory compression," US Patent App. 11/450,214, 2007.
4. J. Basilakis, N. H. Lovell, S. J. Redmond, and B. G. Celler, "Design of a decision-support architecture for the management of remotely monitored patients," *IEEE Trans. Inf. Technol. Biomed.*, vol. 14, no. 5, pp. 1216–1226, 2010.
5. S. Brownsell, D. Bradley, S. Blackburn, F. Cardinaux, and M. S. Hawley, "A systematic review of lifestyle monitoring technologies," *J. Telemed. Telecare*, vol. 17, no. 4, pp. 185–189, 2011.
6. M. Chen, Y. Ma, J. Song, C.-F. Lai, and B. Hu, "Smart clothing: Connecting human with clouds and big data for sustainable health monitoring," *Mobile Netw. Appl.*, vol. 21, no. 5, pp. 825–845, 2016.
7. H. T. Cheng and W. Zhuang, "Bluetooth-enabled in-home patient monitoring system: Early detection of Alzheimer's disease," *IEEE Wireless Commun.*, vol. 17, no. 1, pp. 74–79, 2010.

8. S. D. Choudhari and V. Giripunje, "Remote healthcare monitoring system for drivers community based on IoT," *Int. J. Emerg. Technol. Eng. Res.*, vol. 4, no. 7, pp. 117–121, 2016.
9. B. Chowdhury and R. Khosla, "RFID-based hospital real-time patient management system," in *Proc. 6th IEEE/ACIS Int. Conf. Comput. Inf. Sci.*, 2007, pp. 363–368.
10. D. C. Classen, S. L. Pestotnik, R. S. Evans, and J. P. Burke, "Computerized surveillance of adverse drug events in hospital patients," *JAMA*, vol. 266, no. 20, pp. 2847–2851, 1991.
11. H. Costin, C. Rotariu, I. Alexa, G. Constantinescu, V. Cehan, B. Dionisie, G. Andruseac, V. Felea, E. Crauciuc, and M. Scutariu, "Telemon—a complex system for real-time medical telemonitoring," in *Proc. World Congr. Med. Phys. Biomed. Eng.*, 2009, pp. 92–95.
12. *Adxl335: Small, Low Power, 3-Axis±3 g Accelerometer*, Analog Devices, Norwood, MA, USA, 2012.
13. *Ad8232: Single-Lead, Heart Rate Monitor Front End. Rev. A*, Analog Devices, Norwood, MA, USA, 2013.
14. R. Hillestad, J. Bigelow, A. Bower, F. Girosi, R. Meili, R. Scoville, and R. Taylor, "Can electronic medical record systems transform health care? Potential health benefits, savings, and costs," *Health Affairs*, vol. 24, no. 5, pp. 1103–1117, 2005.
15. J. J. Hopfield, "Neural networks and physical systems with emergent collective computational abilities," *Proc. Nat. Acad. Sci.*, vol. 79, no. 8, pp. 2554–2558, 1982.
16. M. Imhoff and S. Kuhls, "Alarm algorithms in critical care monitoring," *Anesth. Analg.*, vol. 102, no. 5, pp. 1525–1537, 2006.
17. R. S. Istepanian, S. Hu, N. Y. Philip, and A. Sungoor, "The potential of the Internet of m-health things "m-IoT" for non-invasive glucose level sensing," in *Proc. Annu. Int. Conf. IEEE Eng. Med. Biol. Soc.*, 2011, pp. 5264–5266.
18. L. Jing-Jian and X. Hai-Qiao,, "Embedded system in the 21 century," *Semicond. Technol.*, vol. 1, 2001.
19. R. Kamal, *Embedded Systems: Architecture, Programming, and Design*. Noida, India: Tata McGraw-Hill Education, 2011.
20. M. Kikuya, T. Ohkubo, K. Asayama, H. Metoki, T. Obara, S. Saito, J. Hashimoto, K. Totsune, H. Hoshi, and H. Satoh, "Ambulatory blood pressure and 10-year risk of cardiovascular and non-cardiovascular mortality: The Ohasama study," *Hypertension*, vol. 45, no. 2, pp. 240–245, 2005.
21. G. J. Klir, *The Architecture of Systems Problem-Solving*. New York, NY, USA: Springer Science & Business Media, 2013.
22. S. Kumar, W. J. Nilsen, A. Abernethy, A. Atienza, K. Patrick, M. Pavel, W. T. Riley, A. Shar, B. Spring, and D. Spruijt-Metz, "Mobile health technology evaluation: The mhealth evidence workshop," *Am. J. Prev. Med.*, vol. 45, no. 2, pp. 228–236, 2013.
23. F. Lau, C. Kuziemsky, M. Price, and J. Gardner, "A review of systematic reviews of health information system studies," *J. Am. Med. Inform. Assoc.*, vol. 17, no. 6, pp. 637–645, 2010.
24. A. M. Law, W. D. Kelton, and W. D. Kelton, *Simulation Modeling and Analysis*, vol. 2. New York, NY, USA: McGraw-Hill, 1991.
25. C.-C. Lin, M.-J. Chiu, C.-C. Hsiao, R.-G. Lee, and Y.-S. Tsai, "Wireless health care service system for the elderly with dementia," *IEEE Trans. Inf. Technol. Biomed.*, vol. 10, no. 4, pp. 696–704, 2006.
26. G. López, V. Custodio, and J. I. Moreno, "LOBIN: E-textile and wireless-sensor-network-based platform for healthcare monitoring in future hospital environments," *IEEE Trans. Inf. Technol. Biomed.*, vol. 14, no. 6, pp. 1446–1458, 2010.

27. M. Mehta, “Esp 8266: A breakthrough in wireless sensor networks and the Internet of things,” *Int. J. Electron. Commun. Eng. Technol.*, vol. 6, no. 1, pp. 07–11, 2015.
28. N. Mohammadzadeh and R. Safdari, “Patient monitoring in mobile health: Opportunities and challenges,” *Med Arch.*, vol. 68, no. 1, pp. 57–60, 2014.
29. S. G. Mougiakakou, C. S. Bartsocas, E. Bozas, N. Chaniotakis, D. Iliopoulou, I. Kouris, S. Pavlopoulos, A. Prountzou, M. Skevofilakas, and A. Tsoukalis, “Smartdiab: A communication and information technology approach for the intelligent monitoring, management, and follow-up of type 1 diabetes patients,” *IEEE Trans. Inf. Technol. Biomed.*, vol. 14, no. 3, pp. 622–633, 2010.
30. S. Patel, B.-R. Chen, T. Buckley, R. Rednic, D. McClure, D. Tarsy, L. Shih, J. Dy, M. Welsh, and P. Bonato, “Home monitoring of patients with Parkinson's disease via wearable technology and a web-based application,” in *Proc. Annu. Int. Conf. IEEE Eng. Med. Biol. Soc.*, 2010, pp. 4411–4414.
31. M. Pavier and T. Sammon, “Embedded power management control circuit,” US Patent App. 11/078:807, 2005.
32. R. A. Rahman, N. S. A. Aziz, M. Kassim, and, M. I. Yusof, “IoT-based personal health care monitoring device for diabetic patients,” in *Proc. IEEE Symp. Comput. Appl.: Comput. Appl. Ind. Electron.*, 2017, pp. 168–173.
33. *Programmable Resolution 1-Wire Digital Thermometer Datasheet*, Maxim integrated, San Jose, CA, USA, 2008.
34. S. Robinson, *Simulation: The Practice of Model Development and Use*. Inglaterra, U.K.: Wiley, 2004.
35. R. Roine, A. Ohinmaa, and D. Hailey, “Assessing telemedicine: A systematic review of the literature,” *Can. Med. Assoc. J.*, vol. 165, no. 6, pp. 765–771, 2001.
36. A. Sangiovanni-Vincentelli, and G. Martin, “Platform-based design and software design methodology for embedded systems,” *IEEE Des. Test Comput.*, vol. 18, no. 6, pp. 23–33, 2001.
37. D. Sciuto, F. Salice, L. Pomante, and W. Fornaciari, “Metrics for design space exploration of heterogeneous multiprocessor embedded systems,” in *Proc. 10th Int. Symp. Hardw./Softw. Codes.*, 2002, pp. 55–60.
38. A. M. Shams, T. K. Darwish, and M. A. Bayoumi, “Performance analysis of low-power 1-bit CMOS full adder cells,” *IEEE Trans. Extensive Scale Integr. Syst.*, vol. 10, no. 1, pp. 20–29, 2002.
39. Y. Shim, “Exercise systems in virtual environment,” US Patent App. 12/216,540, 2009.
40. S. Siebig, S. Kuhls, M. Imhoff, U. Gather, J. Schölmerich, and C. E. Wrede, “Intensive care unit alarms—How many do we need?” *Crit. Care Med.*, vol. 38, no. 2, pp. 451–456, 2010.
41. M. Singh, M. Rajan, V. Shivraj, and P. Balamuralidhar, “Secure MQTT for the Internet of things (IoT),” in Proc. *5th IEEE Int. Conf. Commun. Commun. Syst. Netw. Technol.*, 2015, 746–751.
42. P. Swathy, C. Periasamy, “Modular health care monitoring for patients using IoT,” *Indian J. Appl. Res.*, vol. 8, no. 4, pp. 21–25, 2018.
43. R. P. Tripathi, G. Mishra, D. Bhatia, and T. K. Sinha, “Classification of cardiac arrhythmia using hybrid technology of fast discrete Stockwell-transform (FDST) and self-organizing map,” 2018, doi: 10.20944/preprints201806.0321.v1
44. T. Yokotani and Y. Sasaki, “Comparison with HTTP and MQTT on required network resources for IoT,” in *Proc. Int. Conf. Control, Electron., Renew. Energy Commun.*, 2016, pp. 1–6.

CHAPTER 8

IoT-Based Smart Agriculture in India

AKANKSHA GUPTA[1] and UMANG SINGH[2*]

[1]*Department of Computer Science, Swami Shraddhanand college, University of Delhi, New Delhi, Delhi 110036, India*

[2]*Institute of Technology and Science, Mohan Nagar, Ghaziabad, Uttar Pradesh 201007, India*

[*]*Corresponding author. E-mail: singh.umang@rediffmail.com*

ABSTRACT

Proliferation of technologies can strengthen agriculture field for assessment of agriculture-related information such as water level, productivity of crops, soil quality, and proper fertilizers as per soil type. In fact, farmers can remotely access, receive updates of weather forecasts, and monitor their land through mobiles and computers. Furthermore, there is also a need to enhance skills and knowledge of farmers in harvesting so that excessive use of pesticides and fertilizers should not affect natural ecosystems and quality of food products as well. Thus, proper awareness of important information in agriculture, such as soil forecasting and weather forecasting, is one of few common problems for nationwide Indian farmers. For transformation of traditional agriculture to smart agriculture, the Internet of things (IoT) plays a crucial role in providing information to farmers about their agriculture fields. Monitoring of environmental factors can be done with the help of IoT-based devices and related environment. This chapter initially critically analyzes, assesses, and addresses the existing problems related to agriculture, and then, integration of sensor technology and its integration with the IoT have been studied and reviewed based on real-life existing problems. Furthermore, this attempt presents a solution for sustaining soil quality in varying weather conditions so that issues related to sufficient knowledge about the soil can be optimized.

8.1 INTRODUCTION

Agriculture is the back bone of India as it provides employment to over 60% of the population. In addition to this, food products such as pulses, rice, spices, wheat, etc., are also outsourced from India to worldwide. Therefore, agriculture plays a vital role in the economic development of the nation and contributes to making a country from developing to a developed nation. Furthermore, India is well known for its diversity in culture, religions, caste, traditions, and beliefs with a mixed compound of 28 states, nine union territories, geographical location 8° 4′ and 37° 6′ North and longitudes 68° 7′ and 97°25 East within an area of 3,287,263 km^2, where approximately 1,210,569,573 people are living. Due to different weather conditions (temperature, humidity, rain, etc.) and soil conditions (moisture, pH level, and nutrition level), Indian farmers are facing many problems, which is a very important and serious concern in today's scenario [1–3].

In the past few decades, numerous technological transformations have been observed in the area of farming, which are based on a technology-driven approach as compared to the old manual techniques. This scenario has led to smart farming—an emerging concept of integrating technology with agriculture in order to increase output with minimum efforts.

With the help of this new emerging concept, information such as soil, temperature, humidity level, rail fall status, fertilizers, storage capacity of water tanks, etc., can be easily obtained. In addition to this, farmers have control over livestock monitoring and growing crops performance. However, in traditional agriculture, extensive farming, use of indigenous tools, cattle raising, lack in harvest production, and use of the slash and burn method are important characteristics. Due to the use of the slash and burn method in the traditional approach, the organic matter from soil is reduced very rapidly.

There is a need to motivate small farmers and provide ways to boost crop productivity such as rice. Table 8.1 shows the current rice productivity in various states of India.

Most of the farmers here possess less than 2 hectares of land. Therefore, there is need of employing more technological tools to improve the crop productivity within limited resources, knowledge, and awareness.

8.2 LITERATURE SURVEY

For agriculture, soil, light, and water are important sources of productivity of crops. However, soil erosion continues to be a major environmental problem with regard to land use in India. In fact, soil erosion is a big problem in India. Zaimes et al. [13] have discussed about soil conversation issues in India. The authors have focused on soil degradation in Himalaya region, Indo-Gangetic plains, dry and arid regions, and coastal lands in India.

TABLE 8.1 Top 10 Rice-Producing States in India (2016)

Sr. No.	State	Rice Productivity (in million tons)
1.	West Bengal	15.75
2.	Uttar Pradesh	12.5
3.	Punjab	11.82
4.	Tamil Nadu	7.98
5.	Andhra Pradesh	7.49
6.	Bihar	6.5
7.	Chhattisgarh	6.09
8.	Odisha	5.87
9.	Assam	5.14
10.	Karnataka	3.95

Soil erosion on hill slopes, shifting cultivation, sheet erosion, ravine lands and floods, shifting sand dunes, wind erosion, and improper land management are major challenges in various agroclimatic areas of India. Thus, there is a requirement for various reclamation programs for efficient learning and training [14] and land use planning to improve productivity in limited lands. In 2016, India's total geographical area was 328.73 million hectares (mha), reporting area was 304.89 mha, and area used for agriculture purpose was 264.5 mha. The authors discussed that important combination of factors for soil degradation depends on various factors and equals 147 million hectares of land (94 mha from water erosion, 16 mha from acidification, 14 mha from flooding, 9 mha from wind erosion, and 6 mha from salinity). To enhance the production capacity of the ecosystem, there is a need to monitor the domain to restore vibrancy. For this, farmers require thorough information of farming cycle, market price, and current production level statistics along with available basic knowledge of each crop for in depth understanding. Mohanraj et al. [15] have discussed the benefits of information and communication technology (ICT) in agriculture sector and have presented the path of rural farmers to replace traditional techniques. In this chapter, comparative analysis between the developed system and the existing systems is discussed. Gubbi et al. [16] have focused on various Internet of things (IoT) devices, which will enable farmers to enhance food production by 70% by 2050.

An automated soil erosion monitoring system that works on the principle of measuring the soil erosion on the surface as well as the factors that affect the erosion process has been proposed in [15]. It works by measuring the ground-level changes, rainfall patterns, air quality, weather temperature, soil temperature, and soil moisture through remote communication (using ultrasound waves) and then analyzes the data for measuring the soil erosion. The authors have presented this automated system in Thasos Island, Northern Greece, where they have studied the impacted areas for erosion and collected data for environmental quantities to study soil erosion. They have used solar power supply, a special-purpose data logger and a remote communication unit to collect data. Such a system can highly benefit by incorporating the IoT system and practices. IoT devices such as sensors and microcontrollers can sense the surrounding data and communicate it over a network to the cloud. Thereon, the collected data can be analyzed easily for environmental factors and the soil erosion process. The IoT can provide a more robust and automated system to study the erosion factors. Such a system, if used in India, could prove to be an asset in increasing the productivity of agriculture to a large extent.

8.3 CURRENT PROBLEMS FACED BY FARMERS

In India, agriculture is the major source of income for about 47% of the population, contributing 16% to the national GDP. The economy of the country is highly dependent on the agricultural yield. In addition, it is also responsible for fulfilling the basic needs of food, raw materials, nourishment, and strengthened environment for the people. Even though agriculture is the main source of income for a vast majority of people in our country, there are many problems that are faced by the farmers on a daily basis leading to bad crop and eventually bad turnover for these people. A few of these problems can be observed in Figure 8.1.

A major reason for failure to produce a crop good in quality and quantity is soil erosion. Either the soil selected is not according to the expected output or the good soil gets eroded by climactic conditions such as rain or winds. Bad soil often leads to a low yield. Seeds are also the most essential input for attaining a high yield, but uneven distribution of good seeds in the market and lack of high quality seeds due to their cost often leads to poor input and hence equally poor yield for the farmer. In the current conditions, the change in climate over the years has led to an unstable and nonstatic environment for the agricultural field. Extreme heat across the world has led to a large fall

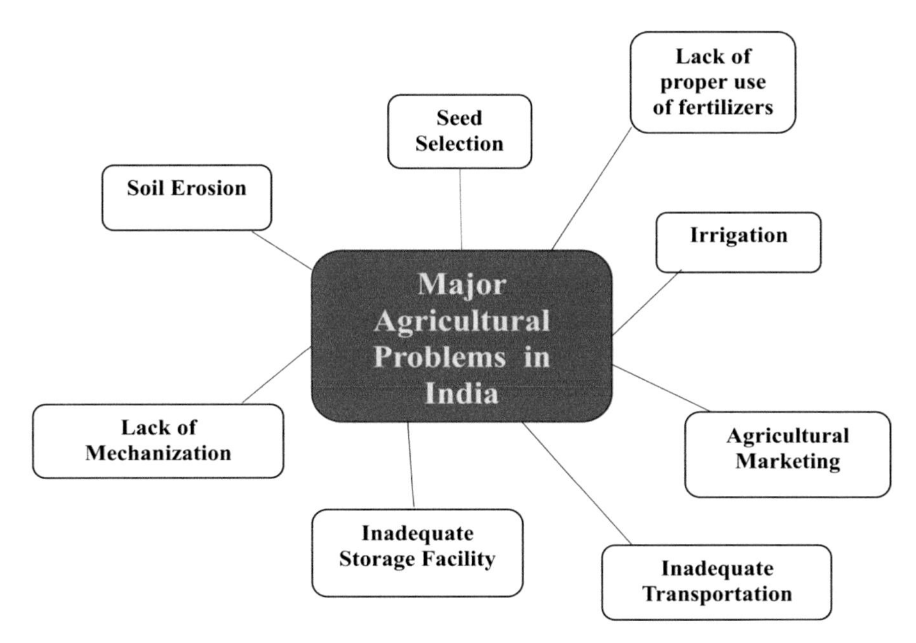

FIGURE 8.1 Major agricultural problems in India.

in the yield of the farmers. Erratic rainfall also causes major losses and crop damage. Irrigation is another most important aspect of farming, but despite the fact that India is the second largest irrigated country of the world after China, the irrigation facilities are not up to the mark, and only a small amount of the total cropped area is covered. Due to varied climate conditions and uncertain rainfall, irrigation becomes more important and should be taken care of more prominently if we want to reach high standards of farming and agriculture in our country. Most of the work in farming such as irrigation, crop sowing, soil monitoring, weeding, and transportation of crops is still done manually in most parts of India. Carrying out such tasks manually, using tools such as wooden ploughs, leads to a more time-consuming and difficult-to-implement process. There is also the added lack of precision, which leads to low yields for the farmers and wastage of human resources as well. Hence, there is a requirement to automate basic tasks in agriculture for maximizing output and minimizing human efforts. Employing machinery and related tools can highly benefit to help in avoiding wasting of labor force and add convenience and efficiency to farming methods to facilitate increased production. There is also a lack of storage facilities available to the farmers in India, which often leads to selling of the crops immediately after harvesting and mostly at lower than expected prices. The income of a farmer could be highly increased if adequate storage is made available to them so that there is no rush to sell the crops fearing damage. The Parse Committee estimated the postharvest losses at 9.3% of which nearly 6.6% occurred due to poor storage conditions alone [10]. Transportation of crops is also a major concern due to the lack of good connectivity of cities to the village areas in the country. The existent roads are also not durable enough in changing weather conditions and hence are unusable many times, leading to failure of transporting of the crops on time without spoilage. In conclusion, we can say that the agricultural sector in our country suffers from a large number of basic problems, which should be rectified and dealt with on an immediate basis. One of the best solutions to counter these problems would be to deploy IoT methods in smart farming.

8.4 INCORPORATION OF IOT IN AGRICULTURE

In India, agricultural practices are generally carried out manually by the farmers, which have often led to low yield of crops. To resolve this, we are going to introduce the IoT in the field of agriculture to automate the process of crop monitoring and prediction.

The IoT is an enabling technology that interconnects various devices, sensors, and objects to transfer data within each other over a network with minimum human interference. It is used to build up a smart environment such as smart agriculture or smart transport or smart health, by utilizing physical devices such as sensors, microprocessor, controls, and various communication protocols. These physical devices are not enabled to interact with the Internet directly but can do so by using an IoT gateway. An IoT system, hence, is considered to be a network system by interfacing its environment with the Internet by using sensors, global positioning system (GPS), lasers, scanners, and other information sensing devices available.

Smart solutions to agriculture issues have been deployed in farming at various levels, and the IoT has already brought revolutionary changes in agriculture. Intelligent agriculture systems that deploy IoT practices can be depicted in Figure 8.2

An intelligent agriculture system first deploys a platform where the production of crops is maximized, along with an expert service providing platform for integrating IoT with farming practices. Finally, an online trading platform is used to advertise market and sell the yield at adequate prices leading to high profits.

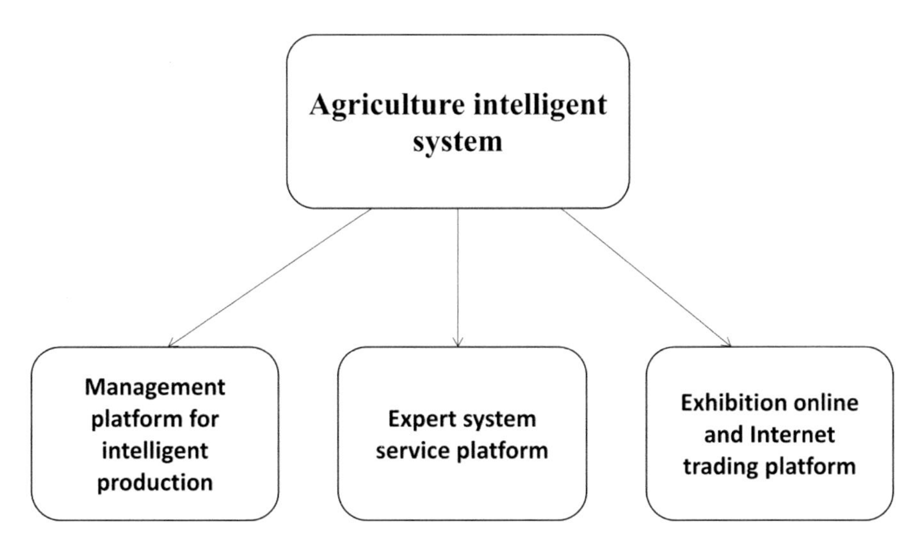

FIGURE 8.2 The modules of an agriculture intelligent system.

Agricultural sector is plagued by a variety of problems such as investment costs, lack of lands, limited knowledge regarding best practices, improper use of fertilizers, lack of good quality seeds and soil, limited storage facilities, etc. The IoT, if deployed in conjunction, can highly help the farmers tackle these issues to a large extent. The innovative methods of farming with the IoT can address the issues of irrigation, soil monitoring, climate change, etc., and increase the output, food safety, storage, and sustainability of the production. To accommodate a larger population, the demand for food will increase, and new techniques need to be designed to create more efficient agricultural production methods. Moreover, global climate is changing, and the growing conditions of the agricultural goods are being affected as well, so there is a need to create new agricultural production models with the focus on productivity and rational usage of environmental resources. In order to overcome variability, one of the biggest challenges to agricultural productivity, farmers need a broader understanding about the characteristics of the field and the development of the crops. Crop investigation and prediction is a time-consuming process, which, if implemented using IoT and data analytics, can be carried out in a much effective and precise manner. This shall lead to optimized yield, minimum costs, and reduction of environmental effect on crops.

Important parameters needed for such a system, such as soil moisture, pH values, weather conditions, etc., can be easily collected using IoT, and an effective strenuous system is created for a better crop monitoring system that maximizes crop yield on the basis of the farm field. The main objective of the system is to improve the quality of life for farm workers by reducing high-labor tasks. Replacing human labor with help of IoT is an emerging trend across the world.

8.5 SMART FARMING SOLUTIONS

As we have already discussed, integrating the IoT with agriculture leads to a better crop production and lesser human intervention in the process, leading to an efficient and highly profitable area for the farmers of the country. Precision agriculture deploys IoT practices in the field of farming in order to ensure optimum growth, health, and sustainability (as shown in Figure 8.3). There are many ways in which IoT could be deployed in this field, some of which we discuss in the following.

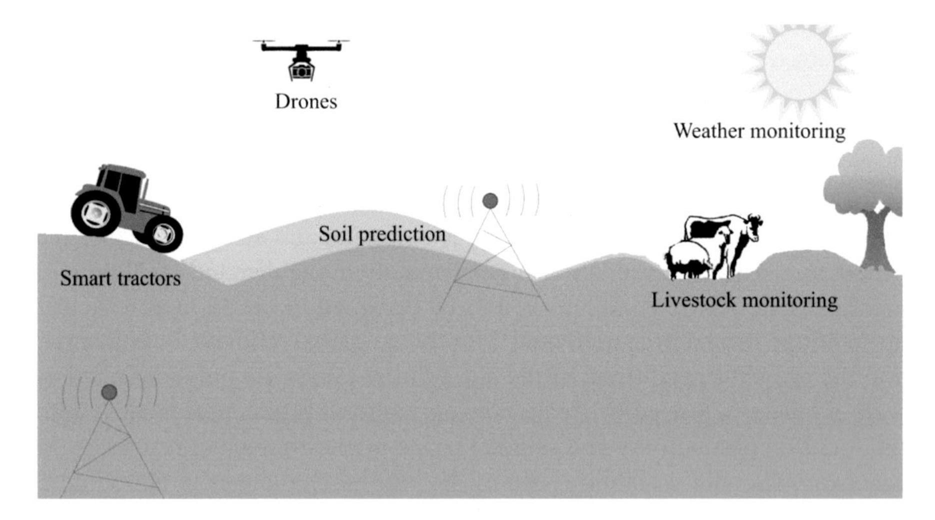

FIGURE 8.3 Applications of precision farming.

8.5.1 SOIL PREDICTION

Soil is the most important part of any agricultural practice, and proper use of soil for specific crop growth is a critical process that determines the actual yield of crops for a farmer. In the older farming methods, soil was planted following the traditional approach of selection and sowing. Not choosing the appropriate soil for the intended crops and the actual weather conditions leads to bad crop growth and losses in the turnover eventually. The IoT can greatly help in predicting the soil by studying the climate of growth and the kind of crops required on the field. Data collected over a period of time can be accessed from over the cloud and used to predict what kind of soil would lead to maximum crop growth in the region. Precise soil prediction is an essential part of precision farming and smart agriculture.

8.5.2 WEATHER MONITORING

Climate is a crucial aspect for agricultural crop growth, and therefore, there is a need to keep track of the surrounding weather conditions, according to which the soil, water level, and crop types can be updated without relying on the imprecise meteorology predictions or manually checking the field for rains. Smart agriculture deploys various sensors across the farmer's field to collect data from the environment and stores it over a cloud. This data can then be used to study the weather conditions and select appropriate crops, soil, and irrigation needs of the farm.

8.5.3 GREENHOUSE MONITORING

Since greenhouse monitoring requires manual intervention for keeping track of the irrigation process, temperature, light, and humidity, IoT technology could be deployed for automating majority of this process [2]. Such a system can monitor the greenhouse atmosphere and alter the surrounding conditions accordingly to form an automated greenhouse environment for the crops. The greenhouse can be segregated into multiple areas that are managed through a base station. Sensor nodes are placed in these measurement areas to collect relevant information, which is then passed onto the controlling sensors to control or change the in-house environment parameters. This arrangement leads to an effective and precisely controlled greenhouse system.

8.5.4 LIVESTOCK MONITORING

Monitoring the hundreds of livestock on a farm can be a tedious process with keeping track of their location as well as looking out for any health issues. This process is an important function to be taken care of, in order to carry out the proper management and growth of farming itself. Sensors can be attached to the livestock to track their location, grazing patterns, bodily functions, etc., which are then sent over to the cloud, from where the farmers can monitor and take appropriate steps to maintain their animals.

8.5.5 SMART TRACTORS

Smart farming is evolving to use more artificial intelligent hardware for agricultural purposes. The tractors on a farm are important for carrying out various tasks on the field based on the kind of equipment used. Artificial-intelligence-powered tractors can highly reduce the manual labor involved with farming by introducing driverless tractors involving minimum human intervention. Inclusion of technologies such as GPS, camera, and IoT connectivity would enable these machines to be autonomous to a large extent and diminish active human control required.

8.5.6 DRONES

Surveillance of the farm field is one of the most valuable pieces of information for precision program. Any kind of problems can be detected early and taken care of, before they lead to more serious issues. However, the traditional approach using helicopters, etc., does not guarantee accurate data collection across very large fields, and hence, smart farming uses drones—unmanned aerial devices—with sensors and built-in digital cameras, giving farmers a better view and more accurate representation of their fields. These drones can be used at on-demand basis and are easy to use and deploy providing real-time data with low investment. They are also a safe and reliable solution to farm management.

8.6 SOLUTION FOR SUSTAINING SOIL QUALITY IN VARYING WEATHER CONDITIONS

In varying weather conditions, soil can be protected by adding organic matter and careful management of fertilizers and appropriate pesticides [4–6]. Table 8.2 lists products that fulfill the criteria of soil protection.

TABLE 8.2 Products Fulfilling Soil Protection

Ref	Product Name	Product Feature	Product Solution	Facilities
[17]	Yuktix Products ColdSense	Monitoring Weather conditions & Asset, managing Pesticides	It conserves resources and enhanced yield and quality	Email Notification Text Message Maintains Dash Board, Live Display and sensor nodes are connected with Yuktix cloud through gateway via Ethernet
[18]	Weeding Robot	It uses the concept of digital image processing to extract the images from databases of related crops so that similar crops can be pesticides timely with robot arms	It reduces the cost of spraying pesticides in the cops.	Autonomous electric robot Dino with precision guidance(GPS and camera)

8.7 LIMITATIONS OF SMART AGRICULTURE

Although the IoT in agriculture would accomplice a great deal of comfort and increase in productivity on a whole, but there are several challenges that arise in implementing technologies in the field [7,8]. First, the lack of knowledge in this field would cause major apprehension in deploying such a system by a farmer on his land. The initial cost of implementation would also be a concern for them. If the system does work, the devices used in smart agriculture will have to be exposed to the harsh environmental conditions on a daily basis, leading to wear and tear and also needing high source of power to keep them in use all through the time. These devices will have to be programmed in a way that they can work on low power for long amounts of time without needing a reset or battery change, since such frequent changes would require shut downs of the stations entirely leading to loss of data. Most often than not, the farmers do not have access to a good-quality network in order for the interconnectivity to work as intended. Even if there is, the link quality at the networking level has also to be maintained in order to assure seamless transfers to and from the cloud storage for the sensors. The security, authenticity, and privacy of the involved have also to be maintained from any kind of internal or external attacks. The hardware for a secure system could cost even more to the meagerly earning farmers. Another consideration to develop IoT solutions in smart farming is to make the farmers aware of and comfortable with using new technologies without any fear or apprehension. Due to lack of education and access to technology, this could be a major hindrance in implementing precision agriculture. In addition, all agricultural fields are different in nature and environmental conditions around them. Since our country is vast in geographical area, similar digitization of things is not feasible. Methodology and development would differ for different fields of study.

8.8 FUTURE SCOPE

Smart farming is the future of all agricultural practices, which would result in maximum yield and minimum effort. Farmers would be in better control over the process of growing crops, livestock management, irrigation, and soil monitoring. Although India has not yet implemented smart solutions in the agricultural sector extensively, many countries have already developed IoT-based solutions in the field of farming which India can also benefit from by incorporating such practices. The Kaa IoT platform, built in Miami, is an enterprise-grade IoT enablement technology that allows walking safely into the agriculture IoT field [9]. It works on the principle of tying different sensors and devices together and provides a variety of IoT-based services such as remote crop monitoring, predictive analytics for crops and livestock, climate monitoring, livestock tracking, and many other facilities. Dashboard is another agricultural monitoring tool using which a farmer can operate satellite imagery and weather data for a better visualization of his practices. LoRa is another Australian revolutionary technical advancement used for real-world smart agriculture deployment such as smart cattle ranching, ingestible cattle health tracker, soil moisture monitoring, cattle health monitoring, autonomous irrigation, and others. ThingsBoard, Inc., is a US corporation founded in 2016 with RnDcentre in Kyiv, Ukraine, that delivers robust and affordable IoT platform with out-of-the-box components and application programming interfaces for smart solutions. They have devised an interactive dashboard that can represent smart farming IoT data visualization embedded in the agricultural project as a smart farm solution. Other interesting IoT agriculture applications are also being used in various countries, such as CROPX's soil monitoring system to indicate the level of irrigation required by measuring moisture, temperature, and electrical conductivity in the soil, TempuTech's wireless sensors to ascertain safety in agriculture storage, CLASS's smart equipment for crop flow management, PRECISIONHAWK's drone for data collection, surveying and imaging in the field, JMB's monitoring solution, in North America, to monitor pregnant cows, and many more. We can take inspiration from these smart solution applications in the future in order to effectively incorporate smart agriculture in our country.

8.9 CONCLUSION

The importance of agriculture in a developing nation, such as India, is high, and incorporating the concepts of IoT in this field can result in increase of agricultural yield to a large extent. Smart farming solutions such as drones, automated tractors, and soil moisture monitoring can ease the problems faced by farmers in the current manual setup [11,12]. We have also seen how the IoT is being used extensively for precision farming in various countries other than India and their applications in implementing solutions for precision agriculture. Although there are limitations of cost, resources, knowledge, and technicalities in putting up these concepts in practice, the overall benefits and ease of work for the farmers shall result in high yields, and hence, deployment of IoT in farming should be done in the majority of regions in our country. Such a scenario shall have a positive impact on the growth of the nation and economy as a whole.

KEYWORDS

- **Internet of things (IoT)**
- **precision farming**
- **smart agriculture**
- **weather monitoring**
- **smart irrigation**
- **soil erosion.**

REFERENCES

1. J. A. Stankovic, "Research directions for the Internet of things," *Internet Things J.*, vol. 1, no. 1, pp. 3–9, 2014.

2. Z. Li, J. Wang, R. Higgs, L. Zhou, and W. Yuan, "Design of an intelligent management system for agricultural greenhouses based on the Internet of things," in *Proc. IEEE Int. Conf. Comput. Sci. Eng./IEEE Int. Conf. Embedded Ubiquitous Comput.*, 2017, pp. 154–160.
3. P. A. H. Williams and V. Mccauley, "Always connected: The security challenges in Internet of things," in *Proc. IEEE 3rd World Forum Internet Things*, 2016, pp. 30–35.
4. T. Min, X. W. Fei, and D. Z. Zhao, "Facility agriculture monitoring system based on Internet of things," *Wireless Int. Technol.*, 2016.
5. X. Hu and S. Qian, "IOT application system with crop growth models in facility agriculture," in *Proc. 6th Int. Conf. Comput. Sci. Convergence Inf. Technol.*, 2011, pp. 129–133.
6. Z. Lin, X. Mo, and X. Yueqin, "Research advances on crop growth models," *Acta AgronomicaSinica*, vol. 29, no. 5, pp.750–758, 2003.
7. M. M. Hossain, M. Fotouhi, and R. Hasan, "Towards an analysis of security issues, challenges, and open problems in the Internet of things," in *Proc. IEEE World Congr. Services*, 2015, pp. 21–28
8. M. Shereef and P. Viswanathan, "Smart agriculture IoT with cloud computing," in *Proc. Int. Conf. Microelectron. Devices, Circuits Syst.*, 2017, pp. 1–7.
9. J. Gutiérrez, J. F. Villa-Medina, A. Nieto-Garibay, and M. Á. Porta- Gándara, "Automated irrigation system using a wireless sensor network and GPRS module," *IEEE Trans. Instrum. Meas.*, vol. 63, no. 1, pp. 166–176, Jan. 2014.
10. [Online]. Available: http://www.yourarticlelibrary.com/agriculture/10-major-agricultural-problems-of-india-and-their-possible-solutions/20988
11. [Online]. Available: https://www.kaaproject.org/smart-farming/
12. R. Bhattacharyya, B. N. Ghosh, P. Dogra, P. K. Mishra, P. Santra, S. Kumar, M. A. Fullen, U. K. Mandal, K. S. Anil, M. Lalitha, D. Sarkar, D. Mukhopadhyay, K. Das, M. Pal, R. Yadav, and V. P. Chaudhary "Soil conservation issues in India," *Sustainability*, vol. 8, 2016, Art. no. 565
13. G. N. Zaimes, V. Iakovoglou, P. Koutalakis, *et al*, "The automated soil erosion monitoring system (ASEMS)," *Int. J. Geol. Environ. Eng.*, vol. 9, no. 10, 2015.
14. D. Kurrey, R. S. Rajput, and R. K. Singh, "Status of soil erosion or land degradation in India" in *Proc. Nat. Conf. Manag. Soil Resources Environ. Sustain.: Challenges Perspectives*, December 9–10, 2016.
15. I. Mohanraj, K. Ashokumar, and J. Naren, "Field monitoring and automation using IoT in agriculture domain," *Proc. Comput. Sci.*, vol. 93, pp. 931–939, 2016
16. J. Gubbi, R. Buyya, S. Marusic, and M. Palaniswami, "Internet of things (IoT): A vision, architectural elements, and future directions," *Future Gener. Comput. Syst.*, vol. 29, no. 7, pp. 1645–1660, 2013.
17. [Online]. Available: https://www.sigfox.com/en/iot-soil-condition-monitoring-sensors-will-optimize-agriculture-through-data-2
18. [Online].Available: https://www.cropin.com/iot-internet-of-things-applications-agriculture/

PART III

Statistical and Mathematical Model-Based Solutions

CHAPTER 9

Analysis of Epidemiology: Integrating Computational Models

DEEPAK KUMAR[1*], VINOD KUMAR[2], and POOJA KHURANA[1]

[1]*Department of Mathematics, Manav Rachna International Institute of Research and Studies, Faridabad, Haryana 121004, India*

[2]*Department of Mathematics, University College of Basic Science and Humanities, Guru Kashi University, Talwandi Sabo, Punjab 151302, India*

[*]*Corresponding author. E-mail: deepakman12@gmail.com*

ABSTRACT

An epidemiology is very serious situation, in which epidemic disease spreads animal to animal, person to person, animal to person through inhaling of the virus liable to spread it to healthy person. The human society has suffering from epidemic diseases for thousands of years because of viral infection. Viral infection transmits from birds or animals like hens, ducks, pigs, swine, turkeys, and many other species of warm-blooded vertebrates. In the study of epidemic diseases, computational modeling approaches to be progressively abundant in epidemiology research. The complex nature of the study of epidemics is appropriate to quantitative techniques as it gives challenges and opportunities to new advancements. The practical implementation of the computational modeling must rely on an epidemiological model. Computational models can supplement test and clinical investigations, social media users, and yet additionally challenge current paradigms, redefine our understanding of mechanism driving epidemic and shape future research in public health policies, and to provide the number of techniques to control all kinds of epidemic diseases.

9.1 INTRODUCTION

Epidemiology is a very dire situation, in which epidemic disease spreads from animal to animal, person to person, and animal to person through inhaling of the virus. The human society has suffered from epidemic diseases for thousands of years because of viral infection. Viral infections transmit from birds or animals such as hens, ducks, pigs, swine, turkeys, and many other species of warm-blooded vertebrates. In the study of epidemic diseases, computational modeling approaches to be progressively abundant in epidemiology research. The complex nature of the study of epidemics is appropriate for quantitative techniques as it gives challenges and opportunities to new advancements. The practical implementation of computational modeling must rely on an epidemiological model. Computational models can supplement test and clinical investigations, social media users, and yet additionally challenge current paradigms, redefine our understanding of mechanism driving epidemic, and shape future research in public health policies, and provide the number of techniques to control all kinds of epidemic diseases.

The study of epidemic disease is the investigation and examination of the distribution determining the factors of epidemic diseases in a population. This way, computational modeling demonstrates to epidemiology explore by helping to elucidate the mechanism and by giving quantitative predictions that can be validated.

9.1.1 INTRODUCTION TO COMPUTATIONAL MODELING

GTA, Need for Speed, Counter-Strike, PUBG, Call of Duty; Don't these names sound familiar to you? Yes, these are among the top-most games played in the world, which almost everyone knows. These games somewhat give us the "Reality Experience!" We all enjoy playing these, don't we, but have you ever pondered on this fact, that 'How?' How are these games so well built? How is everything made in an organized manner? How is the reflex action taking place when you enter a command? Well, these games use computational modeling and are examples of the same. Not just these, there are a lot many things that use this method. So, what is computational modeling?

Well, these games use computational modeling and are perfect examples of the same. Not just these, there are a lot of many things that use this method. So, what is computational modeling? A computational model, computer-based system that forms a direct link or a joint between mathematics and computer science, used to carry out the complex operations of mathematics with the help of computer simulation. Computational modeling is about two critical topics, which are vital for the functioning of the computational model: mathematics and computer.

"For the things of this world cannot be made without a knowledge of mathematics"—Roger Bacon Mathematics, very well known, an integral part of modern civilization, is a subject that nurtures a lot, many qualities of a person like a way of thinking, understanding, logical reasoning, creativity, problem-solving skills, spatial thinking, social experiences, and many more.

In simple terms, representation of the problems in mathematics to know the behavior of that problem quantitatively is mathematical modeling.

"The computer was born to solve problems that did not exist before"—Bill Gates. Computer technology makes business and other fields faster and easier. It also provided all the students (needy ones and the disabled ones) a better tool to become familiar with the various skills (basic, master, and advanced) required for performing the tasks. Due to this, computational science has reached its height where understanding and solution of complex problems become very easy with the help of advanced computing capabilities. It is an interdisciplinary approach, including mathematics, computer science, biology, and psychology, forming a core link between the model development and simulation to understand the behavior of natural systems. Now, coming back, as said earlier, the computational modeling method is a way we interconnect technology with mathematics to solve real-life problems.

9.1.2 INTRODUCTION TO EPIDEMIOLOGY AND IMPORTANCE

Epidemiology is a Greek word derived from "epi," which means on or upon, "demos," which means people, and "logos," which means the study of, provided the meaning of the word as the study of study of the distribution and determinants of health-related states or events in specified populations, and the application of this study to the control of health problems [1]. To determine the number of disease cases in a particular area during a specific period is the main aim of epidemiology [2]. Frequency and the health events pattern in a population are the main determinants of epidemiology. It means how often disease occurs in a community to that of the size of the people.

Pattern refers to the happening of a series of health-related events by time, person, and place. Epidemiology is an essential part of the fundamental illustration of a particular disease. Epidemiological studies are used as a template description to prevent illness. It also helps in the management of a patient, who has already developed an infection. Risk factor identification in the food production system by representing critical control points is the importance of epidemiology [3]. One more impact factor of epidemiology is high accuracy in the diagnostic procedures helpful to both the patient for the reduction in risk factors and also to the physician for providing the appropriate medication [4].

9.1.3 INTRODUCTION TO SOCIAL NETWORKS

Social networking is a relatively modern advancement in science. A social network is a chain of entities (such as friends, colleagues, and partners) connected by antisocial connections. Social networking is an excellent form of fun, is a place for finding and meeting human beings with similar interests and ideas, and is useful for connecting and staying in touch with old friends/colleagues. It is also a useful promotional tool because of businesspeople, entrepreneurs, writers, actors, musicians, and artists. It is a committed Internet site, which permits users after talking together with every other by way of posting information, comments, messages, images, etc. It is the use of Internet-based social media applications to rearrange connections with friends, family, classmates, and clients. Social networking can occur for social purposes, business purposes or both through sites such as Facebook, Twitter, LinkedIn, Bebo, Classmates, Instagram, MySpace, Path, Pinterest, Reddit, Stumble Upon, Tumblr, Yik Yak, YouTube, and Yelp. Perhaps, the easiest way to understand social networking is to think of it like high school. An online social network is a meeting place for individuals to expand their range and remain in contact with their associations. Person-to-person communication is not a static thing. Systems are developing and constantly changing, with new ones flying up at a quick rate. Many systems' administration sites are outfitted toward users with particular interests and needs, while others wish everybody to join. Many individuals join a social network as long as their loved ones are utilizing the application, and they need to remain in contact. Once you have been using a social network site for some time, you will undoubtedly interact with other individuals you know or knew long before. These networks are fantastic spots to make up for lost time with old companions and to share current and old photographs, and find different companions whom you may have lost contact with a journey. Experts have dependably organized somehow. Regardless of whether it is a conference, a gathering, or a significant industry occasion, meeting other individuals facing a similar situation is a need. Social networks, particularly those like LinkedIn that take into account organizations and experts, give another stage to meet carriers and compelling individuals in the business. Given the fast growth over conventional networks within the advent years, the gradual increase regarding the Internet is now forming itself around the characteristics of associative networks.

9.2 LITERATURE REVIEW

In today's world, the healthcare environment needs experimentally proved evidence for better decision making process. The literature review provides a fundamental background knowledge that answers many health-related questions. It includes the prevalence of particular in the next upcoming years, cause of illness, risk factors associated with the disease, treatment patterns, and the effectiveness of treatment.

Kumar et al. [53], done a meta-analysis of 233 studies on Dengue and reported 180 reviews as a confirmed Dengue infection and 77 as a fatal case. Harder et al. [54] performed 12 studies, of which six were applicable for meta-analysis, and the conclusion drawn was the significant increase in the chickenpox of those cases that have not done any prior vaccination of varicella. Likewise, many more diseases were reviewed and analyzed. Even the process of study of disease is continuing, the prevention and cure of different types of conditions, including rare diseases, makes the entire world full of healthy people.

Thakare et al. [44] explained the improved SIR model for epidemic control in a social network. It gives better realistic simulation results by considering crowding or protective effect. The efficiency of the model can be analyzed in social subnetworks with some potential immunization strategies. It includes a random set immunization, dominating set immunization, and high degree set immunization. Woo et al. [40] presented the SIR model of diffusion in web forums. It is used to analyze disease outbreaks. Evaluation of model occurs on a large longitudinal dataset from the web forum of a major retail company. The fitting results showed that the SIR model is a plausible model to describe the diffusion process of a topic. The research showed that epidemic models could expand their application areas to topic discussion on the web, particularly social media such as web forums. Wang and Wang [41] proposed a novel SIR model to study rumor spreading. They did by taking the influence of the social network into consideration. They found that the influence of the network medium on homogeneous networks is greater than on inhomogeneous networks. They performed numerical simulations which showed that rumor spreading accelerates with an increase of the infectivity between persons and the network medium. Cannarella and Spechler [55] applied a modified epidemiological model to outline the acceptance and disused progress of online social networks by active users. They validated the proposed infectious recovery SIR model (irSIR model). In addition, the usage of publicly available Google enquires question facts for MySpace. Then, that applies the irSIR model in conformity with ask query data because of "Facebook," which confirmed the instance regarding a disused phase. Extrapolating the superior in shape model predicts a rapid decline in Facebook recreation into the next few years. Sotoodeh et al. [46] proposed a general compartmental information diffusion model and extracted some of the parameters which are beneficial to analyze the model. To the acceptance of a deterministic manner to stochastic one, the Markovian property has been used to find out transition probability. Then, the probability obtained has been applied to get the mean value of the population per each group. Wei et al. [47] proposed a general compartmental information diffusion model and extracted some of the parameters that are beneficial to analyze the model. To the acceptance of a deterministic manner to stochastic one, the Markovian property has been used to find out transition probability. Then, the probability obtained has been applied to get the mean value of the population per each group. Medianama [56] calculated that WhatsApp customers in India use the video calling feature for a complete of 50 million minutes per day, the highest aggregate usage inside the global, in keeping with the enterprise. Singh [57] told about WhatsApp users in India that about 200 million people are actively using this app in India.

9.3 RESEARCH GAP IN THE STUDY

In the literature review, the significant work is done on computational modeling. Here, it is showing that there is a gap in a different area of computational modeling on epidemiology. Previous research investigates the board area of modeling, which needs to be accomplished with so far not explored this matter to improve the computational modeling on epidemic diseases. In recent years, maximum research work on different diseases with mathematical modeling has been present in a different manner. Current research work is a unique approach to fill the gap with the help of computational modeling on epidemiology. It may be used to reduce the effect of epidemic diseases on humans and creating awareness among human beings to be protected and future medication as well as future predictions. The gapes noted passed up a major opportunity in prior research amid the work for displaying for transmittable illnesses should be filled. An epidemic disease (influenza A H1N1, Swine Flu) study has been presented [5]. There should be consideration of removal rate and the disease-free equilibrium. Discussion of contagious diseases can happen with the help of the *SIR* model by which a new compartment like the influence of treatment can add. There is a conversion of the *SIR* model to a new model named *SITR* model. Occasionally, the fixed population infected by influenza produces bronchitis, the minor infection, and then applies the SI_1TI_2RS Model. Very few transmittable diseases are there in a long history on which mathematical modeling can be applied to develop a model.

9.4 PROBLEM IDENTIFICATION

The world in which we live today is a place very different from the one that existed 50 years ago. The modern life is full of facilities and technology (everything at one click today) due to which there is a significant change observed in the lifestyle of the people. Change in lifestyle such as advancement in technology and uptake of fast food in excess provides many factors that have a high potential of developing diseases in an individual. Risk factors associated with the environment such as pollution in the air, water, and soil, change in climate, and exposure to chemicals contribute significantly toward the development of the disease. In addition, genetic factors such as a change in the single or multiple base-pair by mutation due to environmental and lifestyle factors are responsible for disease development. Although technology in medicine has reached its height, it helps in finding the cure and methods for the prevention of many diseases. However, the cure for all is very challenging. Therefore, epidemiology came into existence that provides better decision making to improved accuracy in the diagnostic processes.

9.5 PROPOSED SOLUTION

The solution to the problem in this chapter has inspired by ethnographies from different computational techniques present cultures. The concept focused on human health-related issues.

9.5.1 DETERMINISTIC MODELS

In order to find the population size as a uniquely determined function of time "t" without randomness but by probability distribution are called deterministic models.

9.5.1.1 SIR MODEL

In the simple deterministic model, we consider that the total population, say N, at any time "t" is taken to be constant. If a small set of infected persons is introduced into a big population, the fundamental problem is to explain the spread of the infection within the population. It depends on a variety of conditions, which includes the particular disease concerned. We consider a disease in which removal is also included recovery after taking any drug or death or loss of interest. Consider the disease is such that the population can be divided into three different classes: the susceptible, S, who are prone to disease; the invectives, I, who are having the disease and able to transmit it; and the removed class, R, namely, those who are removed from the population by recovery, death, hospitalization or by any other means. The structure of the above model is represented in Figure 9.1.

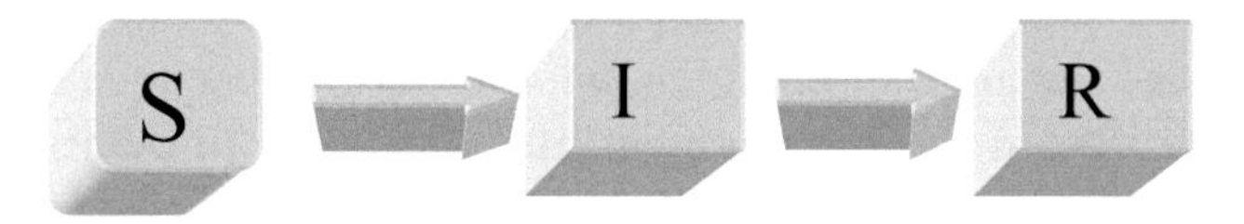

FIGURE 9.1 Presentation of the SIR model.

Such type of models is called *SIR* models. Let n be the initial number of susceptible persons in the total population in which there is only one infected person has introduced. As a result, the number of susceptible starts decreasing. At the same time, the number of infected persons increases. We now consider the different classes when mixed uniformly with the condition that every individual has an equal probability of coming in contact with each other. Basic assumptions for the mathematical model are the following.

- The total number of the population is fixed.
- The infectious disease is transmitted by direct individual contact.
- The recovery from an infectious disease will vary.

$$\frac{dS}{dt} = -\beta SI$$

$$\frac{dI}{dt} = \beta SI - \gamma I$$

$$\frac{dR}{dt} = \gamma I$$

Adding the above equations, we get $\frac{dS}{dt} + \frac{dI}{dt} + \frac{dR}{dt} = 0$, which implies that $S + I + R = N$, at times with the initial conditions that $S(0) > 0$, $I(0) > 0$, and $R(0) = 0$.

Let $S(t)$ denote the number of persons who can be infected from any given set of population, called susceptible class. The standard epidemic model was first invented in 1927 by Kermack and McKendrick [5] and has played a significant role in mathematical epidemiology. We use probability to simulate the variability of the total number of the infected population in the compartmental model. The newly infected population caused by the infected population at each stage is obtained by the numerical value of the probability of an event, while the infection rate is not fixed. Therefore, here, we have two cases to investigate the status of the disease in society.

Case 1. If the infection rate is higher than one or
Case 2. If the infection rate is less than one.

Case 1:

Probability of event	1	2	3	4	5
Number of the infected population	0	3	4	7	8

The expected rate of infection $= \frac{1}{5}*0 + \frac{1}{5}*3 + \frac{1}{5}*4 + \frac{1}{5}*7 + \frac{1}{5}*8 = \frac{22}{5} = 4.4$

Which is higher than one, so the epidemic will spread-out in the population?

Case 2:

Probability of event	1	2	3	4	5
Number of the infected population	3	1	0	0	0

The expected rate of infection $= \frac{1}{5}*3 + \frac{1}{5}*1 + \frac{1}{5}*0 + \frac{1}{5}*0 + \frac{1}{5}*0 = \frac{4}{5} = 0.8$

which is less than 1, so the epidemic will die-out in the population. Different types of epidemiological improved SIR model are as follows:

Stability Analysis for the System of Differential Equations (SIR) Model

Jacobian matrix of the governing equation (SIR model) is given as

$$J = \begin{bmatrix} -\beta I & -\beta S & 0 \\ \beta I & \beta S - \alpha & 0 \\ 0 & \alpha & \mu \end{bmatrix} . \begin{bmatrix} S \\ I \\ R \end{bmatrix}.$$

$$\text{Now, Det } (J - \lambda I) = \begin{vmatrix} -\beta I - \lambda & -\beta S & 0 \\ \beta I & \beta S - \alpha - \lambda & 0 \\ 0 & \alpha & \mu - \lambda \end{vmatrix} = 0$$

$$(\mu - \lambda)\left[\lambda^2 + (\beta I - \beta S + \alpha)\lambda^1 + \beta\alpha I\right] = 0$$

$$(\mu - \lambda)\left[\lambda = \frac{-(\beta I - \beta S + \alpha) \pm \sqrt{(\beta I - \beta S + \alpha)^2 - 4\beta\alpha I}}{2}\right] = 0$$

i.e., $\lambda_1 < 0$, $\lambda_2 < 0$, and $\lambda_2 < 0$; if $-(\beta I - \beta S + \alpha) > \sqrt{(\beta I - \beta S + \alpha)^2 - 4\beta\alpha I}$.

Since all the eigenvalues are negative, then the given model is steady (stable), otherwise nonsteady (unstable).

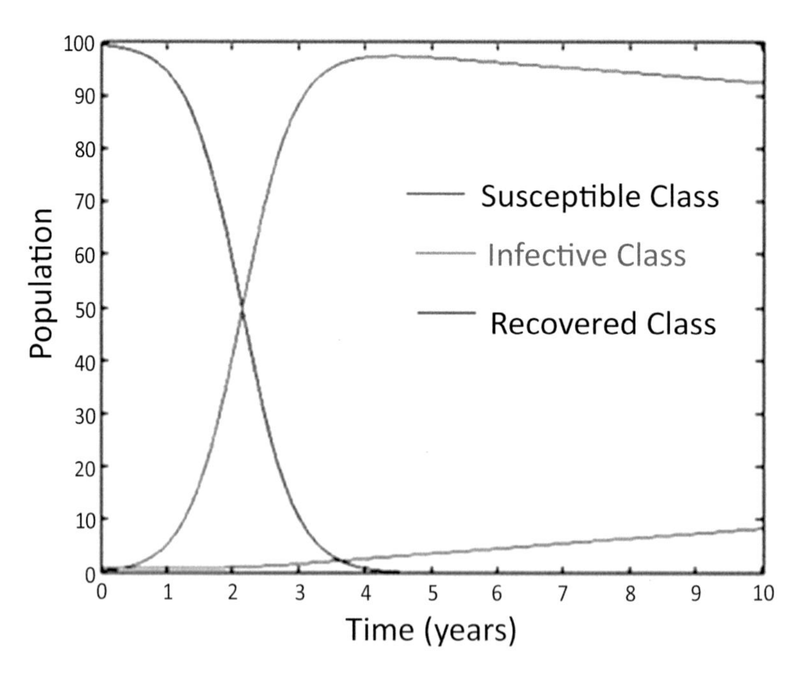

FIGURE 9.2 The graphical relationship of SIR.

In the above graph, the susceptible class (*S*) shows that the persons, who are prone to get any communicable disease, go on decreasing as they are getting infected when coming in contact with infectives. The infected persons (*I*) show that the number of infective goes on increasing and then become stable as at the same time people are getting recovered with some vaccinations and getting into recovered class. It can be seen from the graph that the recovered class (*R*) is increasing but with less rate as it includes death cases also.

In a social network, the user's dynamicity is a significant feature, which can impact the user's behaviors. The graph shows that the number of possible authors is decreasing concerning time and the range of authors whose posts lose infectivity to others on a subject is increasing concerning time. The infected class is growing initially, but it is decreasing.

9.5.1.2 SITR MODEL

The compartment model SITR is a set of differential equations, which are designed for the susceptible to infection, infection to treatment, and treatment to complete recovery (see Figure 9.3).

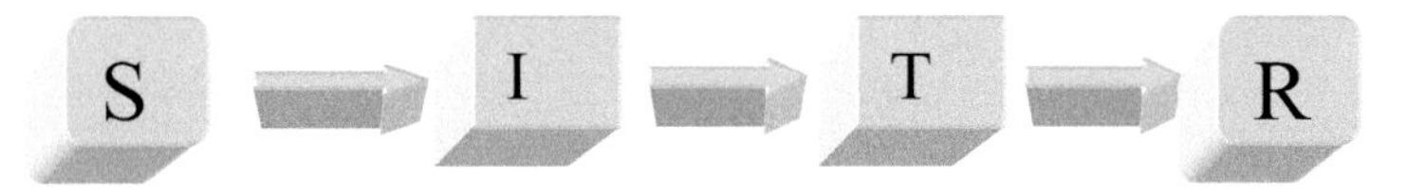

FIGURE 9.3 Presentation of the SITR model.

$$\frac{dS}{dt} = -\beta SI$$

$$\frac{dI}{dt} = \beta SI\text{-}\gamma I\text{-}dI$$

$$\frac{dT}{dt} = \gamma I\text{-}\sigma T$$

$$\frac{dR}{dt} = \sigma T$$

Susceptible class, *S*: susceptible to infection, β transmission rate; Infective class, *I*: infection *I*, γ rate of selection treatment, *d* death rate due to infection; Treatment class, *T*: treatment for infection, σ removal rate from infection due to treatment; Removed class, *R*: complete removal of infection. Table 9.1 represents the estimation of values for the parameters involved in SITR Model.

TABLE 9.1 Estimation of Parameters

Parameters	Values [Reference]
β infection rate	1.28 per year
γ treatment rate	0.70 per year
d death rate	0.20 per year, [51]
σ removal rate due to treatment	0.10 per year [51]
Inchoate value of S_0	1 estimated
Inchoate value of I_0	0.01 estimated
Inchoate value of T_0	0.50 estimated
Inchoate value of R_0	0.20 estimated

Figure 9.4 represents the status of mathematical epidemiology. The result of the above model is verified by the Routh–Hurwitz criterion and provides the future prediction of reverting it [55].

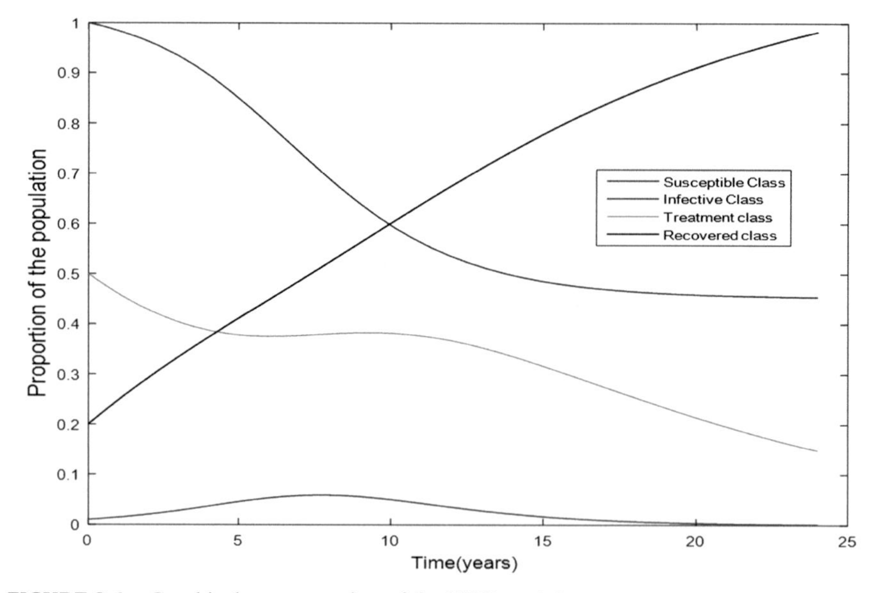

FIGURE 9.4 Graphical representation of the SITR model.

9.5.1.3 SI_1TI_2RS MODEL

In this model, infected individuals become infectious immediately, as shown in Figure 9.5.

S → I_1 → T → I_2 → R → S

FIGURE 9.5 Presentation of the SI_1TI_2RS model.

The governing differential equations of the model are as follows:

$$\frac{dS}{dt} = -\beta_1 S I_1 + \delta R$$

$$\frac{dI_1}{dt} = \beta_1 S I_1 - \gamma_1 I_1$$

$$\frac{dT}{dt} = \gamma_1 I_1 - (\sigma + \beta_2 I_2) T$$

$$\frac{dI_2}{dt} = \beta_2 I_2 S - (\gamma_2 + d_2) I_2$$

$$\frac{dR}{dt} = \gamma_2 I_2 + \sigma T - \delta R$$

TABLE 9.2 Estimation of Parameters Involve in SI_1TI_2RS Model

Sr. No.	Variables/Parameter	Values [Reference no.]
1	S "Susceptible class"	399 per year
2	I_1 "Infective class 1"	23 per year
3	I_2"Infective class 2"	1 per year
4	T "Treatment class"	23 per year
5	R "Recovered class"	0.00125 per year
6	β1 "Transmission rate 1"	0.012531 per year
7	β2"Transmission rate 2"	0.006 per year
8	σ "Rate to secondary infection"	0.0625 per year
9	$d1$ "death rate 1"	0 per year
10	$d2$ "death rate 2"	0 per year, [52]
11	γ1 "Recovery rate 1"	0.2 pear year
12	γ2 "Recovery rate 2"	0.09 per year, [52]

The model provides the information to the human population that they should be more educated to collect more data and attend health awareness camp of the infectious disease transmission and medication for in human population (Table 9.2). In addition, the practitioner can give better attention to successful treatment. Figure 9.6 shows the graphical solution of the SI_1TI_2RS Model.

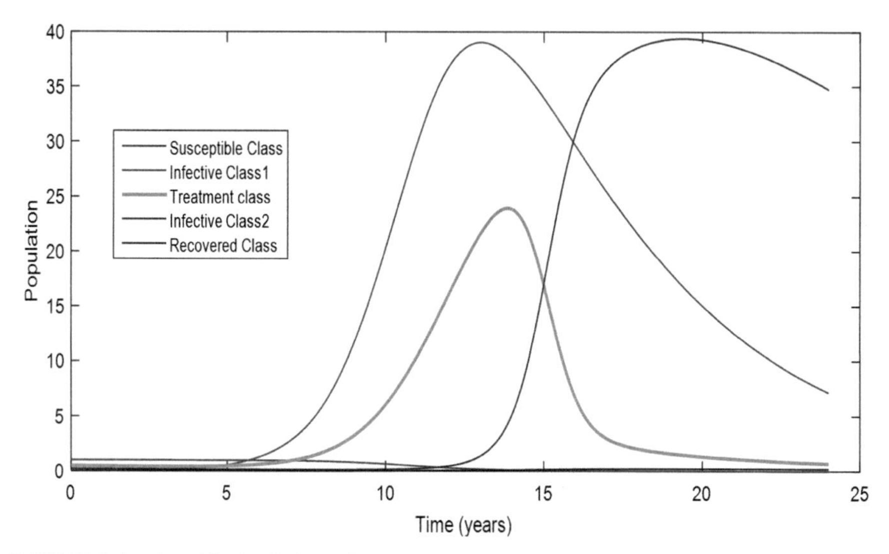

FIGURE 9.6 Graphical solution of the SI_1TI_2R model.

9.5.1.4 SI_1TI_2RV

Figure 9.7 gives the vaccination model.

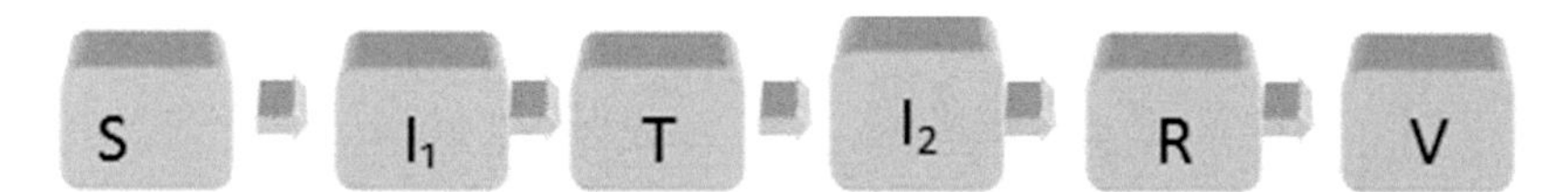

FIGURE 9.7 Presentation of the SI_1TI_2RV model.

The governing differential equations of the model are as follows:

$$\frac{dS}{dt} = -\beta_1 S I_1 - \alpha$$

$$\frac{dI_1}{dt} = \beta_1 S I_1 - \gamma_1 I_1$$

$$\frac{dT}{dt} = \gamma_1 I_1 - (\sigma + \beta_2 I_2) T$$

$$\frac{dI_2}{dt} = \beta_2 I_2 S - (\gamma_2 + d_2) I_2$$

$$\frac{dR}{dt} = \gamma_2 I_2 + \sigma T$$

$$\frac{dV}{dt} = \alpha V$$

This model is helpful to predict the growth of infectious disease and the effect of the vaccination model. Our investigation also provides a considerable role in the correlation between mathematical modeling and dynamical aspects of specific epidemic diseases.

9.5.1.5 SITRS

Figure 9.8 gives the resusceptible model.

FIGURE 9.8 Presentation of the SITRS model.

The governing differential equations of the model are as follows:

$$\frac{dS}{dt} = -\beta S I + \delta R$$

$$\frac{dI}{dt} = \beta S I - \gamma I$$

$$\frac{dT}{dt} = \gamma I - (\sigma + d) T$$

$$\frac{dR}{dt} = \sigma T - \delta R$$

Table 9.3 represents parameter estimations for the parameters involve in SITRS model. The graph shown in Figure 9.9 represents the behavior of susceptible to influenza (*S*), infected with influenza (*I*), treatment for influenza (*T*), and completely recovered from influenza (*R*). The graph represents that the susceptible rate (*S*) decreases concerning time and removal (*R*) from the disease due to treatment. Graphically solutions prove that the variables are not asymptotically stable. Simulation results are showing the trajectories and behavior of SITRS model.

TABLE 9.3 Parameter Estimations for the Parameters Involve in SITRS Model

Parameters and Variables	Values with Reference
β transmission rate	1.30 per year
γ infective is selected for treatment	0.50 per year
d death rate	0.20 per year
α removal rate from the treatment	0.10 per year [51]
μ losing immunity	0.027 per year
Initial value of S_0	1 [51]
Initial value of I_0	0.01 [51]
Initial value of T_0	0.5 [51]
Initial value of R_0	0.2 [51]

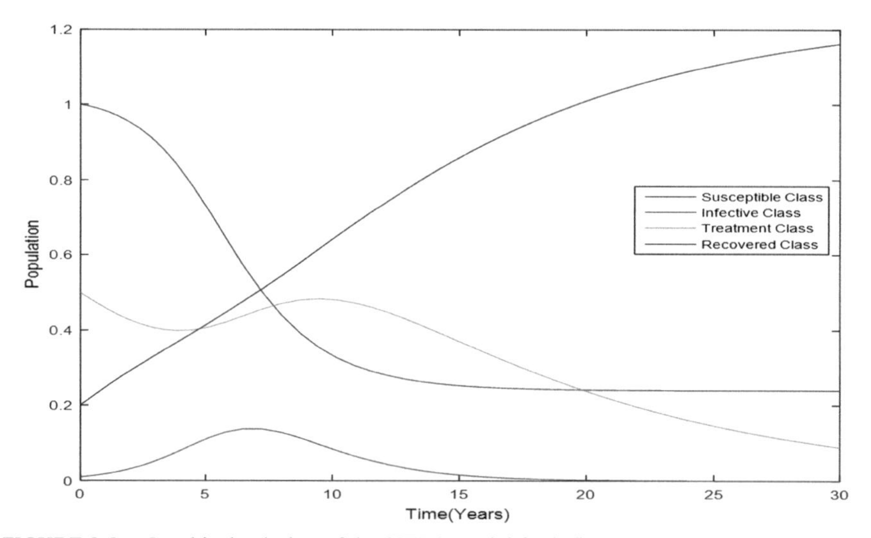

FIGURE 9.9 Graphical solution of the SITRS model for influenza.

Result and Discussion

- The number of new infection increases in the population because the basic reproduction number is positive and greater than one.
- The force of infection is also increased in the population, and the number of the suspected class will decrease with concerning time.
- The numerical data are verified and analyzed by MATLAB Graph analysis.

This work represents simultaneous differential equations which are formed for the Susceptible Class (S), Infective Class (I), Recovered Class (R), Infected with Influenza Virus (I_1), Recovered from Infection (T), Infected with secondary pneumonia (I_2), Transmission rate of Influenza (β_1), Recovery Rate of Infection (β), Rate at which an individual loses susceptibility (σ), Rate of Re-susceptibility (δ), Transmission Rate of Infection (β), Excess death rate due to Infection (d), Transmission Rate of Bacterial Infection (β_2), Recovery Rate of Bacterial Infection (γ_2), Excess Death Rate due to Bacterial Infection (d_2), and vaccination rate (v) represents the parameters and variables used/ involved in SITRS Model. In the mathematical model, each compartment can compute the number of effective people due to infectious disease at any time. In mathematical models, it is also essential to keep a list of variables, which can be changed according to situations of the model. Each variable has the properties to express the epidemic situation in the human population.

9.5.1.6 STOCHASTIC MODEL

If we conduct a large number of experiments or carry out a significant number of surveys under the same initial conditions, we find that the population size at a time t is not uniquely determined and keeps on fluctuating randomly. It means that we require probability distributions are called stochastic models.

Statistical methods in various sorts of epidemiological studies are shown through examples with real or imaginary ("fictive") data. Essential measures of recurrence and impact will be presented. Distinctive relapse models will be introduced as instances of sophisticated systematic strategies.

Basic Equation: $\frac{dp_n}{dt} = -p_n \sum_{j \neq 0} f_j(n) + \sum_{j \neq 0} p_{n-j} f_j(n-j).$

Multiplying by x^n, summing for all n, and using the definition of probability generating function, we obtain $\phi(x,t) = \sum_{n=0}^{\infty} p_n(t) x^n$.

Here, p_n is the probability of n susceptible persons and $f_j(n)$ is the probability that the number changes to $n + j$ in the time interval $t + \Delta t$. n and j-integer.

9.5.1.6.1 The Stochastic SIR Model

This model consists of susceptible, infected, and removed as a population.

Example: In this model, the compartment of individuals of three classes are considered, which are refer to as class S (susceptible), class I

(infective), and removed (either the individuals recovered or died). The cause of infection's probability during the time interval $t, t+\Delta t$, where the removal of one individual from class S and addition of that individual to class I is $BSI/N\ \Delta t + o(\Delta t)$. Consider γ to be the rate of recovery of infected individuals; then, the probability for the state of recovery where one individual from class I has been removed and added to class R during the time interval of $[t, t+\Delta t]$ is $\gamma I\ \Delta t + o\ (\Delta t)$. The cause of infection's and recovery probabilities are $((S_{t+\Delta t}, I_{t+\Delta t})-(S_t, I_t) = (-1,1)) = \beta S_t I_t /N\ \Delta t + o(\Delta t)$. $P((S_{t+\Delta t}, I_{t+\Delta t})-(S_t, I_t) = (0,-1)) = \gamma It\Delta t + o(\Delta t)$ with the complementary probability $P((S_{t+\Delta t}, I_{t+\Delta t})-(S_t, I_t) = (0,0)) = 1-(\beta S_t /N+\gamma)I_{t+\Delta t} + o\ (\Delta t)$ in the time interval [53].

9.5.1.6.2 Discrete-Time Stochastic Models

It is assumed that stochastic variables depict the number of susceptible and infective. In a simple case, the number at t depends only on the numbers at the previous time. So susceptible maybe, considering controlled population.

$$\Pr(i_1,.......y_T, s_1...........s_T \mid i_0) = \prod_{t=1}^{T} \Pr(i_t \mid i_{t-1}, s_{t-1}) \times \Pr(s_t \mid i_t, s_{t-1}).$$

9.5.1.6.3 Continuous-Time Stochastic Models

To build a continuous-time stochastic model, this considers the epidemics (infection) and recoveries on a continuous time scale. It is a continuous-time stochastic (Markov chain) for frequency-dependent transmission. The probabilities of susceptible approaches to infective are

$$\Pr\left(\begin{bmatrix} S(t+\Delta t) \\ I(t+\Delta t) \end{bmatrix} = \begin{bmatrix} s-1 \\ i+1 \end{bmatrix} \middle| \begin{bmatrix} S(t) \\ I(t) \end{bmatrix} = \begin{bmatrix} s \\ i \end{bmatrix}\right) = \frac{\beta si}{N} \Delta t$$

$$\Pr\left(\begin{bmatrix} S(t+\Delta t) \\ I(t+\Delta t) \end{bmatrix} = \begin{bmatrix} s \\ i-1 \end{bmatrix} \middle| \begin{bmatrix} S(t) \\ I(t) \end{bmatrix} = \begin{bmatrix} s \\ i \end{bmatrix}\right) = \frac{\gamma si}{N} \Delta t.$$

From the standpoint of an infective, each can infect susceptible in time Δt as per rate transmission rate β, where s is the number of susceptible at a time t, and in the perspective of susceptible, each may be infected in time Δt as per rate γ, where i is the number of infective at time t. This approach is not widely used as computationally, where population increases, which is the case with surveillance data.

9.5.2 GAUSSIAN DISTRIBUTIONS

The standard Gaussian distribution has the probability density centered at 0. The probability density at any value z (positive or negative) is given by $0.3989\ e\left[\frac{-1}{2}z^2\right]$.

Example: If the mean and standard deviation of general Gaussian distribution are 100 and 20, respectively, what ranges of values correspond to probabilities of 0.90 and 0.95, respectively?

Similarly, when x has a Gaussian distribution with mean μ and standard deviation σ, then $z = \left(\frac{x-\mu}{\sigma}\right)$ will have a standard Gaussian distribution. This fact can be used to get the probability for a range of values of x using tables of z.

The probability density per unit of x when x has a Gaussian distribution with mean μ and standard deviation σ is $\frac{0.3989}{\sigma}\exp\left[\frac{-1}{2}\left(\frac{x-\mu}{\sigma}\right)^2\right]$.

9.5.3 APPLICATIONS OF SIR MODELS

The SIR model can be applied to any disease, whether acute or chronic with the condition that the mass is prone to get that disease and then can spread the disease. Some common applications are as follows.

9.5.3.1 HUMAN/ANIMAL RELATED: EPIDEMICS

The SIR model is formulated on any disease, which flows through any medium when a susceptible person meets with an infected person and possibly becomes infected, which includes the parameters like genetics or contagious.

S(t) represents the number of feasible people or animal, who can be in contact with infected person/animal and prone to be infected at any time t.

I(t) represents a wide range of people/animal who has the disease and likely to spread the infection to other persons/animals.

R(t) represents a range of people/animal that has recovered from that disease and has lost power to infect other persons/animals.

Some common diseases are as follows:

Hypertension, a complex issue, described by the dimension of systolic and diastolic blood pressure more than 140 and 90 mm Hg, respectively [6]. It is uncovered to be the most significant risk factor for individuals who have influenced the way of life. Reports from World Health Organization (WHO), Global Burden of Disease, and Noncommunicable Disease Risk Factor Collaboration demonstrate the high predominance of hypertension in India and detailed more than 1 million individuals that are experiencing this ailment (disease) [7]. India has turned into the third driving position of causing demise in India, adding to about 10.8% of all passing in India. 29% of the whole stroke and 24% of all heart attacks are connected with hypertension [8].

Diabetes, a metabolic disorder characterized by the increased level of glucose levels (more than 180 mg/dL) in the blood plasma, causing a condition of hyperglycemia [9–11]. According to the WHO, it has been estimated that sound 422 million adults have diabetes, and around 1.6 million deaths are directly associated with diabetes each year [12].

Alzheimer's disease, a type of dementia, is a neurodegenerative disorder, characterized by the accumulation of beta-amyloid protein (plaques) and the tau proteins (neurofibrillary tangles) [13]. It is a stepwise progressive disorder affecting the memory, thinking, and behavior of an individual, resulted from neuronal cell death [14], contributing 60–70% of all dementia cases [15]. It has been revised that around 5.7 million Americans have Alzheimer, and this number can be increased to 13.8 million. Deaths caused due to Alzheimer's have increased by 123% [16].

The parasite of protozoa causes Malaria, a common, parasitic, and life-threatening disease that belongs to the genus Plasmodium [16]. It remains one of the world's most devastating infectious diseases found in the tropical and subtropical regions of the world. Plasmodium falciparum, plasmodium vivax, and plasmodium malaria are the primary parasites responsible for causing malaria and sometimes plasmodium knowlesi in human infections, but mortality rates are usually low. It is transmitted by female infected Anopheles mosquito [17]. Plasmodium falciparum is known to be a fatal parasite that is responsible for most of the mortality rates [18]. The WHO has reported about 219 million cases of malaria in 2018, but somehow less than 2010, where about 239 million cases were reported [19]. The frequency of occurrence of Malaria is more in the African region that is about 92%,

5% in the southeast Asian area, and about 2% in the Eastern Mediterranean region [29].

Aedes mosquitoes transmit Dengue, a vector-borne viral disease. Dengue virus belongs to the genus Flavivirus and family Flaviviridae and consists of DEN 1, DEN 2, DEN 3, and DEN 4 serotypes [22,23]. About 100 countries are suffering from Dengue, including the USA and Europe [24]. It has become a significant health challenge in the tropical and subtropical regions. An increase in the growth rate of population, lack of mosquitoes repellant, global warming, water and air pollution, lack of awareness, and health facilities cause a 30-fold increase in Dengue infection worldwide [25,26]. In addition, it has been reported that there may be 40% upsurge in the spread of Dengue infection by 2080 due to change in climate [27].

Contagious diseases are as follows:

Swine Flu, also known as Swine-Influenza A (H1N1) flu is the infectious and respiratory illness of the pigs, spread by strains of Influenza a virus [28]. The WHO reported the pandemic nature of the H1N1 virus in 2009 that affected most of the world's population. In 1918, approximately 50 million people were infected with H1N1 Influenza virus around the world, with 50–100 million people deaths reported [30]. Around 43–89 million cases were reported in 2009 by the Centre for Disease and Control and Prevention and about 1799 deaths in 178 countries worldwide [31]. In 2019, 10,000 cases of swine flu have been reported across India with 774 deaths [30].

Chickenpox, a highly contagious disease, causes infection in more than 90% of the people that are not vaccinated during their lifetime [32]. It is transmitted by Varicella Zoster Virus, signalized by a pruritic vesicular eruption associated with fever and malaise [33]. Varicella (chickenpox) is more prominent in temperate regions as compared to tropical regions. The WHO estimated the burden of chickenpox to about 140 million cases with 4.2 million severe complications and 4200 deaths in 2014 [34]. The varicella-zoster-virus-infected 16 cases per 1000 people in developed countries annually [34]. The more burden of disease has been reported in children with 90% of cases [35].

Whooping Cough, also called Pertussis, is an acute illness of the respiratory system. Bordetella pertussis is responsible for causing whooping cough [32]. In 2008, the WHO reported 16 million cases, of which 90% were in developing countries with 195,000 deaths [33]. In 2009, 40% of the cases have been noticed in the USA. Children of age group between 7 and 10 years suffered from whooping cough from an estimate of around 9%, 13%, 23.5%, and 23% in 2003, 2007, 2008, and 2009, respectively [34].

9.5.3.2 STANDARD EPIDEMIC SIR MODEL IN WEB FORUM

The SIR model is formulated on an idea or topic diffusion on social networks as social media has become one of the easiest ways to share any information within a flying quick. It has influenced many streams such as business, politics, and marketing. *SIR* is the standard epidemic model, which was first invented in 1927 by W. O. Kermack and A. G. Mckendrick and has played a significant role in mathematical epidemiology. In the web forum context:

$S(t)$ represents the number of feasible authors, who might have a hobby in an issue;

$I(t)$ represents a wide range of authors who write down posts on any issue throughout the identical duration;

$R(t)$ represents a range of recovered authors whose posts not affecting others on a subject.

The infection rate, α, indicates how many possible authors will be infected per contact between an infective and a susceptible. The recovery rate, β, indicates how many infective authors per infective recover during a unit time.

9.5.3.2.1 Some Other Simple Examples of SIR Models through Social Media Are Rumour/Fake News Spreading on WhatsApp

Fake news in India is a growing problem. The practice of the use of social media platforms such as WhatsApp to popularize false information is leading in a dangerous direction. Two main aspects are urging the fake news surge: first, the trend of declining smartphone prices over the last couple of years, and, second, the fall in Internet data prices. Therefore, the problem of fake news is not going to leave soon, and knowing the truth is not going to be easy. We heard fake news at the time of demonetization of currency notes in India during November 2017. Lots of fake news, videos, and photos on the conflicts of Hindus and Muslims are shared through WhatsApp.

Viral marketing

Product information can be popularizing through social networks. This gives a positive impact on the customer, and as we see, lots of people are purchasing things through online shopping sites. It is gaining more fans, and these fans are spreading information and more, and more people are getting involved in every social network.

Audience applause

The widespread example is audience applause. We can see when sitting in a group of people and one person starts to clap and immediately the person sitting in the crowd starts clapping. This is the psychology of the persons in the crowd that they change their behavior quickly in response to others.

The spread of personal computer virus

Nowadays, with the increasing rate of IT (information technology) development, the security of networks is a significant issue in our daily routine. A computer virus is one of the threatening aspects nowadays. It has a high impact on the computer world. The virus attacks the computer system and damages the software.

Academic spreading

Here, the susceptible are research scholars, infected are the research scientists, and the carriers are journals and conferences. There is immigration into research scholars from the educational stream whose strength is assumed to be proportional to the strength of research scholars' class. The increase in research journals and conferences depends on the number of research scientists.

9.6 PROS AND CONS OF SOLUTION

The primary thought behind building up a computational model is to make things simpler to clarify the application of science and humanism related hypotheses.

9.6.1 EPIDEMIOLOGY

Epidemiological studies can prevent future outbreaks. Epidemiological studies obtain relevant data. Data collected are more reliable and precise as compared to animal-based studies. It helps clinicians in better decision making for their patient's welfare that was uncertain before the introduction of epidemiological studies. Epidemiological studies also play a significant role in determining the cause of illness, disability so that the research and action can be performed according to the determined purpose. Chronic diseases can also be prevented to a great extent by epidemiological studies.

More time is required to create a database and for the conduction of research. It does not take into account the health of the environment. Many pollutants are present in the atmosphere responsible for causing health-related diseases in people. Difficulty in the collection of data due to restricted access and lack of resources and communication and transport problems to some of the areas is one of the disadvantages of epidemiological studies. Some diseases lead to rapid changes in the population, and by the time data collection and analysis are done, the conclusions drawn from that studies become out of date, which is a considerable disadvantage of epidemiological studies.

9.6.2 SOCIAL NETWORKS

The main advantage of the social community is that it allows you to stay linked to friends and families in today's busy paced world. One is capable of antique flash friendships, proportion family pictures, and special occasions on your lifestyles with just about anyone you know, at the equal time. Social networking is additionally a significant way to pair those humans who are kind of minded. You enter the above groups up to expectations that are centered closer to the one-of-a-kind interests, ideas, and interests. Companies, artists, or musicians can attain an impossibly widespread yet numerous aggregate about people the usage of neighborly media sites. This lets in them in imitation of promoting yet need themselves then their products in access so much has by no means been viewed before. Any information can keep spreader as wildfire over social media sites. Important matters kind of recalls, reminders, storm facts is wholly communicated quickly. People fast operate now not think about the penalties regarding where he submit concerning these social sites. Just like talked about above, anything may spread to thousands and thousands regarding people inside hours and days concerning social media. This also, unfortunately, consists of things so are fake then performed up. These data perform motive panic and then extreme misinformation between societies. Online social interactions with social networking hold now not solely were beginning modern relationships, however closing deep others. This contemporary invitation has been driving wedges within people's actual life, offline relationships, frequently age last to them because of good. Social networking places have faith to the limit.

One of the most critical dangers of overusing social networking websites are the college students arrive addicted in imitation of it. They spend hours on these popular networking websites, which may degrade the student's performance. A few people may in general use these informal communication sites till midnight or much more, which can prompt wellbeing-related issues. Many people may supply data, such as phone numbers and address, which is entirely unsafe due to the fact that anyone may be, without difficulty, tracked down by way of strangers.

9.7 CONCLUSION AND FUTURE SCOPE

This chapter focused on mathematical and statistical aspects of the models, yet maybe the most important part is creating models and interpreting results to respond to real-life problems/situations. Such stays to be done on time-dependent models.

Epidemiology is a fundamental pillar for answering many health-related queries such as the prevalence of a particular disease in the next upcoming years, cause of illness, risk factors associated with the disease, and treatment patterns. It provides experimental-based evidence for better decision making and prevention interventions. Future epidemiology will be based upon communicating the translation process in a better way, meaning that epidemiologist must develop the procedures for generating rapid-fire hypothesis and their testing and make aware people how to live with uncertainty. One future aspect of epidemiological studies will include the examination of the factors or a real factor (high-calorie food, lack of physical activity, and spending more time on computers or watching TV) responsible for causing obesity.

Influence of change in the global environment on human health is one of the future aspects of epidemiology in the form of ecoepidemiology. To find the route of curing health-related challenges such as rare diseases, infectious diseases, obesity, aging, and environment health, epidemiology is like a panacea.

Social media incorporate several personal information and may be used as additional records supply for evaluation of social strategies in the actual real world. The SIR model can be described for the analysis of social media user growth, Internet user growth, mobile user growth, and mobile social media user growth for prediction. Prediction is based on factors that are estimated on the online data sources. The SIR model can be applied for the prediction of the number of people that have the same level of interests on any unique topic. The behavior of social users with the help of epidemiology SIR model can be studied and observed that these networks are not being used positively. In addition, the improved epidemiological SIR model can be used to describe the dynamics of user activity of online social networks. Utilizing publicly accessible information on social network websites, we can demonstrate that the SIR model for displaying dynamics provides a short description of the data. At the first step, we have a deterministic model based on differential equations that need to be examined and then examine the possibility of applying the SIR model on social media users. For future research, this model can be applied to different social networks such as Facebook, WhatsApp, LinkedIn, etc., in India or anywhere in the world. An epidemiological SIR model to describe the dynamics of spreading of fake news through social media platform named WhatsApp. The rate of growth of misinformation through the social media platform is increasing every year by the age group 19–24 very rapidly that is also shown in the graph, whereas the recovery rate is comparatively low in the age group 40–60. It can be concluded that the fake news spreading on WhatsApp is increasing day by day. Mathematics result of the model represents that the situation is being stable using different methods. Society needs to think to stop the fake news spreading and explain the demerits of such communication. We will advise readers to not believe all the news on this app, because some of the discussion may be fake.

There are various types of discovered mathematical models that have been used in mathematical epidemiology as well as in another field. By modifying the basic SIR model, we have proposed different compartmental models for further development in the future of analysis of epidemiology.

KEYWORDS

- **epidemiology**
- **computational modeling**
- **stochastic modeling**
- **deterministic modeling**
- **SIR models**

REFERENCES

1. Kaufman, J. S.; Cooper, R. S.; Commentary: Considerations for use of racial/ethnic classification in etiologic research. *American Journal of Epidemiology*. **2001**, 154, 291–298.
2. Gehlbach, S. H.; Interpreting the medical literature: Practical epidemiology for clinicians. New York, NY, USA: Macmillan. **1998**.

3. Bartlett, P. C.; Judge, L. J.; The role of epidemiology in public health. *Revue Scientifiqueet Technique—Office International des Epizooties*. **1997**, 16, 331–336.
4. Cimmino, M. A.; Hazes, J. M.;. Introduction: Value of epidemiological research for clinical practice. *Best Practice & Research Clinical Rheumatology*. **2002**, 16, 7–12.
5. Kermack, W. O.; McKendrick, A. G.; A contribution to the mathematical theory of epidemics. *Proceedings of the Royal Society of London. Series A, Containing Papers of a Mathematical and Physical Character*. **1927**, 115, 700–721.
6. Kokubo, Y.; Iwashima, Y.; Kamide, K.; Hypertension: Introduction, types, causes, and complications. *In Pathophysiology and Pharmacotherapy of Cardiovascular Disease.* New York, NY, USA: Springer, **2015**, 635–653.
7. Gupta, R., Gaur, K., & Ram, C. V. S.. Emerging trends in hypertension epidemiology in India. *Journal of Human Hypertension*. **2019**, 33, 575–587.
8. Commission on Health Research for Development. *Health Research: Essential Link to Equity in Development*. New York, NY, USA: Oxford University Press, **1990**.
9. Modak, M.; Dixit, P.; Londhe, J., Ghaskadbi, S.; Devasagayam, T. P. A.; Recent advances in Indian herbal drug research guest editor: Thomas Paul Asir Devasagayam Indian herbs and herbal drugs used for the treatment of diabetes. *Journal of Clinical Biochemistry and Nutrition*. **2007**, 40, 163–173.
10. Kooti, W.; Farokhipour, M.; Asadzadeh, Z.; Ashtary-Larky, D.; Asadi-Samani, M.; The role of medicinal plants in the treatment of diabetes: A systematic review. *Electronic Physician*. **2016**, 8(1), 18–32.
11. Forlenza, O. V.; de Paula, V. J.; Machado-Vieira, R.; Diniz, B. S.; Gattaz, W. F.; Does lithium prevent Alzheimer's disease? *Drugs & Aging*. **2012**, 29, 335–342.
12. Macchi, B.; Marino-Merlo, F.; Frezza, C.; Cuzzocrea, S.; Mastino, A.; Inflammation and programmed cell death in Alzheimer's disease: Comparison of the central nervous system and peripheral blood. *Molecular Neurobiology.* **2014**, 50, 463–472.
13. Duthey, B.; Background paper 6.11: Alzheimer disease and other dementias. *A Public Health Approach to Innovation*. **2013**, 1–74.
14. Alzheimer's, A.; 2015 Alzheimer's disease facts and figures. Alzheimer's & dementia. *The Journal of the Alzheimer's Association*. **2015**, 11, 332.
15. Tangpukdee, N.; Duangdee, C.; Wilairatana, P.; Krudsood, S.; Malaria diagnosis: A brief review. *The Korean Journal of Parasitology*. **2009**, 47, 93–102.
16. Cox-Singh, J.; Singh, B.; Knowlesi malaria: Newly emergent and of public health importance? *Trends in Parasitology*. **2008**, 24, 406–410.
17. Perkins, D. J.; Were, T.; Davenport, G. C.; Kempaiah, P.; Hittner, J. B.; Ong'echa, J. M.; Severe malarial anemia: Innate immunity and pathogenesis. *International Journal of Biological Sciences*. **2011**, 7(9), 1427.
18. World Health Organization. *World Malaria Report.* **2018**. [Online]. Available: http://www.who.int/malaria/publications/world-malaria-report-2018/report/en/ (Accessed on December 27, 2018).
19. Joel G Breman. *Malaria: Epidemiology, Prevention, and Control*. Philadelphia, PA, USA: Wolters Kluwer Health. **2019**.
20. Hasan, S.; Jamdar, S. F.; Alalowi, M.; Beaiji, S. M. A. A.; Dengue virus: A global human threat: Review of literature. *Journal of International Society of Preventive & Community Dentistry*, **2016**, 6(1), 1–1.
21. Halstead, S. B.; Pathogenesis of dengue: challenges to molecular biology. *Science.* **1998**, 239, 476–481.

22. San Martín, J. L.; Brathwaite, O.; Zambrano, B.; Solórzano, J. O.; Bouckenooghe, A.; Dayan, G. H.; Guzmán, M. G.; The epidemiology of dengue in the Americas over the last three decades: A worrisome reality. *The American Journal of Tropical Medicine and Hygiene.* **2010**, 82(1), 128–135.
23. World Health Organization, Special Programme for Research, Training in Tropical Diseases, World Health Organization. Department of Control of Neglected Tropical Diseases, World Health Organization. Epidemic and Pandemic Alert. Dengue: Guidelines for diagnosis, treatment, prevention and control. *World Health Organization.* **2009**
24. Guzman, M. G.; Halstead, S. B.; Artsob, H.; Buchy, P.; Farrar, J.; Gubler, D. J.; Nathan, M. B.; Dengue: A continuing global threat. *Nature Reviews Microbiology.* **2010**, 8(12 suppl), S7–S16.
25. Colón-González, F. J.; Fezzi, C.; Lake, I. R.; Hunter, P. R.; The effects of weather and climate change on dengue. *PLoS Neglected Tropical Diseases.* **2013**, 7, e2503.
26. Dandagi, G. L.; Byahatti, S. M.; An insight into the swine-influenza A (H1N1) virus infection in humans. *Lung India: Official Organ of Indian Chest Society.* **2011**, 28, 34.
27. Jilani T. N.; Siddiqui A. H.; *H1N1 Influenza (Swine Flu).* Treasure Island, FL, USA: StatPearls. **2019**.
28. Hasan, F.; Khan, M. O.; & Ali, M; Swine Flu: Knowledge, attitude, and practices survey of medical and dental students of Karachi. *Cureus.* **2018**, 10 e2048.
29. Cohen, J.; & Breuer, J; Chickenpox: treatment. *BMJ Clinical Evidence*, **2015**, 2015, 0912.
30. Skull, S. A., & Wang, E. E. L. Varicella vaccination—A critical review of the evidence. *Archives of Disease in Childhood.* **2001**, 85, 83–90.
31. WHO Varicella and herpes zoster vaccination position paper. *World Health Organization*, Genève, Switzerland. **June 2014**
32. Seward, J. F.; Marin, M.; Sage V. Z. V.; Working Group, April. Varicella disease burden and varicella vaccines. *In WHO Sage Meeting.* **2014**.
33. Scanlon, K. M.; Skerry, C.; Carbonetti, N. H.; Novel therapies for the treatment of pertussis disease. *FEMS Pathogens and Disease.* **2015**, 73, ftv074.
34. WHO. Pertussis vaccines: WHO position paper. Weekly Epidemiological Record. 2010, 85, 385–400.
35. Harvey, K.; Esposito, D.H.; Han, P.; Kozarsky, P.; Freedman, D. O.; Plier, D. A.; Sotir, M. J.; Surveillance for travel-related disease—GeoSentinel surveillance system, United States, 1997–2011. Morbidity and *Mortality Weekly Report: Surveillance Summaries.* **2013**, 62, pp.1–23.
36. Calderon, R. L.; Measuring risks in humans: the promise and practice of epidemiology. *Food and Chemical Toxicology*, **2000**, 38, S59–S63.
37. Fowkes, F. G.; Dobson, A. J.; Hensley, M. J.; Leeder, S. R.; The role of clinical epidemiology in medical practice. *Effective Health Care.* **1984**, 1, 259–265.
38. Brownson, R. C.; Epidemiology in public health practice: By A. Haveman-Nies, SC Jansen, JAM van Oers, and P. van't Veer. *American Journal of Epidemiology.* **2011**, 174, 871–872.
39. Remington, P. L.; Brownson, S. E.; Siegel, C. R.; The role of epidemiology in chronic disease prevention and health promotion programs. *Journal of Public Health Management & Practice.* **2003**, 9, 258–265.
40. Woo, J.; Chen, H.; Epidemic model for information diffusion in web forums: experiments in marketing exchange and political dialog. *Springer Plus.* **2016**, 5, 66.
41. Wang, J.; Wang, Y. Q.; SIR rumor spreading model with network medium in complex social networks. *Chinese Journal of Physics.* **2015**, 53, 020702-1.

42. Kumar, V., & Kumar, D.; SIR model of Swine Flu in Shimla. In Advanced Computing and Communication Technologies (pp. 297–303). Singapore: Springer. **2016**.
43. Cannarella, J.; Spechler, J. A.; Epidemiological modeling of online social network dynamics. *arXiv preprint arXiv:1401.4208*. **2014**.
44. Thakare, P.; Mathurkar, S. A.; Review on modeling of epidemics spreading in social interactions, *International Journal of Science, Engineering and Technology Research.* **2016**, 5, 192–195.
45. Opuszko, M.; Ruhland, J.; Impact of the network structure on the SIR model spreading phenomena in online networks. In *Proceedings of the 8th International Multi-Conference on Computing in the Global Information Technology (ICCGI'13)*. **2013**.
46. Sotoodeh, H.; Safaei, F.; Sanei, A.; Daei, E.; A general stochastic information diffusion model in social networks based on epidemic diseases. *arXivpreprint arXiv:1309.7289*. **2013**.
47. Wei, Z.; Yanqing, Y.; Hanlin, T.; Qiwei, D.; Taowei, L.; Information diffusion model based on social network. In *Proceedings of the 2012 International Conference of Modern Computer Science and Applications*. **2013**, 145–150.
48. Cooper, B.; *We Are Socials Global Digistatshot: Facebook & Smartphones Still Rule, Technology*, Surry Hills NSW, Australia: The Misfits Media Company Pty Limited, **2015**.
49. Fierro, R.; Discrete-time stochastic epidemic models and their statistical inference. *In Stochastic Modeling and Control. Intech Open*. **2012**. [Online]. Available: http://dx.doi.org/10.5772/39271
50. Wakefield, J.; Dong, T. Q.; Minin, V. N.; Spatio-temporal analysis of surveillance data. *arXiv preprint arXiv:1711.00555*. **2017**.
51. Bais, K.V.; Kumar, D.; Mathematical analysis on bronchitis infection. In *Proceeding 2016 International Conference on Computing for Sustainable Global Development*. **2016**, 1861–1864.
52. Kumar D.; Bais K. V.; Mathematical model on influenza disease with re-susceptibility. *Australian Journal of Basic and Applied Sciences*. **2016**, 10(15), 177–182.
53. Ganesh Kumar, P.; Murhekar, M. V.; Poornima, V.; Saravanakumar, V.; Sukumaran, K.; Anandaselvasankar, A.; John, D. Mehendale, S. M.; Dengue infection in India: A systematic review and meta-analysis. *PLoS Neglected Tropical Diseases.* **2018**, 12(7), p.e0006618.
54. Harder, T.; Siedler, A.; Systematic review and meta-analysis of chickenpox vaccination and risk of herpes zoster: A quantitative view on the "exogenous boosting hypothesis". *Clinical Infectious Diseases*. **2018**, 69, 1329–1338.
55. Cannarella J, Spechler JA. Epidemiological modeling of online social network dynamics. *arXiv preprint arXiv:1401.4208*. **2014**.
56. Medianama, M.; Indian; WhatsApp user's video calling for 50 million minutes per day; Calculations, [Online]. Available: *https://www.medianama.com/2017/05/223-indian-WhatsAppusers-video-call-for-50-million-minutes-per-day/.* (URL) **2017**.
57. Singh M.; Mashable India, Guess WhatsApp's biggest market? India. [Online]. Available: *http://mashable.com/2017/02/24/WhatsAppindia-200-million-active-users/#yUVHWR mJGsqK* (URL). **2017**.

PART IV

Applying Neural Networks and Deep Learning Models for Cognitive Problems

CHAPTER 10

Recent Issues with Machine Vision Applications for Deep Network Architectures

NARESH KUMAR

Department of Mathematics. Indian Institute of Technology, Roorkee, Uttarakhand 247667, India, E-mail: naresh.csiitr@gmail.com, nkumar123@ma.iitr.ac.in

ABSTRACT

In this chapter, we put an effort to focus on the processing and storage issues that can be carried out from the updates of unstructured data in the unconstraint environment. In the technological aspects, we discuss the existing intelligent deep network architectures along with traditional machine learning approaches. The outcomes of the analytical discussion presented in this chapter enable to picturize a generalized intelligent neural network for processing highly complex visual data with graphs and manifold structures. The essential issues with updating the hidden layers and some fast optimization techniques are also introduced. Finally, this can be concluded that the presented work reflects the sounding challenges to process and extract qualitative information from the densely unstructured tremendous amount of training data. In case of video processing, we need to frame out various deep learning aspects that can lead the research work to highly resourceful scope for deep data analytics and many problems that require high-performance computing in visual media.

10.1 INTRODUCTION

In a real-life scenario, every small activity is inherently structured with a massive amount of visual context. The variations in visual information can be easily pointed out by human and machine vision. Making records and processing of continuously generated visual information for any specific objective can create a significant challenge to the traditional storage capacity and processing issues with machines intelligence system. Such problems are widespread in several domains such as particular human activity recognition at public places, road traffic monitoring, chemical process, and financial updates in the share market. Graphs have been a ubiquitous computational entity from many decades to deal with a greater extent of computational work in many disciplines of applied science and engineering. A graph is a simple network of nodes that can be used collectively to process, control, and communicate the flow of information between the nodes through its edges. Therefore, a graph can be considered a computational structure, which creates the scope of the large computational framework in many disciplines of science and engineering applications. The detailed characteristic and attributes of the components of graphs vary with the specific domain of the network. In the computational views, the nodes in any graphs correspond to the random variables, and the edges between the nodes reflect statistical operations between the variables. To develop an efficient graphical model for the evaluation of the system in any engineering or scientific discipline requires detailed practical knowledge of graph theory and statistical mechanics. Hidden Markov model, Markov random fields, and Kalman filter Bayesian network are famous examples of the graphical model. In applied mathematics and engineering problems, these models are primarily used to deal with uncertainty and complexity, which are known as classification score or categorization of the objects in the spatiotemporal scenario. All the very specific classical multivariate probabilistic systems come in the categories of the phenomena that are frequently studied in pattern recognition, information theory, and statistical mechanics.

Semantically visual key information is ever hardly fixed from the perspective several perceptual vision, that is, the information related to any event due to human actions varies with video understating [25]. Real-world things are realized to the computers by artificial intelligence (AI) from the data captured from perceptual vision. Jordan [1] proposed that the general graphical model that can efficiently formulate and design a new framework for a system. AI techniques can produce the design and development of any complex system for many disciplines in a very smooth fashion. The processing criterion covers logic programming, reinforcement learning, expert system, neural network, cognitive science, swarm intelligence, and fuzzy logic. For achieving more exceptional data analytics, AI is promoted to machine learning and deep learning. All the building blocks of AI can be broadly exploited in sounding research domain such as natural language processing (NLP), medical imaging, game playing, computer vision, and robotics.

Regarding data generation, a report from the National Security Agency presented that the whole universe exerts the total amount of 1826 Petabytes data per day. This data statistics represent the entire amount of energy consumed and stored per day in the world. In 2009, it was reported that the data generation rate had become nine times more than the data generated in the last five years from 2004. From similar statistics, it is predicted that this amount will be 35 trillion gigabytes of the world in 2020 [6,7]. This is amazing and revolutionary information for data scientists. On the positive side, this information can meet the dream of data scientists to develop the several smart technologies such as automated healthcare diagnoses, safety and security, and an intelligent system for education and psychological training. However, from the development aspects, the processing issues of such huge amount of data to get trained require to develop a very big network. This will be the biggest hurdle to resolve the issues created due to big data characteristics. However, this opens a very good research domain of developing an intelligent online system that can reduce the overhead due to big data issues. Here, instead of focusing on big data analytics, we have chosen the prime objective to represent the issues with real-world data analytics and to get trained a deep neural network model of graph-based real-world information [31–34]. The deep neural network is expected to resourceful to solve the challenging many challenging issues in the visual domain. In fact, this is a very broad area and a generalized topic in computer vision problem. Apart from this, the graphical evolution to develop a deep network is also depicted along with several operations in hidden layers. This analytical study will help to develop a highly optimized and fast deep network model.

10.2 REAL-WORLD STRUCTURE

Evidence from neurology and psychological studies represents the fact that understating the things of the real world varies with the various perspectives of human nature such as locality, age factor, etc. In contrast, the machines remain constant and act in all the conditions accordingly once the instructions are set to perform any specific job related to visual media. This process requires to put the majority of efforts to achieve learning of the things with the desired accuracy. Zero-shot learning or unsupervised learning becomes the necessity in very common cases of real-life problem where extracting specific ground truth for a particular research problem is extensively burdensome. This means that real-world structure creates unconstraint issues to any machine learning problem. For instance, this fact can be realized as recognizing frontal face can give far better accuracy than the face in wild environments. Therefore, several such complex research problems in computer vision such as human–computer interface, scene understating, and human brain interface is referred to as hot research problems due to unstructured real-world visual media. This needs to develop an efficient machine intelligence system (MIS), which can efficiently provide an interface to solve the many physical world problems.

10.2.1 DEEP NETWORK ARCHITECTURE OF MIS

Non-Euclidean structures are vital substances in real world that create hard issues in processing and analysis the specified information using machines. This happens because of the variations in the features of similar objects and similarity of features in different objects or overlapping the coincident features. The example of the chair is shown in Figure 10.1 to explain this concept in static vision. In contrast, different objects may have a color similarity or may overlap one to another for a particular moment in motion. This problem is referred to as occlusion.

FIGURE 10.1 Visionary issues with random features of real-world structures (e.g., chair).

In Figure 10.2, a very general overview is presented to reflect the requirements of the vision based deep architecture of an intelligent system. This includes three components: (1) heterogeneous system real-world objects with random features, (2) a neural network system that required to develop by training with the specific ground from the physical datasets, and (3) the specific library to work as a back-end for the front-end network.

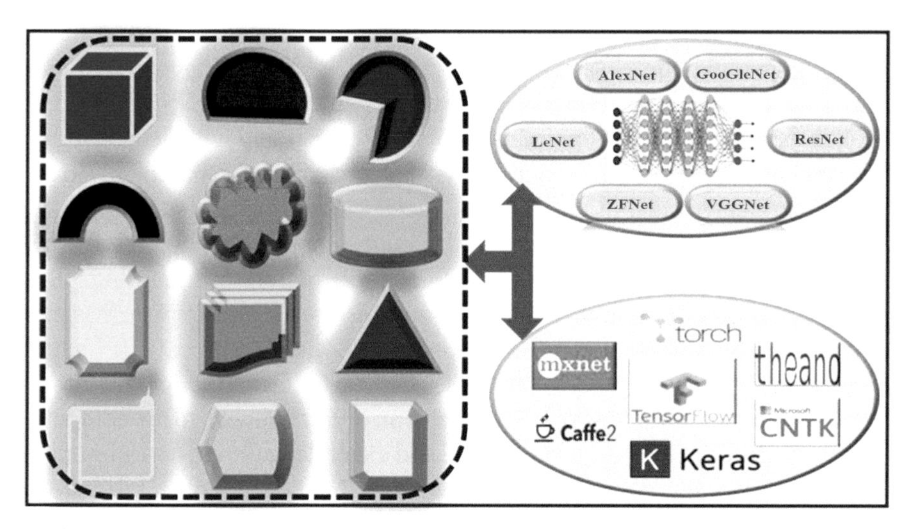

FIGURE 10.2 A general overview of visual media to develop an intelligent system.

This is not very necessary to use exactly the famous deep architectures such as AlexNet [14], GoogleNet [35], and VGGNet [36]. In addition, developing a new network is not very difficult because the concept of deep network architecture allows updating the hidden layers accordingly the specified objective. There are several factors that can affect the structure of a deep network that may include a number of layers and parameters selected to process the data, size of stride, and type of layers introduced.

10.3 VISUAL INTELLIGENCE WITH REAL-WORLD GRAPHS

Acoustic data consider 1D signal processing of sound, whereas images and videos processing take account for 2D and 3D signal processing. Machine intelligence is characterized by supervised and unsupervised learning of deep geometrical structures of the real-world objects. Geometrical deep knowledge can extract important features, from which an optimized and faster model can be built. Therefore, the analysis and evaluations of all the non-Euclidean structures can be performed faster. Several applications of computational science and domain network analysis can be found to exploit the geometrical deep learning detection and tracking of the occluded objects in video sequences. Furthermore, NLP, video captioning, and descriptor-based research work for features analysis in recognition phases are supposed to consider the importance of deep network architecture for real-world geometry.

A convolutional neural network (CNN) [37] has been proved to be an outstanding backbone of several deep networks, which is characterized by a cluster of matrices multiplication. For graphs and manifolds, a generalized CNN is represented to exploit image and video processing with deep learning [2]. In addition, it highlighted that the data related to graphs and manifold are most prominently used in the network social media, transportation, and the sensor-based anatomical structure of the human brain. Graphs are efficient tools to represent any complex real-world information, but learning them with the machine is quite difficult. In this case, deep learning is very useful to automate graph-based representation of real-world entities [3]. Graph-based kernel, indexing, and hashing use map classification and recognition. Detecting an object in video or still image is an open challenge to the computer vision community. This requires precise and exact detection in less amount of time. Several variations of the CNN are reported to develop object detectors, single-shot detection [12], the real-time object detection scheme "you look only once" (YOLO), and its series YOLOv2 and YOLOv3 [39]. All the methods outperformed the region proposal network, which is simply a faster recurrent neural network (RNN), faster-R-CNN [40]. Deep learning in graph clustering is outshined the spectral in terms of computational complexity [17]. In this, graph clustering with deep learning, spectral clustering, and *k*-means clustering are utilized to implement the graph encoder with sparse autoencoder. Sparse encoder controls reconstruction and sparsity errors. Processing the whole data as a graph is the toughest problem, but this can provide efficient analytical solutions in several disciplines such as molecular biology, pattern recognition, and astronomy with the study of the geospatial satellite. The graphs with the CNN are combined as a graph convolution network (GNN) [20,29]. In the same tune, a graph coevolution network is used to develop graph autoencoders for unsupervised learning [21]. As a vision perspective, the worldly information is updated with variations in the characteristics of nodes, which means any specific worldly event is happened due to the changes in the functions of a particular object. Considering the physical world as a graph becomes an issue of high-performance computing. This approach can provide alternative solutions of many problems like suspicious event detection, undirected human actions on the basis of nodes, and edges information in the connected graph [22]. This concludes that the possibility of success of such experiments can be expected only with the support of deep learning methodology. Motivated by this fact, the spatial-temporal GNN is proposed for human action recognition from depth information [18,30]. In the proposed work, the human body is assumed a graph of joints as nodes and bones as the edges connecting to nodes. The experiments were performed on NTU RGBD human action benchmarks. Several such experiments ensure that deep learning is remarkably capable of filling the gaps of spatiotemporal events by developing a deep model of the action sequence. Such experiments can be successful in providing desired results by jointly exploiting sequence learning networks such as RNN and spatiotemporal graphs. With the help of graphs of the depth information collected from the human body and structural-RNN, the important experiments to model the human motion and interaction with machines are performed [19]. Practically, lack of ground truth may be noticed to raise the training issues with the model for many existing or new problems in the machine learning domain. Leveraging the lack of exact training semisupervised learning is a better option. For this case, fast approximation convolution is performed, and the comparison processor was shown on central processing unit and graphics processing unit (GPU) [23]. Furthermore, the label propagation jointly with semisupervised learning utilized a large number of facts from unlabeled data with neural networks [26]. Machine learning with graphs and deep network opens a ubiquitously high range of solutions.

In the review work presented in [24], the main problem for machine learning with graphs is highlighted as the information association between the nodes. The encoding and decoding scheme with the embedding of graphs is helpful for informatics. Recently observed state-of-the-art presents scalability and interpretability in temporal graphs as open issues [26].

10.3.1 GRAPH- AND MANIFOLD-BASED LEARNING SCENARIO

Graphs provide a mechanism of lucid representation to the processing components in a network. Computer vision literature is full with plenty of research on 1D, 2D, and 3D data representation, which is referred as analysis of acoustic signal, image, and video processing, respectively. All these representations are termed Euclidean structures. Non-Euclidean structures include graphs and manifolds. Recently, the success of deep learning has been reported as an interesting work with non-Euclidean geometry, but several techniques for graph-based image processing such as normalized cut-based image segmentation remain ever sounding [41–43]. With the advent of deep learning, they constructed a deformation invariant model of manifolds and graphs in 3D spatial domain. Achieving deformation invariant features for non-Euclidean objects in frequency domain is an open challenge to the computer vision community. On these structures, convolution is not supposed easily applicable since recovery of such lower dimensional structures manifolds referred as nonlinear dimensionality reduction, which can be consider as instance of unsupervised deep learning. Being very specific for graph-based deep learning application, the experiments on human skeleton are performed for action recognition [44]. They utilized lie group features on the graphs of skeleton joints, which can be easily aligned with temporal features due to rotation in joints.

10.4 STANDARD NEURAL NETWORK AND CLASSIFICATION CRITERION

All vision problems generally require many complex phases up to final solution of any particular research work. Table 10.1 presents basic classification strategies that are referred as core machine learning. The structure of the basic neural network topology is represented in Figure 10.3. Kumar [27,28] presented detail architecture of the important deep networks architecture and large-scale data analytics issues, which are shown in the next subsections.

10.4.1 DEEP LEARNING SPECIFIC COMPONENTS

In this section, the specific components of deep network structure are described. Basic features of these components discriminate the conventional neural network to deeply learning neural network in the context of exploiting the complex feature space of the visual objects and resolve the severe computational issues with such feature space.

The main component of a deep neural network is the CNN, which focus on convolution as major operation of a large number of matrices. The intelligent system with the CNN can efficiently process complex features. In this reference, apart from the CNN, the vital technology includes the RNN, long short-term memory, autoencoders, and restricted Boltzmann machine. In this chapter, we primarily focus on the brief architecture of the CNN and describe its performance to solve video processing tasks. In general, every CNN model contains four basic building blocks of convolution, sampling, nonlinearity unit used in CNN, and fully connected layers used for feature classification.

TABLE 10.1 Machine Learning Algorithms and Its Feature Processing

Machine Learning Schemes	Feature Description
Supervised (labeling based) Learning methods	• A training model is developed with training labeled datasets and tested against the samples. For example, classification and regression
Unsupervised (unlabeled) learning	• Missing the training labeled samples. For example, clustering and noise reduction etc.
Reinforce learning	• Based on penalty and reward function for input sample Data. For example, Markov decision process
Active machine learning	• Sampling is performed based on query on selective data
Representation-based learning	• Includes feature engineering of its selection, extraction, and reduction.
Transfer-based learning	• Domain-invariant-based transfer learning
Kernel-based machine learning	• Nonlinear multimedia processing to reduce high dimensional, e.g., LDA, PCA, support vector machine, etc.

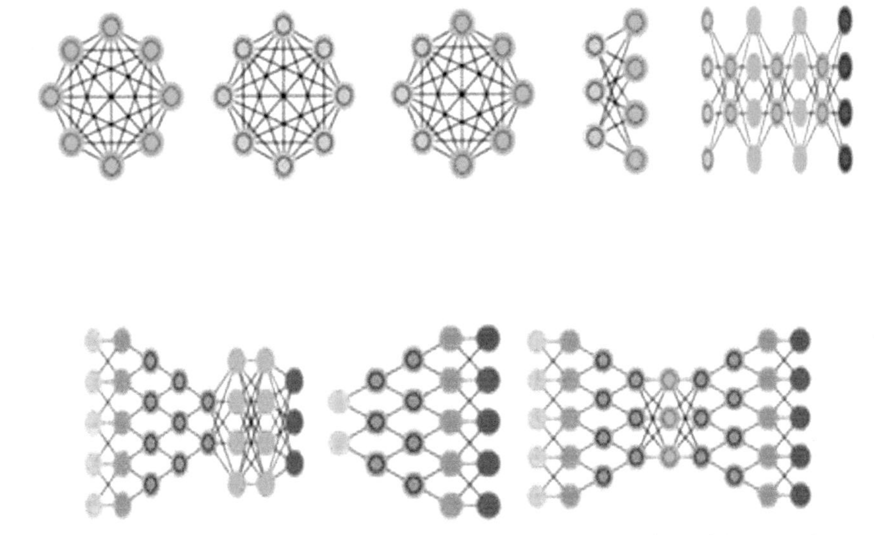

FIGURE 10.3 (a) Markov chain model. (b) Hopfield of network model. (c) Boltzmann machine structure. (d) Restricted Boltzmann machine. (e) Deep belief network. (f) Deep convolution network (DCN). (g) Deconvolution network. (h) DCN graphic network.

10.4.2 NONLINEARITY IN DEEP NETWORK

This refers to introduction of a class of thresholding in the layers of traditional neural network. The layers of nonlinearity are introduced by utilizing the activation or sigmoid function. In a real-life scenario, almost the problems are nonlinear, which are incorporated by introducing by the same nature of activation function to get threshold for the CNN. The most basic thresholding functions are rectified linear unit (ReLU) [8,9], logistic function, and exponential linear unit [10]. In general, all the activation functions work on pixelwise operations. The literature of deep learning evidences that several advanced versions of ReLU are more powerful. The examples include modified ReLU, leaky ReLU, parametric ReLU, and randomized leaky ReLU represented in (10.2); (10.3) outperformed the state of the art

$$\begin{cases} f(x) = x, \; x \geq 0 \\ 0, \; x < 0 \end{cases} \tag{10.1}$$

$$\begin{cases} f(y) = y, \; y \geq 0 \\ ay, \; y < 0 \end{cases} \tag{10.2}$$

$$\begin{cases} z, \; z \geq 0 \\ f(z) = \underline{z}, \; z < 0. \end{cases} \tag{10.3}$$

10.4.3 CONVOLUTION FILTERS

The activation of neural network from convolution layers mitigates fine-tuning issues and maps the higher dimensional data space to lower dimensional data space. The convolution process by a 3´3 kernel with image is basically characterized by three parameters. The number of convolution filters used in the CNN determines the depth of CNN. Other basic components of the CNN include stride and zero padding. Stride refers to the number of pixels that jump during one convolution. In the third component, zero padding is the mechanism that provides the boundary pixel to involve in the convolution process. The basic convolution filters include Gaussian, Laplacian, Sobel, and box filter for the basic building block of filters. It is customary that selecting more number of filters gives better training to the deep network provided that computational tools are sufficient to take care the processing overhead. Apart from convolution, there are many more conceptual layers are referred to introduce for developing a deep network. These layers may be pooling layers, batch normalization layers, drop out, and fully connected layers. All the layers correspond to different operation in deep network, and introducing these layers in the network depends on the particular objectives.

10.5 STANDARD DEEP NETWORKS

In this section, we discuss some standard deep networks, which ensure the success of deep learning methods applicable to many engineering and science discipline.

10.5.1 LENET (1990)

In 1990, when there was no sound in research community for deep learning, Yann LeCun developed a very first convolution neural network. This is edge of deep learning get break though with specific domain of optical character recognition. The basic LeNet architecture is given in Figure 10.4, which presents the discrimination of visual objects at the prediction layer. This network provides the base to all the modern deep neural networks, which need to have cascading of convolution and pooling and nonlinearity layers. ReLU [15,16] is generally applied before pooling and fully connected layers of the network.

10.5.2 ALEXNET (2012)

The architecture of AlexNet exploits details of convolution neural network blocks developed in LeNet. Noticeably, only difference is the number of filters used for reducing dimensionality between various pooling layers. The details of the network are represented in [3]. The evidence from ImageNet challenge reported that the network is trained on two NVidia graphics card "GTX 580," with over 1.2 million sample images of the large dataset. For purely classification, the training of such data sample takes five to six days. The network uses five coevolutions and trains 60 million parameters and 6.5 lakh neurons.

10.5.3 ALEXNET (2012)

Zeiler and Fergus developed a deep network with their name (ZFNet) in 2013 that exploits the intermediate functionality of classification methodology inside the deep network layers. They tweaked the complete architecture proposed in AlexNet convolution neural network. ZFNet demonstrated the state of the art on Caltech 101 and Caltech-256 benchmarks. By training with ImageNet [14] for classification on GTX 580 for 12 days, we developed features of pixel maps as opposing the convolution layers. In the experiential phase of the network, activation and error operations are performed by the ReLU and the cross-entropy loss function. In this network, error computation and action operations are performed by cross-entropy loss and ReLU, respectively. During the classification process on the ImageNet benchmark, the drop-out approach is utilized to achieve regularization.

10.5.4 GOOGLENET (2014)

The massive data of real-world scenario pertain huge numbers of parameters, which mark a black spot at the success of earlier deep network models. The number of parameters used in GoogleNet was only 4 million, whereas Alexnet used 60 million parameters. The reduction of such a large of number parameters was possible by introducing an inception model [35]. The architecture of the inception model is represented in Figure 10.4.

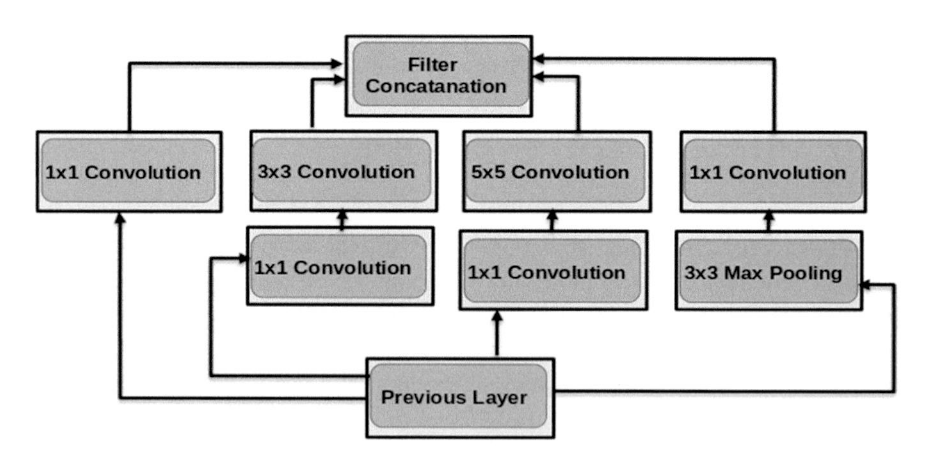

FIGURE 10.4 Inception module of the GoogleNet architecture [35].

The inception model of the GoogleNet set benchmark in detection and visual recognition literature. From the evidences of several experiments on ImageNet14, this is concluded that the inception model with its successive versions, that is, inception 3 and inception 4, is a fast and suitable model to resolve the issues complex visual analytics. In this model, the Hebbian principle is adopted to get the better optimization control and results in a model with only 22 layers. The mechanism of the inception model is motivated by selection of the size of convolution filters and concatenation them. The mechanism of convolution used in inception model gets the CNN free from overfitting. The overfitting layers are renamed as global average pooling layers.

10.5.5 VGGNET (2014)

The issue of a large number of layers as required for handling huge number of parameters in large-scale data models was considered a severe problem till 2014. Therefore, this model is highly sounded to deal with large-scale data statistics without bothering about the number of layers used in the network.

The experimental setup for training VGGNet is developed with four Titan Black GPUs and took three to four weeks to accomplish the training phase. This network takes 21–28 days to train with two GPUs "Titan Black." The model used Caffe toolbox as background and training data is optimized by utilizing stochastic gradient descent (SGD) scheme. The experimental observation [11,13] reports presented in Table 10.2 show that there is only 7.5% error in the validation set on the model with 19 layers and 7.3% error in the test set on top five classification layers.

10.5.6 RESNET(2015)

After the classical performance of VGGNet on large-scale data, it is assumed to develop big network results in higher performance. Such deeper neural networks outperform better with extensive data, but training such a network is a highly tedious job. The main credit is given to Kaiming H, for resolving the processing issue with the deeper network, in ILSVRC 2015 challenge.

The layers used in the network were modified by learning residual functions [11]. This function optimizes the computations and achieves higher accuracy. From the experiments, it is observed that batch normalization also fails to reduce validation and training error while introducing the extra layer to the network. In the ResNet inception model presented in Figure 10.5, this problem is solved by introducing bypass to the summing up the layer with the CNN. The model took 21 days to process 152 layers with the ImageNet benchmark by utilizing eight GPUs.

TABLE 10.2 The Comparative Classification Performance of the ResNet Model

Deep Network Model	Error Analysis (%)	Computation Time (ms)
VGG Model-A	28.90	381
ResNet Model-34	26.70	230
BN-Inception Model	26.10	192
ResNet Model-50	24.00	402
ResNet Model-101	22.10	651
Inception-v3 Model	20.90	493

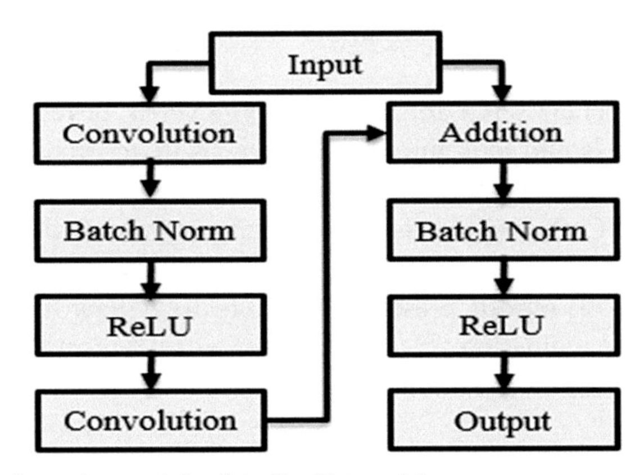

FIGURE 10.5 Inception module of the ResNet model.

10.6 CONCLUSION

In this chapter, the prime concern is represented to the processing issues with heterogeneous contents in the visual media of real life. For resolving the highlighted issues, deep data analytics and graphical processing algorithms are needed to develop for extracting and evaluating the information from real-life visual media. Training an optimized deep network with graph-based pixel-level processing is a better choice than handicraft feature engineering for local and global feature extraction. A visual domain generally consists of real-time multimedia activities, which are still an open research problem regarding the semantic understanding of poorly retrieved information from a complex environment. Another issue with the processing of complex information is reported as big data evolution from live streaming of video sequences. In such a case, without introducing big data analytics, the experimental issues can degrade the performance of conventional machine algorithms. Furthermore, five components of big data issues can cause to fail every conventional machine learning technique to retrieve 100% exact information. Therefore, to exploit deep network architecture, it is very customary to develop optimized deep data analytics system and high-performance algorithms. In brief, this work raises the future scope of developing deep neural networks with some optimized algorithms that can give better performance to compute the unstructured visual data with large scale analytics.

10.6.1 FUTURE RECOMMENDATIONS

The CNN has been performed successfully for solving an image classification problem. However, in the case of video-based experiments on big data, developing a big network does not guarantee higher performance. This motivates that processing of large-scale visual media requires to build up the mechanism for every layer in the network. Such high-level processing cannot be expected without a rich source of processing devices. Therefore, to compromise with costly computational resources and time economy, this is necessarily required to develop an optimized deep network for extracting information from unconstraint visual media. The conclusive remark for future issues this sounds the absence of an adaptively fast and small deep network. Although stacking with more layers and effective subsampling can increase the size of receptive field but central receptive field of each neuron does not participate equally. The throughout study of this work concludes that many issues such as SGD, graphs, and Riemannian-manifold-based deep learning can still be needed to practice for a better solution to open challenging problems in social media communication, sensor network of neuron for analysis of the human brain, stock market for financial evaluations, and geographical study based on a network of satellites. In the review work presented in [24], the main problem for machine learning with graphs is highlighted as the information association between the nodes. The encoding and decoding scheme with the embedding of graphs is helpful for informatics. Recently observed state of the art presents scalability and interpretability in temporal graphs as open issues [26].

KEYWORDS

- **big data analytics**
- **convolution neural network**
- **long short-term memory**
- **deep learning**
- **graphs and manifolds**
- **unstructured complex datasets**
- **visual media**

REFERENCES

1. Jordan, M. I. (Ed.) (1998). *Learning in Graphical Model*, Vol. 89. New York, NY, USA: Springer Science & Business Media.
2. Monti, F., Boscaini, D., Masci, J., Rodola, E., Svoboda, J., & Bronstein, M. M. (2017, July). Geometric deep learning on graphs and manifolds using mixture model CNNs. In *Conference on Computer Vision and Pattern Recognition*, p. 3.
3. Riba, P., Dutta, A., Lladós, J., & Fornés, A. (2017, November). Graph-based deep learning for graphics classification. In *14th IAPR International Conference on Document Analysis and Recognition* (pp. 29–30).
4. K. Tombre and B. Lamiroy (2003), "Graphics recognition-from re-engineering to retrieval," in *International Conference on Document Analysis and Recognition*, pp. 148–155.
5. N. Nayef and T. M. Breuel (2010), "A branch and bound algorithm for graphical symbol recognition in document images," in *International Workshop on Document Analysis Systems*, pp. 543–545.
6. J. Gantz and D. Reinsel, *Extracting Value from Chaos*. Hopkinton, MA, USA: EMC, Jun. 2011.
7. J. Gantz and D. Reinsel, *The Digital Universe Decade—Are You Ready*. Hopkinton, MA, USA: EMC, May 2010.
8. Naresh Babu, K. V., & Edla, D. R. (2017), "New algebraic activation function for multi-layered feed forward neural networks," *IETE Journal of Research*, 63(1), 71–79.
9. Xu, B., Wang, N., Chen, T., & Li, M. (2015). Empirical evaluation of rectified activations in convolutional network. *arXiv preprint arXiv:1505.00853*
10. Clevert, D. A., Unterthiner, T., & Hochreiter, S. (2015). Fast and accurate deep network learning by exponential linear units (ELUs). *arXiv preprint arXiv:1511.07289*.
11. Huang, F., Ash, J., Langford, J., & Schapire, R. (2017). Learning Deep ResNet Blocks Sequentially using Boosting Theory. *arXiv preprint arXiv:1706.04964*.
12. Liu, W., Anguelov, D., Erhan, D., Szegedy, C., Reed, S., Fu, C. Y., & Berg, A. C. (2016), "SSD: Single shot multibox detector," In *European Conference on Computer Vision* (pp. 21–37.

13. Haung, G., Liu Z., Weinberger, K. Q., & van der Maaten, L. (2016), Densely Connected Convolutional Networks. *arXiv preprint arXiv:1608.06993*.
14. Krizhevsky, A., Sutskever, I., & Hinton, G. E. (2012), "ImageNet classification with deep convolutional neural networks. In *International Conference on Neural Information Processing Systems* (pp. 1097–1105).
15. Xu, B., Wang, N., Chen, T., & Li, M. (2015) Empirical evaluation of rectified activations in convolutional network. *arXiv preprint arXiv:1505.00853*.
16. Jin, X., Xu, C., Feng, J., Wei, Y., Xiong, J., & Yan, S. (2016), "Deep learning with S-shaped rectified linear activation units," In *30th AAAI Conference on Artificial Intelligence* (pp. 1737–1743).
17. Tian, F., Gao, B., Cui, Q., Chen, E., & Liu, T. Y. (2014, July). Learning deep representations for graph clustering. In *28th AAAI Conference on Artificial Intelligence* (pp. 1293–1299).
18. Yan, S., Xiong, Y., & Lin, D. (2018). Spatial temporal graph convolutional networks for skeleton-based action recognition. *arXiv preprint arXiv:1801.07455*.
19. Jain, A., Zamir, A. R., Savarese, S., & Saxena, A. (2016). Structural-RNN: Deep learning on spatio-temporal graphs. In *IEEE Conference on Computer Vision and Pattern Recognition* (pp. 5308–5317).
20. Scarselli, F., Gori, M., Tsoi, A. C., Hagenbuchner, M., & Monfardini, G. (2009). The graph neural network model. *IEEE Transactions on Neural Networks*, 20(1), 61–80.
21. Kipf, T. N., & Welling, M. (2016). Variational graph auto-encoders. *arXiv preprint arXiv:1611.07308*.
22. Niepert, M., Ahmed, M., & Kutzkov, K. (2016, June). Learning convolutional neural networks for graphs. In *International Conference on Machine Learning* (pp. 2014–2023).
23. Kipf, T. N., & Welling, M. (2016). Semi-supervised classification with graph convolutional networks. *arXiv preprint arXiv:1609.02907*.
24. Hamilton, W. L., Ying, R., & Leskovec, J. (2017). Representation learning on graphs: Methods and applications. *arXiv preprint arXiv:1709.05584*.
25. Baradel, F., Neverova, N., Wolf, C., Mille, J., & Mori, G. (2018). Object Level Visual Reasoning in Videos. *arXiv preprint arXiv:1806.06157*.
26. Bui, T. D., Ravi, S., & Ramavajjala, V. (2018, February). Neural Graph Learning: Training Neural Networks Using Graphs. In *11th ACM International Conference on Web Search and Data Mining* (pp. 64–71).
27. Kumar, N. (2017, December). Large scale deep network architecture of CNN for unconstraint visual activity analytics. In *International Conference on Intelligent Systems Design and Applications* (pp. 251–261).
28. Kumar, N. (2017, December). Machine Intelligence Prospective for Large Scale Video based Visual Activities Analysis. In *Ninth International Conference on Advanced Computing* (pp. 29–34).
29. Gallagher, B. (2006). Matching structure and semantics: A survey on graph-based pattern matching. *AAAI FS*, 6, (45–53).
30. Jargalsaikhan, I., Little, S., & O'Connor, N. E. (2017, August). Action localization in video using a graph-based feature representation. In *14th IEEE International Conference on Advanced Video and Signal Based Surveillance* (pp. 1–6).
31. Zanfir, A., & Sminchisescu, C. (2018). Deep Learning of Graph Matching. In *IEEE Conference on Computer Vision and Pattern Recognition* (pp. 2684–2693).
32. Zhang, M., Cui, Z., Neumann, M., & Chen, Y. (2018). An end-to-end deep learning architecture for graph classification. In *AAAI Conference on Artificial Intelligence*.

33. Duvenaud, D. K., Maclaurin, D., Iparraguirre, J., Bombarell, R., Hirzel, T., Aspuru-Guzik, A., & Adams, R. P. (2015). Convolutional networks on graphs for learning molecular fingerprints. In *International Conference on Neural Information Processing Systems* (pp. 2224–2232).
34. Henaff, M., Bruna, J., & LeCun, Y. (2015). Deep convolutional networks on graph-structured data. *arXiv preprint arXiv:1506.05163*.
35. Szegedy, C., Liu, W., Jia, Y., Sermanet, P., Reed, S., Anguelov, D., & Rabinovich, A. (2015). Going deeper with convolutions. In *IEEE Conference on Computer Vision and Pattern Recognition* (pp. 1–9).
36. Wang, L., Guo, S., Huang, W., & Qiao, Y. (2015). Places205-vggnet models for scene recognition. *arXiv preprint arXiv:1508.01667*.
37. Krizhevsky, A., Sutskever, I., & Hinton, G. E. (2012). ImageNet classification with deep convolutional neural networks. In *International Conference on Neural Information Processing Systems* (pp. 1097–1105).
38. Redmon, J., Divvala, S., Girshick, R., & Farhadi, A. (2016). You only look once: Unified, real-time object detection. In *IEEE Conference on Computer Vision and Pattern Recognition* (pp. 779–788).
39. Redmon, J., & Farhadi, A. (2017). YOLO9000: Better, Faster, Stronger. *arXiv preprint*.
40. Ren, S., He, K., Girshick, R., & Sun, J. (2015). Faster R-CNN: Towards real-time object detection with region proposal networks. In *International Conference on Neural Information Processing Systems* (pp. 91–99).
41. Monti, F., Boscaini, D., Masci, J., Rodola, E., Svoboda, J., & Bronstein, M.M. (2017, July). Geometric deep learning on graphs and manifolds using mixture model CNNs. In *IEEE Conference on Computer Vision and Pattern Recognition*, p. 3.
42. Bronstein, M. M., Bruna, J., LeCun, Y., Szlam, A., & Vandergheynst, P. (2017). Geometric deep learning: Going beyond Euclidean data. *IEEE Signal Processing Magazine*, 34(4), 18–42.
43. Shi J. and Malik J. (2000). Normalized cuts and image segmentation. *IEEE Transactions on Pattern Analysis and Machine Intelligence*, 22(8):888–905.
44. Huang, Z., Wan, C., Probst, T., & Van Gool, L. (2017). Deep learning on lie groups for skeleton-based action recognition. In *IEEE Conference on Computer Vision and Pattern Recognition* (pp. 1243–1252).
45. [Online]. Available: https://gluon-cv.mxnet.io/build/examples_datasets/imagenet.html

CHAPTER 11

Applications of Backpropagation and Conjugate Gradient Neural Networks in Structural Engineering

S. VARADHARAJAN[1*], KIRTHANASHRI S. VASANTHAN[2], SHWETAMBARA VERMA,[3] and PRIYANKA SINGH[4]

[1,4] *Department of Civil Engineering, Amity School of Engineering and Technology, Amity University, Noida, Uttar Pradesh 201301, India*

[2]*Amity Institute of Molecular Medicine and Stem Cell Research, Amity University, Noida, Uttar Pradesh 201301, India*

[3]*Department of Civil Engineering, School of Engineering and Technology, Kaziranga University, Jorhat, Assam.*

[*]*Corresponding author. E-mail: svrajan82@gmail.com*

ABSTRACT

This chapter briefly describes the backpropagation and conjugate networks in structural engineering. The first portion of the section deals with the basic framework of backpropagation and conjugate gradient network. The second portion presents the literature review of research works associated with the application of neural networks in structural engineering. The third section deals with the pros and cons of neural networks. The final part deals with the identification of the problems associated with the neural networks and emphasis on possible solutions to overcome the issues. Finally, conclusions derived from the research study are provided.

11.1 INTRODUCTION

Artificial intelligence is a system specifically designed to exhibit human-like behavior in decision making for solving complex problems. The decisions made by artificial intelligence are quick, efficient, and straightforward. Artificial intelligence is an interdisciplinary subject, which has inherent advantages over traditional modeling and statistical methods. The developed artificial neural networks (ANNs) are useful in the collection, analysis, and processing of data on a large scale. After that, the difference between actual and target values is computed to propose an activation function.

Artificial intelligence combines different fields such as computer science, neuropsychology, psychology, and information technology. The ANN is often used in the field of civil engineering due to its efficiency and lesser time consumption in different aspects such as modeling, design, analysis, and optimization, which are very complex, time-consuming, and prone to error. The field of structural engineering employs techniques of data collection, especially for predicting the strength of concrete. The ANN can be trained based on the experimental dataset for prediction of desired output parameters.

The first section deals with a brief introduction to the ANN, along with a detailed description of its elements. The final portion of this section presents a brief discussion on the backpropagation (BP) network. The second section offers a comprehensive literature review spanning over three decades about the implementation of the ANN in structural engineering problems. Finally, the last portion elucidates the shortcomings of different types of ANNs and provides recommendations to overcome these shortcomings.

11.1.1 ELEMENTS OF NEURON

The biological networks are quite complex and very difficult to explain. A human brain is comprised of hundreds of biological neurons, and it is impossible to represent these neurons using a mathematical model. However, a simple neural network model resembling the functioning of the actual brain can represent complex mathematical problems and can yield accurate results. The artificial neuron receives the input signals from the brain. After that, output data and every information or peripheral data of this neuron are generated, which can be used as the input data for further iterations. The activation function processes the data from surroundings received by the input layer in the hidden layers. The output layer receives these processed data, which forms a solution of the problem. The input layer may comprise of n number of neurons, but every neuron receives one input data only.

11.1.2 COEFFICIENT DESCRIBING WEIGHTS

The weight coefficients are integral elements of the neural network, and weight coefficient is assigned to each neuron depending on its sensitivity to the output parameter. The input signal can be obtained by multiplying the input parameter with its weight coefficient. The weights are described as w1, w2, and w3 and input data are marked as k1, k2, and k3. Therefore, equivalent weights can be calculated as w1k1 + w2k2 + w3k3. The zero weight coefficients show that there is no relationship between input and output data.

11.1.3 BP NETWORKS

The field of structural engineering often employs BP networks and BP algorithm to achieve accuracy. BP networks consist of several hidden layers with different weights. The technique of supervised learning is used in BP networks to calculate the discrepancy between expected and actual outputs.

The functioning of the BP network is as follows:

- Providing the training data to a given sample of ANN.
- The network output is compared with the expected output.
- Estimation of the network error.
- The adjustment of weights to reduce the error.

The multilayer perception is used in conjunction with the BP network. The relationship between weights of individual elements should be accurately determined for the training of multilayer perception. The calculated error depends on the combination of weight coefficients. The main aim of training a perceptron is to determine the minimum failure for any possible combination. The gradient descent technique can be adopted to achieve this purpose.

The BP algorithm refers to a training algorithm of feedforward neural networks or multilayer perceptrons. The network weights are initially set to a random value or an absolute minimum of the error surface. The BP algorithm calculates the gradient of the error surface, and changes in weights are determined for the steepest slope. The weights will converge to minima for small error surface. The BP algorithm is summarized as follows.

- Initialize weights of the network.
- The first input vector is derived from the dataset used for training.

- The output is obtained by propagating the input vector through the network.
- The desired output is compared against actual output to estimate error.
- The error signals are transmitted in the reverse network direction.
- To weights are adjusted to minimize the overall error.
- The repetition of steps is done to reduce the overall error to a permissible limit.

The training can be subclassified into two different types.

- *Online training*: In this training, the corresponding weights of the network are changed after the presentation of each pattern.
- *Batch training*: In this type of training, the error weights are updated at each iteration.

The training iterations are repeated number of times till a satisfactory performance is obtained for a particular problem. The training should terminate when the performance of the individual test data reaches a maximum value. However, it may not result in error minimization necessarily. The counter-propagation algorithm is associated with two parameters that can be adjusted significantly, learning rate and the learning speed, which determines the size of the step adopted during the learning process of the descent of the iterative gradient. A significant value leads to the generation of the network error, but if this step is small and time-consuming

11.2 LITERATURE REVIEW

Adeli and Yeh [4] proposed a new model for engineering design, which employed machine learning technique (Block, 1962). Perception is the sensor that receives an input signal. The perceptron devoid of hidden layers was used to formulate the problem of the structural design. Vanluchene and Sun [95] applied the BP algorithm to the structural engineering problems such as the selection of beam cross section. Design decisions and showed the efficiency of neural network approaches [78].

Hajela and Berke [39] advocated the use of neural networks in the optimization of structures and observed its efficient performance. Hung and Adeli [43] based on perceptron theory developed a new model called as PERHID for machine learning. Yu and Adeli [100] and Hung and Adeli (1994a) utilized object-oriented programming for machine learning and engineering design. Some researchers have introduced an iteration process, which triggered itself in the original perceptron learning model to inhibit the error at the end of each iteration cycle [5]. The authors observed an enhanced performance of new theory in the design of steel beams. Theocaris and Panagiotopoulos [87] applied the BP learning algorithm for parameter identification in fracture mechanics. Gunaratnam and Gero [37] observed better efficiency of neural networks in solving problems of structural engineering. Messner et al. [63] used neural networks for selection of most efficient cross sections for structural members for a given building data like space, site location, budget, and height of the structure.

Yeh [99] used neural networks for the identification of transverse cracking and spalling of concrete joints prestressed concrete piles. Kang and Yoon [50] used neural networks for efficient designing of roof trusses, while others used neural networks for the improvement of wavefront and profile characteristics. Rogers [77] used neural networks in conjunction with the approximate method of structural analysis for structural optimization. Mukherjee and Deshpande [65] used the BP algorithm for the preliminary design of the structures. Alsugaired and Sharar (1995) used neural networks to determine the characteristics of semirigid connections with single-angle plate beam configuration.

Turkkan and Shrivastava [90] conducted wind analysis using neural networks for membranes of different shapes. Mukherjee and Deshpande [65] determined the behavior of axially loaded columns subjected to buckling based using ANNs. Anderson et al. [11] conducted the experimental study on beam-column connections and determined load-carrying capacity of different types of connections. Noor [73] conducted nonlinear and sensitivity analysis using the BP algorithm. Mirmiran and Shahawy [64] used ANNs to accurately predict the behavior of concrete columns with fiber-reinforced composites. Yang et al. [97] determined mechanical properties of lightweight concrete using ANNs. Analysis results elucidated a close agreement between the experimental results and results obtained by ANNs. Konin et al. [56] employed ANNs to determine the penetration of chlorides due to presence of microcracking in the high-performance concrete. Hegazy and Ayed [42] identified selection characteristics, strain in reinforcement bars, distribution of stress in concrete, and crack patterns in concrete using ANNs. Chuang et al. [27] determined the load-carrying capacity in the ultimate range for reinforced concrete columns with pinned supports. Stavroulakis and Antes [85] identified steady-state cracks in the structural members. Cao et al. [21] modeled a cantilever beam subjected to the pair of concentrated loads as an analogy to model the aircraft wings. Mathew et al. [60] employed the ANN to assess the behavior of masonry panels under the influence of different types of bending. Kishi et al. [55] developed an ANN model to assess the alignment of structural members subjected to different end conditions. The authors observed the overestimation of the size of structural members, which yielded a conservative design. Hung and Jan [45] adopted ANNs for determination of the performance of braced and unbraced frame using AISC alignment charts. The global stiffness matrix represents a relation between the load and deflection, but they consume a lot of time. However, the ANN can be effectively used for this purpose with more efficiency and lesser time consumption. Adeli and Kumar [2] adopted the BP algorithm for determining the approximate displacements at the end of each iteration. The analysis results showed that the combination of neural network and BP algorithm improved efficiency. Consolazio [28] presented different combinations of the neural networks in conjunction with iterative equation-solving techniques. William and Shoukry [96] used finite-element analysis in conjunction with neural networks for design and analysis of flat slab highway. The authors observed that adoption of neural networks leads to better convergence. This was later confirmed by Li [58]. Öztaş et al. [74] observed the better performance of the ANN in the prediction of 28 days compressive strength of concrete in comparison to traditional methods. Similar observations were reported by Kewalramani and Gupta [52], Nataraja et al. [70], Adhikary and Mutsuyoshi [7], Narendra et al. [69], El- et al. [32], and Ji et al. [48]. Jeyasehar and Sumangala [47] observed that the ANN yielded close results in comparison to experimental results obtained from nondestructive testing of prestressed concrete. Saini et al. [80] observed the superior performance of the ANN in the analysis of doubly reinforced beams. Neocleous (2006) accurately determined the flexural capacity of concrete with steel fibers using the ANN. Altun et al. [10] adopted BP networks to forecast the characteristic mechanical properties of lightweight concrete. The authors manufactured lightweight concrete with a density ranging from 350 to 450 kg/m^3. The steel fibers (Dramix type) were added in different dosages up to 60%. The input layers for the ANN were the quantity of steel fiber, w/c ratio, superplasticizer, etc. The BP ANN showed better performance as compared to multiple linear regression (MLR) analysis in the estimation of compressive strength of concrete. Topcu and Sarıdemir [89] adopted both ANN and fuzzy logic models for the prediction of mechanical properties associated with the concrete. The authors chose a large amount of experimental data from previous research works for ANN modeling. The ANN model developed adopted raw materials (used in the manufacture of concrete) as input parameters. The compressive strength was chosen as the output parameter. The analysis results showed that the amalgamated system exhibited better performance than the ANN. Demir [30] accurately determined the elastic modulus of concrete using the ANN model. Jung and Kim [49] accurately predicted the mechanical properties of concrete beams without shear reinforcement. Karthikeyan et al. [51] observed the better performance of ANN as compared to conventional methods in predicting durability characteristics such as creep and shrinkage of concrete.

Prasad et al. [75] determined the compressive strength of concrete with high workability employing ANN. The authors used a large number of experimental results to train the developed ANN model. The analysis results showed the superior performance of the ANN as compared to traditional methods. Naderpour et al. [67] observed the efficiency of the ANN in predicting the compressive strength of concrete with fiber-reinforced plastics as reinforcement. Bilgehan and Turgut [16] conducted ultrasonic pulse velocity test in determination of the strength of concrete. The authors used the experimental dataset in developing ANN. The authors observed a close correlation between ANN and experimental results.

Furthermore, the ANN yielded better results as compared to classical methods. Sobhani et al. [84] confirmed the superior performance of the ANN over the ANFIS model in prediction of concrete strength with low workability. Ashrafi et al. [12] employed the ANN to determine the load-displacement behavior of concrete reinforced with composite fibers and observed its efficient performance. Erdem [33] formulated an ANN model to predict the moment-resisting capacity of flat slabs and found close agreement between experimental and ANN results. Basyigit et al. [14] obtained similar results for high-performance concrete using neural network and fuzzy logic models. Uysal and Tanyildizi [92] determined the strength parameters of concrete at 28 days using the ANN. The concrete was manufactured, incorporating different minerals to increase the strength of concrete. The manufactured concrete employed two combinations. The first combination included both limestone and fly ash (FA) in ratios ranging from 15% to 30%. The analysis results showed a good correlation between the results of the experimental study and the ANN. Atici [13] predicted the characteristic compressive strength of blast furnace slag concrete and observed superior performance of ANN as compared to MLR analysis. Hakim et al. [40] found the ANN to predict the mechanical properties of concrete with the superplasticizers accurately. Similar results were observed by Siddique et al. [83] for self-compacting concrete. Nazari and Riahi [71] predicted the split tensile strength and water permeability using ANN. The authors compared the genetic programming technique and the ANN for predicting mechanical and durability properties. The authors conducted the experimental study and collected a dataset of 144 specimens with a different period of curing. The authors adopted 16 different mix proportions with varying proportions of cement, coarse aggregate, fine aggregate, and water. The ANN was developed considering eight different input parameters such as type of aggregate, cement content, water content, nanoparticle content, type of superplasticizer, etc. The analysis results showed comparable performance between neural network and genetic algorithm models.

Uysal [91] used an ANN model for estimation of characteristic compressive strength of the ANN model with polypropylene (PP) fiber under high temperature. In their research study, the authors adopted mineral additives such as FA, granulated blast furnace slag, zeolite, basalt powder, and PP fiber of density 2 kg/m^3. The analysis results showed a loss of compressive strength as the temperature exceeds 600 °C, but the presence of PP fiber prevented the risk of concrete spalling.

Sadrmomtazi et al. [79] predicted mechanical properties of concrete manufactured by use of lightweight expanded polystyrene beads. The authors used a combination of adaptive network-based fuzzy inference system (ANFIS) and ANN models for predicting the compressive strength. The analysis results showed that the ANN is useful in prediction as compared to the prediction capacity of the ANFIS model. In contradiction, Yuan et al. [101] observed close agreement with the BP algorithm and the ANN in evaluating the characteristic compressive strength of high-performance concrete. Chandwani et al. [23] developed a hybrid method by fusion of genetic algorithm and ANN for prediction of mechanical properties of concrete. The authors adopted parameters such as coarse aggregates, FA, fine aggregate, water binder, cement, etc., as input parameters. The analysis results showed the poor performance of linear regression analysis as compared to the ANN. Chithra et al. [25] determined mechanical properties of concrete using ANN and MLR analysis. The authors used copper slag and nanosilica for partial replacement of cement and fine aggregate. The cement was replaced with nanosilica up to 3% at intervals of 0.5%, and fine aggregate was replaced with copper slag up to 50%. In agreement with previous studies, MLR analysis exhibited poor performance as compared to the ANN.

Khademi et al. [54] compared three different techniques, namely, (a) ANN, (b) MLR model, and (c) ANFIS for estimation of characteristic compressive strength of concrete. The analysis results showed superior performance of both ANFIS and ANN models in the determination of mechanical properties of concrete as compared to MLR analysis.

Eskandari-Naddaf and Kazemi [34] developed an ANN model for prediction of characteristic compressive strength of concrete up to 52.5 MPa. The ANN model was trained using an experimental dataset with a different water–cement ratio of up to 0.5%. The analysis results showed that ANN models yielded accurate results. Khademi et al. [53] confirmed the accuracy of the ANN in estimating the compressive strength for concrete of different grades. Naderpour et al. [68] developed an ANN model trained based on 139 experimental datasets for prediction of compressive strength of recycled aggregate concrete. The proposed ANN model consisted of six input parameters such as fine aggregate, coarse aggregate, water–cement ratio, and percentage replacement of recycled coarse aggregate. Bui et al. [19] used a hybrid system with a combination of modified firefly algorithm and ANN for prediction of the tensile strength of concrete. The ANN model was trained from the experimental data, and MFA was used to initialize the weights. The analysis results showed better performance of a hybrid system as compared to ANN in predicting the relationship between compressive and tensile strength. Behnood and Golafshani [15] replaced silica fume by cement to manufacture a silica fume concrete. The authors aimed at investigating the effect of silica fume as a partial substitute of cement to reduce the consumption of cement and inhibit the generation of CO_2. The authors used the ANN and a new methodology called multiobjective gray wolf optimization. The partial replacement of silica fume in concrete yielded good compressive strength, and the ANN predicted the compressive strength accurately. Prasad et al. (2019) determined the efficiency of the ANN model to predict the compressive strength of self-compacting concrete. The authors used 99 datasets from the experimental study to train the neural network. The authors used five input nodes and a three-layered feedforward BP network with 10 hidden layers. The analysis results showed the accurate prediction of compressive strength by the ANN. Rajeshwari and Mandal [76] replaced cement with FA as the binding material to investigate the mechanical properties of concrete. The authors collected 270 datasets from the literature study and 12 from the experimental results to propose an ANN model to predict the compressive strength of concrete. The ANN was developed utilizing eight input parameters associated with raw materials used in the manufacture of concrete. The analysis results showed that proposed ANN accurately predicted the compressive strength with a high coefficient of correlation. It was observed that the compressive strength of concrete decreased with an increase in the percentage of replacement of cement with FA. Hammoudi et al. [41] predicted the compressive strength of concrete at the curing period ranging from 7 to 56 days. The concrete was manufactured by using the industrial byproduct of recycled aggregate and replacing it with coarse aggregate. The authors used response surface methodology and neural networks to predict the compressive strength. The experimental results indicated that strength decreases with the replacement of recycled aggregates was increased from 0% to 100%. The statistical analysis showed an excellent correlation between the response surface method and the ANN method, and both ways accurately predicted the strength of concrete in compression. Hadzima-Nyarko et al. [38] replaced the aggregate in concrete with waste rubber and determined change in the properties of concrete. The authors, based on their experimental results, prepared a dataset of 457 mixes and proposed an ANN model for determination of strength properties. The results of the analytical study showed that strength was strongly influenced by waste rubber percentage. The proposed ANN model exhibited efficiency in the estimation of compressive strength. Similar observations were reported by Yaseen et al. [98] and Mansouri et al. [59].

11.3 PROBLEMS IDENTIFIED

The past literature survey indicates that multilayer perceptrons are widely used in neural networks in comparison to single-layer perceptron. It is due to the inefficiency of single-layer perceptrons in the solution of complex problems. However, multilayer perceptron does present several problems as discussed in the following.

- Multilayer perceptrons are difficult to implement, but the usage of commercially available software can solve these problems to a certain degree (e.g., [17, 81]).
- The second main issue with the multilayer perceptron theory is in the decision of network architecture. The network architecture depends on the number of hidden layers, which, in turn, depends on training data. The training data vary depending on the problem, so there are no specific rules to decide. The BP network may face the convergence issue in case of lesser number of nodes in the hidden layers. However, the large number of nodes cause network overfitting the training data resulting in poor performance.

- The network learns all the training patterns, which leads to overfitting. This overtrained network will lead to an inaccurate prediction of the pattern, which generates clatter. If used for extrapolation, multilayer perceptrons may perform poorly. However, for the interpolation, they may exhibit better performance.

Apart from problems associated with a multilayer perceptron, the BP network has its inherent disadvantages. The first drawback is the slow learning rate. In the case of a BP network, the number of iterations is enormous (Carpenter and Barthelemy, 1994). In addition, the sensitiveness of convergence rate is observed for parameters such as momentum ratio and learning rate. The selection of appropriate values of these parameters is strongly influenced by the nature of the problem [3,99].

11.4 COMPARATIVE STUDY OF ALREADY PROPOSED SOLUTION

Although the reduction in the number of hidden layers may reduce the noise, a network is trained to overlook a small noise in a particular situation. The overfitting of the training data is one of the significant causes of concern for the perceptron. It can be avoided by subdividing the network into several units such as the following.

- Training set
- Test set
- Validation set

The validation set determines the capability of the network to generalize itself during training. The training is discontinued as generalization reaches the maximum value. This process is called as early stopping and is widely adopted in training of multilayer perceptron network. The basic concept of the early stopping technique is as follows.

- The data available for training are divided into data for training and validation in the ratio of 2:1.
- The training set is trained and evaluated to determine the error after the end of each cycle.
- The training process is terminated as soon as the error on the validation set exceeds the previous cycle.
- The weights are kept unchanged at each cycle.

Early stopping is preferred due to its greater efficiency overgeneralization techniques.

The other ways to reduce overfitting are (i) reducing the parameter size which is termed as "greedy constructive learning," and (ii) sharing of weights to the reduced size of each parameter by employing techniques such as regularization, weight decay, etc.

There are several other solutions that can be adopted are the following.

- The functions employed in training the multilayer perceptron should be smooth so that input data become sensitive to the output data. Nevertheless, the used training data should be extensive in number and should cover a wide range of parameters. The adopted training data should represent all the problems to which the multilayer perceptron will be subjected.
- In case the training data adopted are based on extreme or uncommon events, then it would reduce the accuracy of the network. This problem can be rectified by a technique called V-fold cross-validation. In this technique, the data are randomly divided in the form of V-shaped individual subsets. After that, training of multilayer perceptron is done to determine its performance. Several other approaches to overcome this problem can be adopted, such as data preprocessing, adjusting the training algorithm to ignore the well-defined patterns.
- For problems with less complexity, a single hidden layer is required to generate approximate results.

The issues related to the BP networks can be solved as follows.

- The learning rate and momentum parameter need to be carefully chosen based on the problem. LeCun et al. [57] prescribe following thumb rules for the process. The momentum parameter needs to be increased, and the learning rate needs to be decreased for convergence of the network. In addition, the learning rate should be decreased if network fluctuates in the vicinity of the solution.
- More sophisticated algorithms can be adopted so that the necessity of sensitive training parameters is eliminated. Many alternatives such as log sig functions are available, but bounded functions are used for BP networks.
- The rescaling of input data is done between 0 and 1. Then, initial network weights are adopted randomly, but input parameter exhibiting lesser variance is preferred.

- The standardization of inputs is done by division of input data by the respective standard deviation.
- BP is slow, and this problem can be solved by reducing the dimensionality of the input data. This is done employing the process of feature selection, in which the redundant variables are removed from the input data [82].

11.5 PROS AND CONS OF SOLUTION

The solution presented in the previous section has the following advantages as enumerated under

- It improves accuracy.
- Reduces the computation time.
- Less storage space is required.
- It reduces the cost of future measurements.
- It improves data and model understanding.

There is no specific rule to obtain minima in case of early stopping technique. There may be more than one minima in case of real problems, and consequently, no particular criterion is prescribed to judge a minimum of iterations for the desired accuracy. The selection technique requires a large subset of data, which should extensively cover all types of problems; otherwise, it may induce errors in the network. Moreover, the dimensions of the input parameters should also be extensive.

11.6 CONCLUSION

A large number of problems are encountered in the field of structural engineering, and review of past literature study shows that multilayer perceptron is a very useful tool, which can be used to solve these complex problems. The multilayer perceptrons carry a distinct advantage that they are simple, easy to implement, and can be effectively applied to nonlinear problems. These are distinct advantages over nonlinear systems. The multilayer perceptrons are easy to perform with the development of sophisticated software. Nevertheless, multilayer perceptrons do carry some of the inherent disadvantages such as the decision of network architecture and number of layers, which varies for each type of problem. Moreover, overfitting is also one of the major problems encountered in multilayer perceptron theory. These problems can be solved to a large extent to facilitate the easy implementation of multilayer perceptrons.

11.7 FUTURE SCOPE

The other domains of civil engineering such as traffic engineering, disaster mitigation, environmental engineering, etc., can be explored. Nevertheless, more techniques of forecasting can be adopted in conjunction with the ANN. Finally, more case studies can be involved.

KEYWORDS

- **backpropogation network**
- **conjugate gradient network**
- **structural engineering**
- **artificial neural network.**

REFERENCES

1. Adeli, H. & Hung, S. L. (1995). *Machine Learning—Neural Networks, Genetic Algorithms, and Fuzzy Systems*. John Wiley & Sons, Inc., New York, NY, USA.
2. Adeli, H. & Kumar, S. (1999). *Distributed Computer-Aided Engineering for Analysis, Design, and Visualization*. CRC Press, Boca Raton, FL, USA.
3. Adeli, H., & Hung, S. L. (1994). *Machine Learning: Neural Networks, Genetic Algorithms, and Fuzzy Systems*. John Wiley & Sons, Inc., New York, NY, USA.
4. Adeli, H., & Yeh, C. (1989). Perceptron learning in engineering design. *Computer-Aided Civil and Infrastructure Engineering*, *4*(4), 247–256.
5. Adeli, H., & Zhang, J. (1993). An improved perception learning algorithm. *Neural Parallel & Scientific Computations*, *1*, 141–152.
6. Adeli, H., & Zhang, J. (1995). Fully nonlinear analysis of composite girder cable-stayed bridges. *Computers & Structures*, *54*(2), 267–277.
7. Adhikary, B. B., & Mutsuyoshi, H. (2006). Prediction of shear strength of steel fiber RC beams using neural networks. *Construction and Building Materials*, *20*(9), 801–811.
8. Alsugair, A. M., & Sharar, E. A. (1995). A decision support system for pavement maintenance using artificial neural networks technology.
9. Altun, F., Haktanir, T., & Ari, K. (2007). Effects of steel fiber addition on mechanical properties of concrete and RC beams. *Construction and Building Materials*, *21*(3), 654–661.

10. Altun, F., Kişi, Ö., & Aydin, K. (2008). Predicting the compressive strength of steel fiber added lightweight concrete using neural network. *Computational Materials Science*, *42*(2), 259–265.
11. Anderson, D., Hines, E. L., Arthur, S. J. & Eiap, E. L. (1997), Application of artificial neural networks to the prediction of minor axis steel connections, *Computers and Structures*, *63*(4), 685–692.
12. Ashrafi, H. R., Jalal, M., & Garmsiri, K. (2010). Prediction of load–displacement curve of concrete reinforced by composite fibers (steel and polymeric) using artificial neural network. *Expert Systems with Applications*, *37*(12), 7663–7668.
13. Atici, U. (2011). Prediction of the strength of mineral admixture concrete using multivariable regression analysis and an artificial neural network. *Expert Systems with Applications*, *38*(8), 9609–9618.
14. Başyigit, C., Akkurt, I., Kilincarslan, S., & Beycioglu, A. (2010). Prediction of compressive strength of heavyweight concrete by ANN and FL models. *Neural Computing and Applications*, *19*(4), 507–513.
15. Behnood, A., & Golafshani, E. M. (2018). Predicting the compressive strength of silica fume concrete using hybrid artificial neural network with multi-objective grey wolves. *Journal of Cleaner Production*, *202*, 54–64.
16. Bilgehan, M., & Turgut, P. (2010). Artificial neural network approach to predict compressive strength of concrete through ultrasonic pulse velocity. *Research in Nondestructive Evaluation*, *21*(1), 1–17.
17. Bishop, C. M. (1995). *Neural Networks for Pattern Recognition*. Oxford University Press, Oxford, U.K.
18. Block, H. D., Knight Jr., B. W., & Rosenblatt, F. (1962). Analysis of a four-layer series-coupled perceptron. II. *Reviews of Modern Physics*, *34*(1), 135.
19. Bui, D. K., Nguyen, T., Chou, J. S., Nguyen-Xuan, H., & Ngo, T. D. (2018). A modified firefly algorithm-artificial neural network expert system for predicting compressive and tensile strength of high-performance concrete. *Construction and Building Materials*, *180*, 320–333.
20. Cao, J., & Zhou, D. (1998). Stability analysis of delayed cellular neural networks. *Neural Networks*, *11*(9), 1601–1605.
21. Cao, X., Sugiyama, Y. & Mitsui, Y. (1998), Application of artificial neural networks to load identification. *Computers and Structures*, *69*, 63–78.
22. Carpenter, W. C., & Barthelemy, J. F. (1993). A comparison of polynomial approximations and artificial neural nets as response surfaces. *Structural Optimization*, *5*(3), 166–174.
23. Chandwani, V., Agrawal, V., & Nagar, R. (2015). Modeling slump of ready mix concrete using genetic algorithms assisted training of artificial neural networks. *Expert Systems with Applications*, *42*(2), 885–893.
24. Chen, A. M.; Lu, H.-M.; Hecht-Nielsen, R. (1993). On the geometry of feedforward neural network error surfaces. *Proceedings of Neural Computations.*, *5*(6), 910–927.
25. Chithra, S., Kumar, S. S., Chinnaraju, K., & Ashmita, F. A. (2016). A comparative study on the compressive strength prediction models for high performance concrete containing nano silica and copper slag using regression analysis and artificial neural networks. *Construction and Building Materials*, *114*, 528–535.
26. Chopra, P., Sharma, R. K., & Kumar, M. (2016). Prediction of compressive strength of concrete using artificial neural network and genetic programming. *Advances in Materials Science and Engineering*, *2016*, 7648467.

27. Chuang, P. H., Goh, A. T. C. & Wu, X. (1998), Modeling the capacity of pin-ended slender reinforced concrete columns using neural networks, *Journal of Structural Engineering*, *124*(7), 830–838.
28. Consolazio, G. R. (2000). Iterative equation solver for bridge analysis using neural networks. *Computer-Aided Civil and Infrastructure Engineering*, *15*(2), 107–119.
29. Cvetkovska, M. (2002). Nonlinear stress strain behavior of RC elements and plane frame structures exposed to fire. Ph.D. dissertation, Civil Engineering Faculty in Skopje, Ss. Cyril and Methodius University, Skopje, North Macedonia.
30. Demir, F. (2008). Prediction of elastic modulus of normal and high strength concrete by artificial neural networks. *Construction and Building Materials*, *22*(7), 1428–1435.
31. Duan, L., & Chen, W. F. (1989). Effective length factor for columns in unbraced frames. *Journal of Structural Engineering*, *115*(1), 149–165.
32. El-Chabib, H., & Nehdi, M. (2006). Effect of mixture design parameters on segregation of self-consolidating concrete. *ACI Materials Journal*, *103*(5), 374.
33. Erdem, H. (2010). Prediction of the moment capacity of reinforced concrete slabs in fire using artificial neural networks. *Advances in Engineering Software*, *41*(2), 270–276.
34. Eskandari-Naddaf, H., & Kazemi, R. (2017). ANN prediction of cement mortar compressive strength, influence of cement strength class. *Construction and Building Materials*, *138*, 1–11.
35. Flood, I.; Nabil, K. (1994). Neural networks in civil engineering II: Systems and application. *Journal of Computing in Civil Engineering*. *8*(2), 149–162.
36. Flood, I.; Paul, C. (1996). Modeling construction processes using artificial neural networks. *Automation in Construction*. *4* (4), 307–320.
37. Gunaratnam, D. J., & Gero, J. S. (1994). Effect of representation on the performance of neural networks in structural engineering applications. *Computer-Aided Civil and Infrastructure Engineering*, *9*(2), 97–108.
38. Hadzima-Nyarko, M., Nyarko, E. K., Ademović, N., Miličević, I., & Šipoš, T. K. (2019). Modelling the influence of waste rubber on compressive strength of concrete by artificial neural networks. *Materials*, *12*(4), 561.
39. Hajela, P., & Berke, L. (1991). Neurobiological computational models in structural analysis and design. *Computers & Structures*, *41*(4), 657–667.
40. Hakim, S. J. S., Noorzaei, J., Jaafar, M. S., Jameel, M., & Mohammadhassani, M. (2011). Application of artificial neural networks to predict compressive strength of high strength concrete. *International Journal of Physical Sciences*, *6*(5), 975–981.
41. Hammoudi, A., Moussaceb, K., Belebchouche, C., & Dahmoune, F. (2019). Comparison of artificial neural network (ANN) and response surface methodology (RSM) prediction in compressive strength of recycled concrete aggregates. *Construction and Building Materials*, *209*, 425–436.
42. Hegazy, T., & Ayed, A. (1998). Neural network model for parametric cost estimation of highway projects. *Journal of Construction Engineering and Management*, *124*(3), 210–218.
43. Hung, S. L., & Adeli, H. (1991). A model of perceptron learning with a hidden layer for engineering design. *Neurocomputing*, *3*(1), 3–14.
44. Hung, S. L., & Adeli, H. (1991). A model of perceptron learning with a hidden layer for engineering design. *Neurocomputing*, *3*(1), 3–14.
45. Hung, S. L., & Jan, J. C. (1999). Machine learning in engineering analysis and design: An integrated fuzzy neural network learning model. *Computer-Aided Civil and Infrastructure Engineering*, *14*(3), 207–219.

46. Jenkins, W. M. (1999), A neural network for structural re-analysis, *Computers and Structures*, *72*, 687–698.
47. Jeyasehar, C. A., & Sumangala, K. (2006). Nondestructive evaluation of prestressed concrete beams using an artificial neural network (ANN) approach. *Structural Health Monitoring*, *5*(4), 313–323.
48. Ji, T., Lin, T., & Lin, X. (2006). A concrete mix proportion design algorithm based on artificial neural networks. *Cement and Concrete Research*, *36*(7), 1399–1408.
49. Jung, S., & Kim, K. S. (2008). Knowledge-based prediction of shear strength of concrete beams without shear reinforcement. *Engineering Structures*, *30*(6), 1515–1525.
50. Kang, H.-T. & Yoon, C. J. (1994), Neural network approaches to aid simple truss design problems, *Microcomputers in Civil Engineering*, *9*(3), 211–18.
51. Karthikeyan, J., Upadhyay, A., & Bhandari, N. M. (2008). Artificial neural network for predicting creep and shrinkage of high performance concrete. *Journal of Advanced Concrete Technology*, *6*(1), 135–142.
52. Kewalramani, M. A., & Gupta, R. (2006). Concrete compressive strength prediction using ultrasonic pulse velocity through artificial neural networks. *Automation in Construction*, *15*(3), 374–379.
53. Khademi, F., Akbari, M., Jamal, S. M., & Nikoo, M. (2017). Multiple linear regression, artificial neural network, and fuzzy logic prediction of 28 days compressive strength of concrete. *Frontiers of Structural and Civil Engineering*, *11*(1), 90–99.
54. Khademi, F., Jamal, S. M., Deshpande, N., & Londhe, S. (2016). Predicting strength of recycled aggregate concrete using artificial neural network, adaptive neuro-fuzzy inference system and multiple linear regression. *International Journal of Sustainable Built Environment*, *5*(2), 355–369.
55. Kishi, N., Chen, W. F. & Goto, Y. (1997), Effective length factor of columns in semirigid and unbraced frames, *Journal of Structural Engineering*, ASCE, *123*(3), 313–20.
56. Konin, A., Francois, R., & Arliguie, G. (1998). Penetration of chlorides in relation to the microcracking state into reinforced ordinary and high strength concrete. *Materials and Structures*, *31*(5), 310–316.
57. LeCun Y., L. Bottou, Y. Bengio, and P. Haffner (1998). Gradient-based learning applied to document recognition. *Proceedings of the IEEE*, 86(11), 2278–2324.
58. Li, S. (2000), Global flexibility simulation and element stiffness simulation in finite element analysis with neural network, *Computer Methods in Applied Mechanics and Engineering*, *186*, 101–108.
59. Mansouri, I., Gholampour, A., Kisi, O., & Ozbakkaloglu, T. (2018). Evaluation of peak and residual conditions of actively confined concrete using neuro-fuzzy and neural computing techniques. *Neural Computing and Applications*, *29*(3), 873–888.
60. Mathew, A., Kumar, B., Sinha, B. P. & Pedreschi, R. F. (1999), Analysis of masonry panel under biaxial bending using ANNs and CBR, *Journal of Computing in Civil Engineering*, *13*(3), 170–177.
61. McClelland, J. L., Rumelhart, D. E., & PDP Research Group. (1986). Parallel distributed processing. *Explorations in the Microstructure of Cognition*, *2*, 216–271.
62. Meesaraganda, L. P., Saha, P., & Tarafder, N. (2019). Artificial neural network for strength prediction of fibers' self-compacting concrete. In *Soft Computing for Problem Solving* (pp. 15–24). Springer, Singapore.
63. Messner, J. I., Sanvido, V. E. & Kumara, S. R. T. (1994), StructNet: A neural network for structural system selection, *Micro-computers in Civil Engineering*, *9*(2), 109–118.

64. Mirmiran, A., & Shahawy, M. (1997). Behavior of concrete columns confined by fiber composites. *Journal of Structural Engineering*, *123*(5), 583–590.
65. Mukherjee, A., & Deshpande, J. M. (1995). Modeling initial design process using artificial neural networks. *Journal of Computing in Civil Engineering*, *9*(3), 194–200.
66. Mukherjee, A., Deshpande, J. M. & Anmala, J. (1996), Prediction of buckling load of columns using artificial neural networks, *Journal of Structural Engineering*, *122*(11), 1385–1387.
67. Naderpour, H., Kheyroddin, A., & Amiri, G. G. (2010). Prediction of FRP-confined compressive strength of concrete using artificial neural networks. *Composite Structures*, *92*(12), 2817–2829.
68. Naderpour, H., Rafiean, A. H., & Fakharian, P. (2018). Compressive strength prediction of environmentally friendly concrete using artificial neural networks. *Journal of Building Engineering*, *16*, 213–219.
69. Narendra, B. S., Sivapullaiah, P. V., Suresh, S., & Omkar, S. N. (2006). Prediction of unconfined compressive strength of soft grounds using computational intelligence techniques: A comparative study. *Computers and Geotechnics*, *33*(3), 196–208.
70. Nataraja, M. C., Jayaram, M. A., & Ravikumar, C. N. (2006). A fuzzy-neuro model for normal concrete mix design. *Engineering Letters*, *13*(2), 98–107.
71. Nazari, A., & Riahi, S. (2011). Prediction split tensile strength and water permeability of high strength concrete containing TiO2 nanoparticles by artificial neural network and genetic programming. *Composites Part B: Engineering*, *42*(3), 473–488.
72. Neocleous, K., Tlemat, H., & Pilakoutas, K. (2006). Design issues for concrete reinforced with steel fibers, including fibers recovered from used tires. *Journal of Materials in Civil Engineering*, *18*(5), 677–685.
73. Noor, A. K. (1996). Computational intelligence and its impact on future high-performance engineering systems.
74. Öztaş, A., Pala, M., Özbay, E., Kanca, E., Caglar, N., & Bhatti, M. A. (2006). Predicting the compressive strength and slump of high strength concrete using neural network. *Construction and Building Materials*, *20*(9), 769–775.
75. Prasad, B. R., Eskandari, H., & Reddy, B. V. (2009). Prediction of compressive strength of SCC and HPC with high volume fly ash using ANN. *Construction and Building Materials*, *23*(1), 117–128.
76. Rajeshwari, R., & Mandal, S. (2019). Prediction of compressive strength of high-volume fly ash concrete using artificial neural network. In *Sustainable Construction and Building Materials* (pp. 471–483). Springer, Singapore.
77. Rogers, J. L. (1994). Simulating structural analysis with neural network. *Journal of Computing in Civil Engineering*, *8*(2), 252–265.
78. Rumelhart, D. Hinton, G. and Williams, R. (1986). Learning internal representations by error propagation. In *Parallel Distributed Processing*, Rumelhart and McLelland, (Eds). MIT Press, Cambridge, MA, USA.
79. Sadrmomtazi, A., Sobhani, J., & Mirgozar, M. A. (2013). Modeling compressive strength of EPS lightweight concrete using regression, neural network and ANFIS. *Construction and Building Materials*, *42*, 205–216.
80. Saini, B., Sehgal, V. K., & Gambhir, M. L. (2006). Genetically optimized artificial neural network based optimum design of singly and doubly reinforced concrete beams.
81. Sarle, W. S. (1997). Neural Network FAQ, periodic posting to the Usenet newsgroup comp. AI. neural-nets. *URL: ftp://ftp.sas.com/pub/neural/FAQ.html.*

82. Setiono, R., & Liu, H. (1995, August). Understanding neural networks via rule extraction. In *International Joint Conferences on Artificial Intelligence.* (Vol. 1, pp. 480–485).
83. Siddique, R., Aggarwal, P., & Aggarwal, Y. (2011). Prediction of compressive strength of self-compacting concrete containing bottom ash using artificial neural networks. *Advances in Engineering Software*, *42*(10), 780–786.
84. Sobhani, J., Najimi, M., Pourkhorshidi, A. R., & Parhizkar, T. (2010). Prediction of the compressive strength of no-slump concrete: A comparative study of regression, neural network and ANFIS models. *Construction and Building Materials*, *24*(5), 709–718.
85. Stavroulakis, G. E. & Antes, H. (1998), Neural crack identification in steady state elastodynamics, *Computer Methods in Applied Mechanics and Engineering*, *165*, 129–146.
86. *Structures*, 42 (4), 649–659.
87. Theocaris, P. S., & Panagiotopoulos, P. D. (1993). Neural networks for computing in fracture mechanics. Methods and prospects of applications. *Computer Methods in Applied Mechanics and Engineering*, *106*(1–2), 213–228.
88. Theocharis, P. S., & Panagiotopoulos, P. D. (1993). Neural networks for computing in fracture mechanics. Methods and Prospects of applications. *Computer Methods in Applied Mechanics and Engineering*, *106*, 213–228.
89. Topcu, I. B., & Sarıdemir, M. (2008). Prediction of compressive strength of concrete containing fly ash using artificial neural networks and fuzzy logic. *Computational Materials Science*, *41*(3), 305–311.
90. Turkkan, N. & Srivastava, N. K. (1995), Prediction of wind load distribution for air-supported structures using neural networks, *Canadian Journal of Civil Engineering*, *22*, 453–61.
91. Uysal, M. (2012). Self-compacting concrete incorporating filler additives: Performance at high temperatures. *Construction and Building Materials*, *26*(1), 701–706.
92. Uysal, M., & Tanyildizi, H. (2011). Predicting the core compressive strength of self-compacting concrete (SCC) mixtures with mineral additives using artificial neural network. *Construction and Building Materials*, *25*(11), 4105–4111.
93. Uysal, M., & Tanyildizi, H. (2012). Estimation of compressive strength of self compacting concrete containing polypropylene fiber and mineral additives exposed to high temperature using artificial neural network. *Construction and Building Materials*, *27*(1), 404–414.
94. Uysal, M., & Tanyildizi, H. (2012). Estimation of compressive strength of self compacting concrete containing polypropylene fiber and mineral additives exposed to high temperature using artificial neural network. *Construction and Building Materials*, *27*(1), 404–414.
95. Vanluchene, R. D., & Sun, R. (1990). Neural networks in structural engineering. *Computer-Aided Civil and Infrastructure Engineering*, *5*(3), 207–215.
96. William, G. W., & Shoukry, S. N. (2001). 3D finite element analysis of temperature-induced stresses in dowel jointed concrete pavements. *International Journal of Geomechanics*, *1*(3), 291–307.
97. Yang, C. C., Yang, Y. S., & Huang, R. (1997). The effect of aggregate volume ratio on the elastic modulus and compressive strength of lightweight concrete. *Journal of Marine Science and Technology*, *5*(1), 31–38.
98. Yaseen, Z. M., Deo, R. C., Hilal, A., Abd, A. M., Bueno, L. C., Salcedo-Sanz, S., & Nehdi, M. L. (2018). Predicting compressive strength of lightweight foamed concrete using extreme learning machine model. *Advances in Engineering Software*, *115*, 112–125.

99. Yeh, I.-C. (1995), Construction site layout using annealed neural networks, *Journal of Computing in Civil Engineering*, *9*(3), 201–208.
100. Yu, G., & Adeli, H. (1991). Computer-aided design using object-oriented programming paradigm and blackboard architecture. *Computer-Aided Civil and Infrastructure Engineering*, *6*(3), 177–190.
101. Yuan, Z., Peng, H. J., Huang, J. Q., Liu, X. Y., Wang, D. W., Cheng, X. B., & Zhang, Q. (2014). Hierarchical free-standing carbon-nanotube paper electrodes with ultrahigh sulfur-loading for lithium–sulfur batteries. *Advanced Functional Materials*, *24*(39), 6105–6112.

CHAPTER 12

Application of Neural Networks in Construction Management

S. VARADHARAJAN[1*], KIRTHANASHRI S. VASANTHAN[2], SHWETAMBARA VERMA[3], and PRIYANKA SINGH[4]

[1,4] *Department of Civil Engineering, Amity School of Engineering and Technology, Amity University, Noida, Uttar Pradesh 201301, India*

[2]*Amity Institute of Molecular Medicine and Stem Cell Research, Amity University, Noida, Uttar Pradesh 201301, India*

[3]*Department of Civil Engineering, School of Engineering and Technology, Kaziranga University, Jorhat, Assam.*

[*]*Corresponding author. E-mail: svrajan82@gmail.com*

ABSTRACT

The review of past research works pertaining to construction management shows that artificial intelligence is being widely used in this field. The use of artificial intelligence has given numerous advantages to the practicing engineers, which has immensely improved scheduling and management of construction activities resulting in huge economic benefits. This chapter explores use of artificial intelligence in different aspects of construction management and elucidates the problems encountered and presents possible solutions to the problems.

12.1 INTRODUCTION

Artificial intelligence is a system specifically designed to exhibit human-like behavior in decision making for solving complex problems. The decisions made by artificial intelligence are quick, efficient, and straightforward. Artificial intelligence is an interdisciplinary subject, which has inherent advantages over traditional modeling and statistical methods. The developed artificial neural networks (ANNs) are useful in the collection, analysis, and processing the data on a large scale. After that, the difference between actual and target values is computed to propose an activation function.

Artificial intelligence combines different fields such as computer science, neuropsychology, psychology, and information technology. The ANN is often used in the field of civil engineering due to its efficiency and lesser time consumption in different aspects such as modeling, design, analysis, and optimization, which are very complex, time-consuming, and prone to error. The techniques of data collection and analysis are widely applied in the field of structural engineering, especially for predicting the strength of concrete. The ANN can be trained based on the experimental dataset for prediction of desired output parameters.

This chapter mainly focuses on different types of neural networks in cost pricing, scheduling, and estimation of various rent projects related to civil engineering. In the first part, a brief introduction to the concept of activation function has been presented. In the second part, a detailed literature review involving research works of the past two decades has been summarized concerning the utilization of ANN models in cost pricing, construction scheduling, and estimation. The third part deals with the identification of problems involved with currently employed ANN and proposed solutions. The final part deals with suggestions for improvement in the existing solution with scope for future work.

12.1.1 ELEMENTS OF NEURON

The biological networks are quite complex and very difficult to explain. A human brain is comprised of hundreds of biological neurons, and it is impossible to represent these neurons using a mathematical model. However, engineering problems can be simplistically represented by simple neural network models resembling the functioning of the actual brain, which can yield accurate results. The input signals are received by the artificial neuron from the brain. After that, output data and every information or peripheral data of this neuron are generated, which can be used as the input data for further iterations. The input layer receives data from the surroundings and is processed in the hidden layers by the activation function. The output layer receives these processed data, which forms a solution of the problem. The input layer may comprise of n number of neurons, but every neuron receives one input data only.

12.1.2 COEFFICIENT DESCRIBING WEIGHTS

The biological networks are quite complex and very difficult to explain. A human brain is comprised of hundreds of biological neurons, and it is impossible to represent these neurons using a mathematical model. However, engineering problems can be simplistically represented by simple neural network models resembling the functioning of the actual brain, which can yield accurate results. The artificial neuron receives the input signals from the brain. After that, output data and every information or peripheral data of this neuron are generated, which can be used as the input data for further iterations. The data from surroundings received by the input layer are processed in the hidden layers by the activation function. The output layer receives these processed data, which forms a solution of the problem. The input layer may comprise of *n* number of neurons, but every neuron receives one input data only.

12.2 LITERATURE REVIEW

The past literature review shows that the ANN is quite efficient in prediction of tender price and prediction of tender cost [18, 37, 60, 64, 87, 95, 96, 98], cash expenses in a project [25, 75, 99], labor costs [76, 82], prequalification of contractors [58], performance of contract [96], and quantification of risk [64].

12.2.1 COST ESTIMATION

The cost estimation is required for proficient functioning of the construction industry. The past survey of the literature indicates a comprehensive usage of ANN in determination of construction cost [16, 18, 23, 29, 56, 74, 88, 87, 94].

Williams [98] was one of the first researchers who forecasted variation in the construction cost for six months and confirmed the efficiency of the BP algorithm. Hegazy et al. [51] observed that the ANN performs efficiently in different fields of construction management. Hegazy and Ayed [50] adopted the genetic algorithm and optimization techniques for cost optimization of highway projects. Adeli [2] developed an ANN model for optimizing the cost of reinforced concrete (RC) pavements. The analysis results showed that parameters such as atmospheric conditions and human judgment considerably influence the cost forecasting. The authors further observed overfitting to be one of the main problems in the ANN models. Geiger [41] used the ANN network for cost estimation of sheet metals. The authors developed separate models for cost estimation of material, power endurance, and accessories. The results showed that ANN models observed accuracy ranging from 85% to 95%. Elhag and Bossabine [34] proposed the ANN model for cost estimation of buildings.

The results showed an efficient performance of ANN models with high accuracy. Al-Tabtabai [13] determined an incremental increase in the cost of the construction project using ANN and confirmed considerable accuracy of ANN models. Bhokha and Ogunlana [23] developed an ANN model forecasting the building cost by the inclusion of variables such as finishing, structural system, the height of the building, type of decoration, accessibility to the site, etc. The analysis results showed that the proposed model underestimated the values for 42.7% samples and overestimated the results for 57.3% samples. Fang and Froese [38] developed an ANN model and predicted the relationship between parameters such as concrete cost, type of formwork, and quality of concrete used in high-performance concrete. Some researchers advocated the use of hybrid networks in predicting the cost estimate for the buildings. However, the comparison showed the poor performance of the hybrid models as compared to the ANN model. Shtub and Versano [86] used both regression and neural network techniques in determination of steel pipe cost. The authors observed the better performance of the linear regression model as compared to ANN, which was in contradiction to the results of the previous literature study. Assaf [17] observed that ANNs predict the overall effective cost of construction projects in comparison to traditional models. Emsley et al. [37] used a dataset of 288 properties to determine the efficiency of both regression and ANN models. In their research study, the authors used a large number of independent variables, including design and site-related variables. The maximum mean absolute percentage error was observed as 17%, which was very large and cannot be prescribed in practical applications.

Setyawati et al. [83] proposed an ANN model for cost estimation of institutional buildings. The results of the analytical study confirmed a high prediction accuracy of the model. Pathak and Agarwal [73] observed ANN to accurately predict the construction cost of water tanks made of reinforced cement concrete. The authors adopted input parameters like column number, height to diameter ratio, conical wall angle, etc., for modeling ANN. The analysis result showed that the proposed network predicted the cost of the water tank with high accuracy. Günaydın and Doğan [48] advocated the utilization of the ANN model for prediction of the construction cost of buildings. In the proposed ANN model, building height, width, and length were used as the input parameters. The proposed ANN model predicted the results with high accuracy and was easily applicable in the initial project phase. Sonmez [88] studied the influence of parameters such as location, car parking area, and common area as independent variables. The authors compared the linear regression model with a neural network and observed neural network to exhibit a better accuracy of 88%. Günaydın and Doğan [48] developed an ANN model and predicted the cost of building projects with an accuracy of 93%. The authors confirmed that adoption of neural networks reduced the uncertainty in the estimation of building cost. Kim [56] determined the costs of buildings located in Korea and observed the superior performance of the ANN model. Wilmot and Mei [99] reported observed high efficiency of ANN models in separately predicting the labor and construction cost. Sodikov [87] used backpropagation ANN for cost estimation of highway projects and found its accurate prediction. Sayed and Iranmanesh (2008) developed an ANN model to reduce the risk of an increase in project cost and observed its satisfactory performance in comparison to traditional methods. Bouabaz and Hamami [24] observed that the ANN model predicted the maintenance and repair cost of the bridges accurately. Sonmez and Ontepeli [89] observed the superior performance of ANN in comparison to the regression model in the cost estimation of an urban railway system. Wang et al. [94] utilized the BP algorithm for cost estimation of highway projects. The authors trained the model based on a large number of datasets. The authors observed that BP algorithm predicted the cost with considerable accuracy. Arafa and Alqedra [16] developed an ANN model, which was trained based on the experimental dataset of buildings. The analysis results showed that area of the ground floor, type of foundation, and exterior surface area to be the parameters had a strong influence on the preliminary building cost. Elsawy and Higgins [36] developed an ANN model trained based on the experimental dataset of 52 real-life projects constructed from 2002 to 2009 in Egypt. The authors observed that the ANN model yielded an accuracy of 80% in the assessment of the construction cost.

Ahiaga-Dagbui and Smith [9] developed a neural network model using the experimental dataset of building projects in Scotland. The authors developed different models for normalizing the target cost, weights, and cost variable transformations to predict the actual cost. The analysis results showed that the ANN model to be efficient in the prediction of the effective construction cost. Ebrahimnejad et al. [32] proposed a model based on the concept of support vector machine (SVM) for the estimation of construction cost during the preliminary stage. The proposed model exhibited a close correlation in comparison with nonlinear regression and BP algorithm. Alqahtani and Whyte [11] developed an ANN model based on a dataset of 20 building models. The model computed the cost of significant and insignificant items and observed high accuracy. Lyne and Maximino [62] developed an ANN model and used an experimental dataset of a large number of buildings to train the model for predicting the total cost of structural members. The authors observed that results yielded by the proposed model were in close agreement with the experimental results. El-Sawah and Moselhi [35] showed that the mean error of a neural network to range from 17% to 20% for the ANN model used in the proposed study. In addition, the authors observed a linear regression model to be primarily influenced by the training data in comparison to ANN. Yadav et al. [102] proposed an ANN model and trained it based on the experimental dataset of past two decades to predict the cost of the building projects in the preliminary phase. The analysis results showed a high accuracy of the proposed model in comparison to the traditional techniques.

Alshamrani [12] developed a multilinear regression model to compute the initial and sustainable cost for RC and steel framed buildings. The input variables adopted in the model were building area, the height of the floor, and the number of levels. The proposed model was observed to yield accurate results in comparison to traditional techniques. Jeong et al. [53] developed a system to develop building energy efficiency. The authors developed the prediction model using data mining techniques based on a dataset of 437 building models. The authors conducted validation studies on multistory buildings. The results of the analytical research indicated high accuracy of the proposed model.

Bayram [20] conducted a research study to validate "Bromilow's time-cost model" for the estimation of project duration. The ANN model was developed and trained based on the experimental dataset of a large number of buildings. The authors adopted the input data such as the total area of construction, height of the building, gross floor area, actual cost, and project duration. The analytical studies demonstrated the accuracy of the proposed approach. Zhang et al. [105] aimed at improving the efficiency of predictions of cost forecast through a time-series approach. The results indicate the accuracy of the proposed plan in comparison with other methods in project scheduling and cost.

Juszczyk et al. [54] proposed an ANN model for estimation of the building overhead cost. The proposed model was trained based on a dataset of 143 buildings using multilayer perceptron theory with varying activation functions. The analysis results showed that the proposed model exhibited satisfactory results in comparison to the previous models. Kang and Kim [55] used ANN for risk assessment studies. The proposed model was developed in two steps. First, the risk information is collected. Based on the information, a unique software program was developed. The second stage consisted of developing a case study by analyzing the risk information for two plant projects. The analysis results showed that the proposed model exhibited greater accuracy in comparison to the models adopted by the previous researchers.

Xu et al. [101] predicted the cost incurred due to damage incurred during seismic excitation. The proposed model was based on Federal Emergency Management Agency (FEMA) approach, and BIM software was used for the analytical study. In the first stage, time history analysis was conducted, and an algorithm was proposed to identify the damage. After that, fragility curves were developed subsequently. In the second stage, the BIM model was prepared based on the building configuration. In the final step, an algorithm to visualize the component was designed. The proposed algorithm was used to derive results for a six-story building in Beijing. The analysis showed a high accuracy of the proposed algorithm.

12.2.2 CONSTRUCTION SCHEDULING

Construction engineering has integrated components of planning, designing, management, and construction of civil engineering structures such as buildings, bridges, airports, highways, railways, dams, etc. The past literature survey shows applications of ANN in decision making, risk prediction, project scheduling, optimization, resource allocation, etc. Moselhi et al. (1991) have enumerated the applications of ANN in the construction sector. Boussabaine and Kaka [25] conducted a literature survey on the application of ANN in construction management.

Adeli and Karim [5] used ANN in construction scheduling of highway projects. The ANN was employed for solving nonlinear problems of construction scheduling, which involved minimization of the project duration. The proposed modal incorporated the features of both linear scheduling and critical path method for optimization of construction cost. The proposed methodology was observed to be more efficient as compared to traditional methods. Graham and Thomas [47] applied ANN for predicting manufacture time of ready mixed concrete, which was sensitive to the construction operations. The authors adopted the input parameters such as operation time, the month of operation, the total volume of operation, volume of the truck, load arrival, average interval time, etc.

Yahia et al. [103] utilized ANN for prediction of project duration time and observed superior performance and simple application of ANN as compared to traditional methods. Petruseva et al. (2013) proposed a supervised learning algorithm termed as SVM for forecasting the duration of the construction project. The ANN was trained based on the dataset of 75 construction projects completed between 1999 and 2011. The studied projects were located in Bosnia and were obtained through field survey and analysis. The authors also used regression analysis to predict the construction duration and compared it with ANN. The analysis results showed that ANN yielded accurate results as compared to regression analysis.

Maghrebi et al. [63] utilized the ANN in predicting the manufacturing duration of ready mix concrete. The authors mainly focused on supply chain parameters concerned with ready mix concrete. The model was proposed and calibrated based on the experimental dataset of buildings located in Sydney. The comparison of ANN with traditional methods showed its superior performance.

Bhargava et al. [21] proposed a new model to overcome the disadvantages of previously proposed models. The authors developed a technique called a line of balance which was applicable to both linear and repetitive works. Based on the analysis work, the authors suggested that there is a need for robust theory to address the critical aspects of planning and scheduling aspects. El-Gohary et al. [33] introduced a new ANN model to predict the cost of labor in a construction project. In developing the model, a wide range of influencing parameters such as working hours and the number of labors were considered. The proposed approach was applied to model the labor productivity of foundation work. The recommended results showed adequate convergence and accuracy. Nevertheless, the authors proposed new activation functions to achieve better results. Alaloul et al. [10] used multilayered feedforward ANN in predicting the project cost and project duration. The proposed model showed accurate results in comparison with other existing methods, with an average mean square value of 0.0231. Andersen and Findsen [15] developed a pragmatic design strategy based on design and the O and M phases. The proposed approach and implemented and tested for a variety of building models to get accurate results. Ballesteros-Perez et al. [19] proposed two nonlinear theoretical models to estimate the construction cost. The main advantage of this model was that they were both discrete and continues. Moreover, it was observed that the proposed model could be applied to a variety of construction projects. The authors validated the model by testing it on several building models. In addition, the proposed models were very efficient in predicting the crashing cost of the project. Similar observations were reported by Ferreira et al. [39].

12.2.3 UTILIZATION OF ANN IN COST ENGINEERING

The ANN models have been widely utilized in construction management. The neural networks have been increasingly used to determine the construction cost [37, 50, 98] and to predict the tender bids [60, 64], for monitoring the budget of construction [30], construction demand [43], cash flow in a project [25], earthmoving operation (Shi, 1999), labor productivity (Savin and Fazio, 1998; [27, 76], effectiveness (Sinha and Mckim, 2001), new technology development (Chao and Skibniewski, 1995), and organizational behavior. Hegazy et al. [51] developed a neural network for optimization of construction activities and cost. Chua et al. [30] developed an ANN model for identification of the factors affecting the construction budget and situations associated with it. The authors obtained the data from field survey to develop a budget performance model. The proposed approach allowed the model to build the relationship between input and output parameters. The authors used eight input parameters such as project team, planning and control efforts, project manager, hierarchy level in the organization, construction duration, design parameters, and frequency of team meetings. The proposed model was capable of yielding good prediction even with unseen data. The model was observed to perform with excellence in different aspects of construction management.

Assaf [17] proposed an ANN model for determination of overhead project cost in Saudi Arabia. The analysis results showed that the material and project duration significantly influenced the completion time of the project. The authors showed that negative effects of unstable construction on overall project cost.

Emsley et al. [37] proposed an ANN model, which was trained based on the dataset of 300 building projects. The analysis was conducted using both the ANN and the linear regression model. The analysis results showed better performance of ANN in comparison to linear regression analysis. Nevertheless, the ANN was found more efficient in case of nonlinear problems as compared to linear regression analysis. Günaydın and Doğan [48] developed an ANN model for determining construction cost of beams and columns in an RC framed building. The authors trained the ANN building model using an experimental dataset of 30 buildings and adopted eight input parameters, which were a function of building properties. The analysis results showed that the proposed ANN model exhibited a close correlation with the experimental results.

Kim [56] advocated the use of hybrid models by combining the effects of the backpropagation algorithm and genetic algorithms in cost estimation. The authors developed an ANN model and trained it based on data of 530 buildings located in Korea during the period between 1997 and 2000. The analytical study showed better efficiency of the hybrid algorithm as compared to the ANN model. This study paved the way for the adoption of such hybrid algorithms.

Wilmot and Mei [99] developed a cost index describing the overall construction cost of highways. The proposed index included parameters such as cost of construction, cost of labor, and equipment. The analysis results showed that the proposed model yielded accurate results and was very good at forecasting the future prices of the highway projects. Sodikov [87] improved this research work and proposed a new ANN model by including the uncertainties in the project during the design phase. The proposed model was observed to be more accurate in comparison to the previously proposed models, which ignored the uncertainties involved in the project cost.

Golpayegani and Emamizadeh [46] used ANN for the breakdown of the construction activity into several parts and rescheduling it to avoid unnecessary delays in the construction duration. The proposed model was observed to be more accurate in comparison to the existing models in reducing the uncertainties causing project delays. Naderpour et al. [70] developed an ANN model to forecast the project cost. The developed system was based on earned value management systems based on projects selected on a random basis. The proposed model was compared with the previous approach, which showed the efficiency of the previous procedure. Wang et al. (2009) proposed two building models based on bootstrap technology for pre and post planning of buildings. The results demonstrated the efficiency of the proposed approach. Elsawy and Higgins [36] adapted an ANN approach for the development of a cost estimation model to determine overhead site cost for buildings in Egypt. The authors used 52 real cases to train the neural network model.

12.3 PROBLEMS IDENTIFIED

In the previous section, the authors used several types of prediction networks such as BP algorithm, adaptive conjugate gradient network, and radial basis function neural networks (RBFNNs). However, it was observed that each technique has several drawbacks, as discussed in this section.

12.3.1 DEFICIENCIES IN THE BP ALGORITHM

The previous literature shows that BP algorithm has been widely used owing to its simplicity and ease in the application [8,79]. However, this algorithm has a slow rate of learning with the number of iterations exceeding several thousand [26]. Moreover, network convergence is significantly affected by the selection of input parameters and their values along with the momentum ratios. Furthermore, the variety of input parameters has no specific rules and depends upon the type of problem [1]; (Yeh, 1998). Adeli and Hung [1] were one of the first researchers who attempted to remove the deficiencies and developed a new algorithm for training of feedforward networks with a large number of layers. Powell [78] introduced further modifications in the gradient algorithm. The main problem encountered with the BP algorithm about the selection of input variables and momentum ratios has been solved by varying these parameters in a specific range leading to better convergence. The proposed algorithm was tested for image processing, and superior performance was observed. However, this approach had the limitation that it could be applied for specific problems only. Nevertheless, it is more complicated and consumes more significant time and effort to implement it.

In the RBFNN, the mapping of the input–output parameter is performed by using a transfer function surrounding these parameters (Moody and Darken, 1989; Poggio and Girosi, 1990). In the RBFNN, the biological neurons have receptive field such that output is large when input functions are closer to the center. The topology of RBFNN includes input, hidden, and output layers arranged radially. The RRBFFS is more computationally efficient than the BP algorithm but is computationally complex, and its accuracy varies depending upon the number of layers and input parameters.

12.4 SOLUTION TO EXISTING PROBLEMS

The problems with a different type of networks have been identified in the previous section. These problems can be solved by the application of prescribed solutions as follows

12.4.1 INTEGRATING NEURAL NETWORKS WITH OTHER ALGORITHMS

The above algorithms can be effectively integrated by combining their advantages and reducing their disadvantages. The backpropagation network and the genetic algorithm were combined by Adeli and Hung (1991b) to propose a new algorithm. The first stage begins with the acceleration of the learning rate by the adoption of feedforward networks. The corresponding weights of neural networks are decided to reduce the mean square error. The second stage starts with the adoption of the BP algorithm, and the iteration is continued until the end condition is satisfied. The concept of the hybrid algorithm has been used by several authors such as [1,49,68,72], and Topping et al. (1998).

12.4.2 FUZZY LOGIC

The fuzzy logic modeling can be subdivided into three stages. The dataset is subjected to unsupervised learning in the first stage. The second stage precedes with supervised learning and ends with the classification of the training data. The ANN and the genetic algorithm are used in combination for this purpose. After that, the activity of defuzzification and computation of the difference between the initial and final values to compute error is performed.

12.4.3 WAVELETS

The neural networks exhibit poor accuracy in case of extensive data and complicated patterns in case of traffic data collected near a central station. As estimation of construction cost involves a large number of parameters. The neural networks have observed less accuracy in case of noncoincident patterns because of the significant variation in dimensions of the training data. As a solution, one can use a noise reduction technique based on wavelet theory to inhibit unwanted fluctuation in the training data. The wavelets can be effectively used for preprocessing of the data to be fed into the neural networks, which improves the efficiency of these networks. This technique has been effectively used by several researchers such as Adeli and Samant [7], Adeli and Hung [1], and Adeli and Karim [5] for solving traffic-related problems. Several authors, Yu et al. [104] and Zhao et al. [106], determined damage location and identified it using neural networks and wavelet theory.

12.5 PROBLEMS WITH EXISTING SOLUTIONS

The problems with the existing solutions have been presented in the following subsection.

12.5.1 PROS AND CONS OF EXISTING SOLUTIONS

The application of fuzzy logic involves immense technical knowledge of input parameters for the development of fuzzy sets. The second difficulty pertains to the modeling of a fuzzy set by the description of the relationship between input parameters. This theory becomes complicated in case of problems with a large number of input parameters. Moreover, it involves a higher degree of mathematical solutions, which makes it even complex to implement.

The wavelets have a unique advantage that time and frequency of the system can be located more accurately and can be easily adapted to solve the complex engineering problems. The wavelets have distinct advantages over Fourier transforms that they are more simple and easy to implement. Although the implementation of wavelets is simple, its design varies from problem to problem and involves lots of complexities.

However, once fuzzy sets are formulated for a specific design problem. Then, it can be implemented for similar problems with little effort.

12.6 CONCLUSIONS

Artificial intelligence can be defined as a system specially designed to exhibit human-like behavior in the solution of complex issues. The decisions made by artificial intelligence are quick, efficient, and straightforward. Previous literature survey shows that the ANN has been extensively used for problems related to construction scheduling and forecasting. The review of prior literature shows adoption of the BP algorithm in construction management owing to its ease of application and simplicity. However, the BP algorithm had a distinct disadvantage that it was slow and required a large number of iterations to fulfill the convergence criteria. Moreover, the selection of input parameters significantly depended on the type of problem and had no specific rules. Therefore, the formulation of the generalized algorithm for a particular kind of issues becomes difficult. Nevertheless, effects with momentum ratios were also observed to satisfy the convergence criteria. Several solutions to overcome the disadvantages of the BP algorithm were prescribed: (a) integrating the BP algorithms with a genetic algorithm and (b) usage of wavelets and fuzzy logic. The proposed solutions had several disadvantages, which can be overcome using simple measures. Moreover, these techniques can be applied to a variety of problems in construction management. Nevertheless, these measures are quite complex and challenging to implement. However, these techniques can be easily and effectively implemented with the development of new software.

12.7 FUTURE SCOPE

The present study can be extended to more aspects of construction management such as construction scheduling, construction delay, etc. The present study can be conducted with more number of variables, and different activation functions could be adopted.

KEYWORDS

- **neural networks**
- **construction management**
- **artificial intelligence**
- **civil engineering**

REFERENCES

1. Adeli, H., & Hung, S. L. (1995). *Machine Learning—Neural Networks, Genetic Algorithms, and Fuzzy Systems*. John Wiley & Sons, Inc., New York, NY, USA.
2. Adeli, H. (1994). *Advances in Design Optimization*. CRC Press, Boca Raton, FL, USA.

3. Adeli, H., & Hung, S. L. (1994). An adaptive conjugate gradient learning algorithm for efficient training of neural networks. *Applied Mathematics and Computation*, *62*(1), 81–102.
4. Adeli, H., & Karim, A. (1997). Scheduling/cost optimization and neural dynamics model for construction. *Journal of Construction Engineering and Management*, *123*(4), 450–458.
5. Adeli, H., & Karim, A. (2000). Fuzzy-wavelet RBFNN model for freeway incident detection. *Journal of Transportation Engineering*, *126*(6), 464–471.
6. Adeli, H., & Kim, H. (2004). Wavelet-hybrid feedback-least mean square algorithm for robust control of structures. *Journal of Structural Engineering*, *130*(1), 128–137.
7. Adeli, H., & Samant, A. (2000). An adaptive conjugate gradient neural network–Wavelet model for traffic incident detection. *Computer-Aided Civil and Infrastructure Engineering*, *15*(4), 251–260.
8. Adeli, H., & Seon Park, H. (1995). Counterpropagation neural networks in structural engineering. *Journal of Structural Engineering*, *121*(8), 1205–1212.
9. Ahiaga-Dagbui, D. D., & Smith, S. D. (2012). Neural networks for modelling the final target cost of water projects. In *Proceedings of the 28th Association of Researchers in Construction Management Annual Conference* (pp. 307–316).
10. Alaloul, W. S., Liew, M. S., Wan Zawawi, N. A., Mohammed, B. S., & Adamu, M. (2018). An artificial neural networks (ANN) model for evaluating construction project performance based on coordination factors. *Cogent Engineering*, *5*(1), 1507657.
11. Alqahtani, A., & Whyte, A. (2013). Artificial neural networks incorporating cost significant Items towards enhancing estimation for (life-cycle) costing of construction projects. *Construction Economics and Building*, *13*(3), 51–64.
12. Alshamrani, O. S. (2017). Construction cost prediction model for conventional and sustainable college buildings in North America. *Journal of Taibah University for Science*, *11*(2), 315–323.
13. Al-Tabtabai, H. (1998). A framework for developing an expert analysis and forecasting system for construction projects. *Expert Systems with Applications*, *14*(3), 259–273.
14. Amin, A. M., El Korfally, M. I., Sayed, A. A., & Hegazy, O. T. (2009). Efficiency optimization of two-asymmetrical-winding induction motor based on swarm intelligence. *IEEE Transactions on Energy Conversion*, *24*(1), 12–20.
15. Andersen, M. T., & Findsen, A. L. (2019). Exploring the benefits of structured information with the use of virtual design and construction principles in a BIM life-cycle approach. *Architectural Engineering and Design Management*, *15*(2), 83–100.
16. Arafa, M., & Alqedra, M. (2011). Early stage cost estimation of buildings construction projects using artificial neural networks. *Journal of Artificial Intelligence*, *4*(1), 63–75.
17. Assaf, S. A. (2001). Existing and the future planned desalination facilities in the Gaza Strip of Palestine and their socio-economic and environmental impact. *Desalination*, *138*(1–3), 17–28.
18. Bala, B. K., Ashraf, M. A., Uddin, M. A., & Janjai, S. (2005). Experimental and neural network prediction of the performance of a solar tunnel drier for drying jackfruit bulbs and leather. *Journal of Food Process Engineering*, *28*(6), 552–566.
19. Ballesteros-Perez, P., Elamrousy, K. M., & González-Cruz, M. C. (2019). Non-linear time-cost trade-off models of activity crashing: Application to construction scheduling and project compression with fast-tracking. *Automation in Construction*, *97*, 229–240.

20. Bayram, S. (2017). Duration prediction models for construction projects: In terms of cost or physical characteristics? *Journal of Civil Engineering, 21*(6), 2049–2060.
21. Bhargava, P., Phan, T., Zhou, J., & Lee, J. (2015, May). Who, what, when, and where: Multi-dimensional collaborative recommendations using tensor factorization on sparse user-generated data. In *Proceedings of the 24th International Conference on World Wide Web* (pp. 130–140).
22. Bhokha, S., & Ogunlana, S. O. (1999). Application of artificial neural network to forecast construction duration of buildings at the predesign stage. *Engineering, Construction and Architectural Management, 6*(2), 133–144.
23. Bhokha, S., & Ogunlana, S. O. (1999). Application of artificial neural network to forecast construction duration of buildings at the predesign stage. *Engineering, Construction and Architectural Management, 6*(2), 133–144.
24. Bouabaz, M., & Hamami, M. (2008). A cost estimation model for repair bridges based on artificial neural network. *American Journal of Applied Sciences, 5*(4), 334–339.
25. Boussabaine, A. H., & Kaka, A. P. (1998). A neural networks approach for cost flow forecasting. *Construction Management & Economics, 16*(4), 471–479.
26. Carpenter, W. C., & Barthelemy, J. F. (1993). A comparison of polynomial approximations and artificial neural nets as response surfaces. *Structural Optimization, 5*(3), 166–174.
27. Chao, L. C., & Skibniewski, M. J. (1994). Estimating construction productivity: Neural-network-based approach. *Journal of Computing in Civil Engineering, 8*(2), 234–251.
28. Chau, K.W. (1998). The implications of the difference in the growth rates of the prices of building resources and outputs in Hong Kong. *Engineering, Construction and Architectural Management, 5*(1), 38–50.
29. Chen, W., Wilson, J., Tyree, S., Weinberger, K., & Chen, Y. (2015, June). Compressing neural networks with the hashing trick. In *International Conference on Machine Learning* (pp. 2285–2294).
30. Chua, D. K. H., Kog, Y. C., Loh, P. K., & Jaselskis, E. J. (1997). Model for construction budget performance—Neural network approach. Journal of Construction Engineering and Management. 123(3), 214–222.
31. Dave, B., Hämäläinen, J. P., Kemmer, S., Koskela, L., & Koskenvesa, A. (2015). Suggestions to improve lean construction planning. In *Proceedings of the 23rd Annual Conference of the International Group for Lean Construction* (pp. 193–202).
32. Ebrahimnejad, S., Mousavi, S. M., Tavakkoli-Moghaddam, R., Hashemi, H., & Vahdani, B. (2012). A novel two-phase group decision making approach for construction project selection in a fuzzy environment. *Applied Mathematical Modelling, 36*(9), 4197–4217.
33. El-Gohary, K. M., Aziz, R. F., & Abdel-Khalek, H. A. (2017). Engineering approach using ANN to improve and predict construction labor productivity under different influences. *Journal of Construction Engineering and Management, 143*(8), 04017045.
34. Elhag, T. M. S., & Boussabaine, A. H. (1998, September). An artificial neural system for cost estimation of construction projects. In *Proceedings of the 14th ARCOM Annual Conference*.
35. El-Sawah, H., & Moselhi, O. (2014). Comparative study in the use of neural networks for order of magnitude cost estimating in construction. *Journal of Information Technology in Construction, 19*(27), 462–473.
36. Elsawy, B., & Higgins, K. E. (2011). The geriatric assessment. *American Family Physician, 83*(1), 48–56.

37. Emsley, M. W., Lowe, D. J., Duff, A. R., Harding, A., & Hickson, A. (2002). Data modelling and the application of a neural network approach to the prediction of total construction costs. *Construction Management & Economics*, *20*(6), 465–472.
38. Fang, C. F., & Froese, T. (1999). Cost estimation of high performance concrete (HPC) high-rise commercial buildings by neural networks. *Durability of Building Material & Components*, *8*, 2476–2486.
39. [Ferreira, L. B., da Cunha, F. F., de Oliveira, R. A., & Fernandes Filho, E. I. (2019). Estimation of reference evapotranspiration in Brazil with limited meteorological data using ANN and SVM—A new approach. *Journal of Hydrology*, *572*, 556–570.
40. Fortin, J. G., Anctil, F., Parent, L. É., & Bolinder, M. A. (2010). A neural network experiment on the site-specific simulation of potato tuber growth in Eastern Canada. *Computers and Electronics in Agriculture*, *73*(2), 126–132.
41. Geiger, A. R. (1994). *U.S. Patent No. 5,299,054*. Washington, DC, USA: U.S. Patent and Trademark Office.
42. Giles, C. L., Kuhn, G. M., & Williams, R. J. (1994). Dynamic recurrent neural networks: Theory and applications. *IEEE Transactions on Neural Networks*, *5*(2), 153–156.
43. Goh, A. T. (1996). Neural-network modeling of CPT seismic liquefaction data. *Journal of Geotechnical engineering*, *122*(1), 70–73.
44. Goh, S. L., & Mandic, D. P. (2004). A complex-valued RTRL algorithm for recurrent neural networks. *Neural Computation*, *16*(12), 2699–2713.
45. Goh, Y. M., & Chua, D. (2013). Neural network analysis of construction safety management systems: A case study in Singapore. *Construction Management and Economics*, *31*(5), 460–470.
46. Golpayegani, S. A. H., & Emamizadeh, B. (2007). Designing work breakdown structures using modular neural networks. *Decision Support Systems*, *44*(1), 202–222.
47. Graham, B., & Thomas, K. (2006). Knowledge management in Irish construction: The role of CPD accreditation. In *22nd Annual ARCOM Conference*.
48. Günaydın, H. M., & Doğan, S. Z. (2004). A neural network approach for early cost estimation of structural systems of buildings. *International Journal of Project Management*, *22*(7), 595–602.
49. Hajela, P., & Lee, E. (1997). Topological optimization of rotorcraft subfloor structures for crashworthiness considerations. *Computers & Structures*, *64*(1–4), 65–76.
50. Hegazy, T., & Ayed, A. (1998). Neural network model for parametric cost estimation of highway projects. *Journal of Construction Engineering and Management*, *124*(3), 210–218.
51. Hegazy, T., Fazio, P., & Moselhi, O. (1994). Developing practical neural network applications using back-propagation. *Computer-Aided Civil and Infrastructure Engineering*, *9*(2), 145–159.
52. Hung, S. L., & Adeli, H. (1991). A model of perceptron learning with a hidden layer for engineering design. *Neurocomputing*, *3*(1), 3–14.
53. Jeong, J., Hong, T., Ji, C., Kim, J., Lee, M., Jeong, K., & Koo, C. (2017). Development of a prediction model for the cost saving potentials in implementing the building energy efficiency rating certification. *Applied Energy*, *189*, 257–270.
54. Juszczyk, M., Leśniak, A., & Zima, K. (2018). ANN based approach for estimation of construction costs of sports fields. *Complexity*, *2018*, 1–11.
55. Kang, H. W., & Kim, Y. S. (2018). A model for risk cost and bidding price prediction based on risk information in plant construction projects. *Journal of Civil Engineering*, *22*(11), 4215–4229.

56. Kim, Y. (2014). Convolutional neural networks for sentence classification. *arXiv preprint arXiv:1408.5882*.
57. Kim G.-H., Yoon J.-E.; An S. H., & Cho, H.-H. (2004) Neural network model incorporating a genetic algorithm in estimating construction costs. *Building and Environment, 39*, (1333–1340).
58. Lam, K. C., Lam, M. C. K., and Wang, D. (2008). MBNQA-oriented self-assessment quality management system for contractors: Fuzzy AHP approach. *Construction Management and Economics*, 26(5), 447–461.
59. Leśniak, A., & Juszczyk, M. (2018). Prediction of site overhead costs with the use of artificial neural network based model. *Archives of Civil and Mechanical Engineering, 18*(3), 973–982.
60. Li, H. and Love, P. E. D. (1999). Combining rule-based expert systems and artificial neural networks for mark-up estimation. *Construction Management and Economics, 17*(2), 169–176.
61. Li, H., & Love, P. E. (1999). Combining rule-based expert systems and artificial neural networks for mark-up estimation. *Construction Management & Economics, 17*(2), 169–176.
62. Lyne, C. C., & Maximino, C. (2014). An artificial neural network approach to structural cost estimation of building projects in the Philippines.
63. Maghrebi, M., Sammut, C., & Waller, T. S. (2014). Predicting the duration of concrete operations via artificial neural network and by focusing on supply chain parameters. *Building Research Journal, 61*(1), 1–14.
64. McKim, R. A. (1993). Neural network applications to cost engineering. *Cost Engineering, 35*(7), 31.
65. McKim, R. A.; Sinha, S. K. (1999) Condition assessment of underground sewer pipes using a modified digital image processing paradigm. *Tunnelling and Underground Space Technology, 14*, 29–37.
66. Mintorovitch, J., Moseley, M. E., Chileuitt, L., Shimizu, H., Cohen, Y., & Weinstein, P. R. (1991). Comparison of diffusion-and T2-weighted MRI for the early detection of cerebral ischemia and reperfusion in rats. *Magnetic Resonance in Medicine, 18*(1), 39–50.
67. Moghtadaei, M., Golpayegani, M. R. H., & Malekzadeh, R. (2013). A variable structure fuzzy neural network model of squamous dysplasia and esophageal squamous cell carcinoma based on a global chaotic optimization algorithm. *Journal of Theoretical Biology, 318*, 164–172.
68. Moselhi, O., & El-Rayes, K. (1993). Scheduling of repetitive projects with cost optimization. *Journal of Construction Engineering and Management, 119*(4), 681–697.
69. Muniz, A. M. S., Liu, H., Lyons, K. E., Pahwa, R., Liu, W., Nobre, F. F., & Nadal, J. (2010). Comparison among probabilistic neural network, support vector machine and logistic regression for evaluating the effect of subthalamic stimulation in Parkinson disease on ground reaction force during gait. *Journal of Biomechanics, 43*(4), 720–726.
70. Naderpour, H., Rafiean, A. H., & Fakharian, P. (2018). Compressive strength prediction of environmentally friendly concrete using artificial neural networks. *Journal of Building Engineering, 16*, 213–219.
71. Ok, S. C., & Sinha, S. K. (2006). Construction equipment productivity estimation using artificial neural network model. *Construction Management and Economics, 24*(10), 1029–1044.
72. Papadrakakis, M., Lagaros, N. D., & Tsompanakis, Y. (1998). Structural optimization using evolution strategies and neural networks. *Computer Methods in Applied Mechanics and Engineering, 156*(1–4), 309–333.

73. Pathak, K. K. & Agarwal R., (2004). Cost prediction of overhead water tanks using artificial neural networks. *IE (I) Journal. 84* 153–158.
74. Pearce, A. R., Gregory, R. A., & Williams, L. (1999). Range estimating for risk management using artificial neural networks. *Journal of Parametrics*, *19*(1), 3–31.
75. Pewdum, W., Rujirayanyong, T., & Sooksatra, V. (2009). Forecasting final budget and duration of highway construction projects. *Engineering, Construction and Architectural Management*, *16*(6), 544–557.
76. Portas, J., & AbouRizk, S. (1997). Neural network model for estimating construction productivity. *Journal of Construction Engineering and Management*, *123*(4), 399–410.
77. Powell, M. J. (1986). Convergence properties of algorithms for nonlinear optimization. *SIAM Review*, *28*(4), 487–500.
78. Rumelhart, D., Hinton, G., & Williams, R. (1986). Learning internal representations by error propagation. In Parallel Distributed Processing, D. E. Rumelhart and J. L. McLelland, (Eds.), MIT Press, Cambridge, MA, USA.
79. Razavi, S. M.; Iranmanesh, P.; Moeini, A.; Qorani, N.; & Fakhar, A. (2015). Correlation between clinical diagnosis and histopathological diagnosis of oral soft tissue lesions in Isfahan Dental School from 1988 to 2011. Journal of Isfahan Dental School, *Journal of Isfahan Dental School, 11*, 133–141.
80. Savin, D., Alkass, S., & Fazio, P. (1996). Construction resource leveling using neural networks. *Canadian Journal of Civil Engineering*, *23*(4), 917–925.
81. Savin, D., Alkass, S., & Fazio, P. (1998). Calculating weight matrix of neural network for resource leveling. *Journal of Computing in Civil Engineering*, *12*(4), 241–248.
82. Setyawati, B. R., Sahirman, S., & Creese, R. C. (2002). Neural networks for cost estimation. *AACE International Transactions*, ES131.
83. Seyed, H. I. and Mansoureh, Z. (2008). Application of artificial neural network to forecast actual cost of a project to improve earned value management system. *World Academy of Science, Engineering and Technology. 18*, 210–213.
84. Shi, G., Zhou, X., Zhang, G., Shi, X., & Li, H. (2004). The use of artificial neural network analysis and multiple regression for trap quality evaluation: A case study of the Northern Kuqa Depression of Tarim Basin in western China. *Marine and Petroleum Geology*, *21*(3), 411–420.
85. Shtub, A., & Versano, R. (1999). Estimating the cost of steel pipe bending, a comparison between neural networks and regression analysis. *International Journal of Production Economics*, *62*(3), 201–207.
86. Sodikov, J. (2005). Cost estimation of highway projects in developing countries: artificial neural network approach. *Journal of the Eastern Asia Society for Transportation Studies*, *6*, 1036–1047.
87. Sonmez, R. (2004). Conceptual cost estimation of building projects with regression analysis and neural networks. *Canadian Journal of Civil Engineering*, *31*(4), 677–683.
88. Sonmez, R., & Ontepeli, B. (2009). Predesign cost estimation of urban railway projects with parametric modeling. *Journal of Civil Engineering and Management*, *15*(4), 405–409.
89. Sultan, M. A., Eid, S. Z., & Hegazi, N. H. (1994). Back propagation neural networks for direct adaptive control of dynamic systems. *Advances in Modeling & Analysis*.
90. Topping, B. H., & Papadrakakis, M. (1994). *Advances in Parallel and Vector Processing for Structural Mechanics*. Civil-Comp Press, Stirling, UK.
91. Topping, BHV. & Bahreininejad, A. (1997). *Neural Computing for Structural Mechanics* Saxe Coburg, Gotha, UK.

92. Vahdani, B., Tavakkoli-Moghaddam, R., Modarres, M., & Baboli, A. (2012). Reliable design of a forward/reverse logistics network under uncertainty: A robust-M/M/c queuing model. *Transportation Research Part E: Logistics and Transportation Review*, *48*(6), 1152–1168.
93. Wang, K., Li, M., & Hakonarson, H. (2010). ANNOVAR: Functional annotation of genetic variants from high-throughput sequencing data. *Nucleic Acids Research*, *38*(16), e164–e164.
94. Waziri, B. S. (2010, July). An artificial neural network model for predicting construction costs of institutional building projects in Nigeria'. In *West Africa Built Environment Research (WABER) Conference* (pp. 27–28).
95. Waziri, B. S. (2012). Modelling the performance of traditional contract projects in Nigeria: An artificial neural network approach. In *Proceedings of 4th West Africa Built Environment Research (WABER) Conference, Abuja* (pp. 1383–1391).
96. Waziri, B. S., Bala, K., & Bustani, S. A. (2017). Artificial neural networks in construction engineering and management. *International Journal of Architecture, Engineering and Construction*, *6*(1), 50–60.
97. Williams, T. P. (1994). Predicting changes in construction cost indexes using neural networks. *Journal of Construction Engineering and Management*, *120*(2), 306–320.
98. Wilmot, C. G., & Mei, B. (2005). Neural network modeling of highway construction costs. *Journal of Construction Engineering and Management*, *131*(7), 765–771.
99. Xu, S., & Lam, J. (2006). A new approach to exponential stability analysis of neural networks with time-varying delays. *Neural Networks*, *19*(1), 76–83.
100. Xu, Z., Zhang, H., Lu, X., Xu, Y., Zhang, Z., & Li, Y. (2019). A prediction method of building seismic loss based on BIM and FEMA P-58. *Automation in Construction*, *102*, 245–257.
101. Yadav, R., Vyas, M., Vyas, V., & Agrawal, S. (2016). Cost estimation model (CEM) for residential building using artificial neural network. *International Journal of Engineering Research & Technology*. *5*. DOI: 10.17577/IJERTV5IS010431
102. Yahia, H., Hosny, H., & Razik, M. E. A. (2011). Time contingency assessment in construction projects in Egypt using artificial neural networks model. *International Journal of Computer Science Issues*, *8*(4), 523.
103. Yu, Y., Dackermann, U., Li, J., & Niederleithinger, E. (2019). Wavelet packet energy-based damage identification of wood utility poles using support vector machine multi-classifier and evidence theory. *Structural Health Monitoring*, *18*(1), 123–142.
104. Zhang, R., Ashuri, B., Shyr, Y., & Deng, Y. (2018). Forecasting Construction Cost Index based on visibility graph: A network approach. *Physica A: Statistical Mechanics and its Applications*, *493*, 239–252.
105. Zhao, Y., Noori, M., Altabey, W. A., & Beheshti-Aval, S. B. (2018). Mode shape-based damage identification for a reinforced concrete beam using wavelet coefficient differences and multiresolution analysis. *Structural Control and Health Monitoring*, *25*(1), e2041.

PART V

Speech and NLP Applications in Cognition

CHAPTER 13

Speech Recognition Fundamentals and Features

GURPREET KAUR[1*], MOHIT SRIVASTAVA[2], and AMOD KUMAR[3]

[1]*University Institute of Engineering and Technology, Punjab University, Chandigarh 160025, India*

[2]*Chandigarh Engineering College, Landran, Mohali 140307, India*

[3]*National Institute of Technical Teachers Training and Research, Chandigarh 160026, India*

[*]*Corresponding author. E-mail: regs4gurpreet@yahoo.co.in*

ABSTRACT

Speech communication is an essential part of humans. It is natural in humans but artificial in machines. Speech waveform defines the message communicated along with various parameters regarding the speaker. It may be about his/her age group, gender, and emotions. Speech processing has different areas according to its applications, such as in communication, reproduction of speech signal, and its transmission and recognition of speaker or speech. The speech recognition system means letting a machine to understand voice. In this chapter, the speech recognition fundamentals, features, and difficulties faced are explored.

13.1 INTRODUCTION

Speech is the method to communicate and to tell emotions. It is very easy for human beings but quite difficult with machines. Due to the large dependency of humans on machines, recognition of speech signals by machine is growing area since the last 65 years [12]. To make a good recognition system in any language, any background noise with the large dataset is today's need.

Researchers have developed systems that are available in the market, but the performance of those systems still requires improvement [16,26,35].

Speech processing is a vast area in terms of its purpose and applications. There may be different purposes of speech processing such as:

Understanding of speech signal as a means of communication.

- Reproduction and transmission.
- Automatic recognition of speech signal.
- Characteristics of the speaker.
- Language identification.

Based on the purpose, speech processing has three significant areas:

- Analysis/synthesis
- Recognition
- Coding

Analysis/synthesis of the speech waveform involves characterization of the spectral information of speech signal for transmission and storage. Recognition field is categorized into speech recognition (SR), speaker recognition (SPR), and language recognition (LR). There may be a recognition system, in which the only message is recognized, known as SR, where the only speaker is recognized, known as SPR, and at last, only language is recognized, known as LR. These systems can be developed according to the need of society. Next area is coding of the speech signals. This is an application of data compression of audio signals. The classification of speech processing can be shown in Figure 13.1.

SR, SPR, and integrated speaker and speech recognition (ISSR) systems are explained in the following section.

13.1.1 SPEECH RECOGNITION

SR is the field where words are identified from the spoken entity, that is, what a person is saying? SR systems may be categorized into three types depending upon the speech mode, speaker mode, and vocabulary mode. Types of speech mode can be isolated SR, connected SR, and continuous SR. In isolated SR, the speaker has to take a break between the words. In connected SR, continues speaking is there without taking any pause. Continuous SR is like a person talks. Speaker can speak into the system without prior training, known as speaker-independent systems, and when the speaker take prior training before speaking into the system, it is known as speaker-dependent systems. Types of dataset can be small, medium, and large. The small dataset can be less than a hundred words. The medium dataset can be 1000 of words, and large dataset can be more than 5000 words. The basic building block for the SR system can be shown in Figure 13.2.

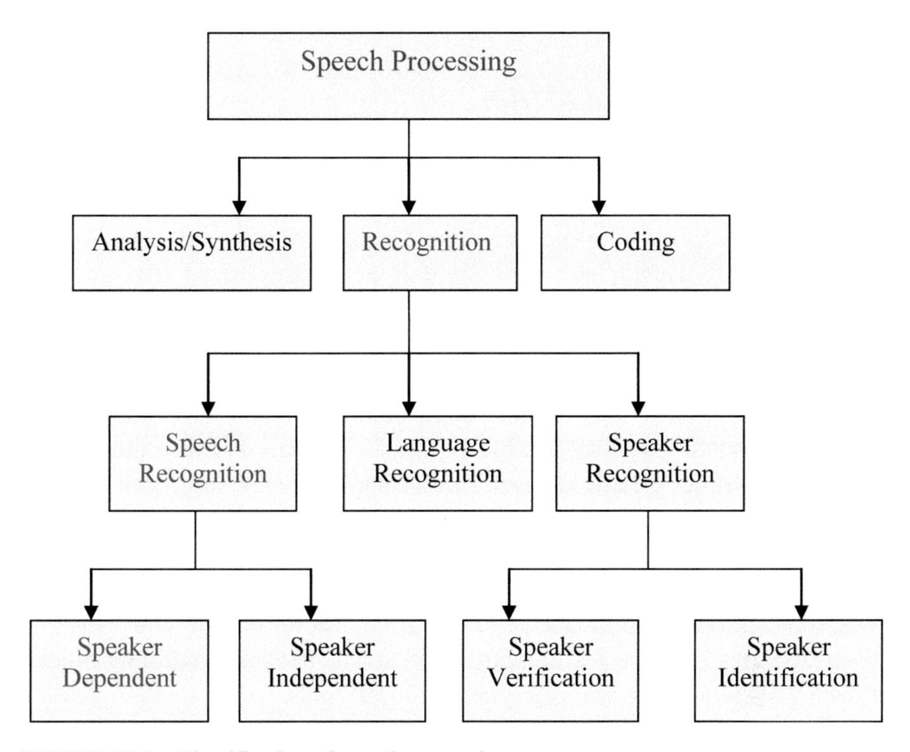

FIGURE 13.1 Classification of speech processing.

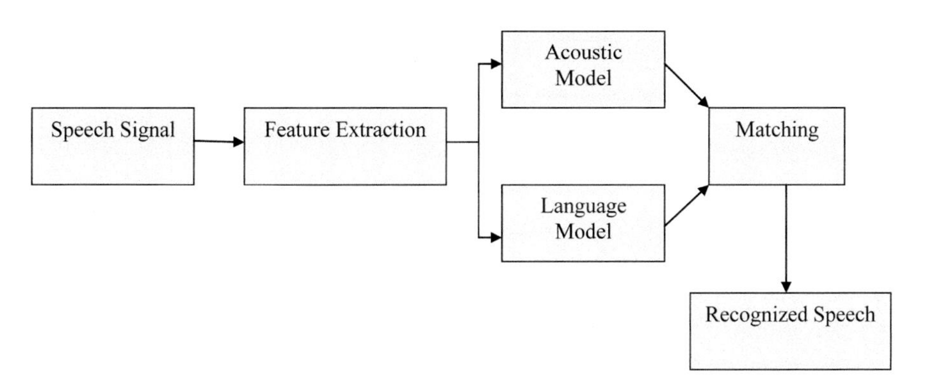

FIGURE 13.2 Block diagram of SR.

Feature extraction is the process for selecting those feature vectors, which are helpful for SR and discarding all speaker-related information. The acoustic model for SR models the context variations, whereas the language model models linguistic information. The performance of SR systems depends upon the feature extraction and classification methods [5].

13.1.2 SPEAKER RECOGNITION

SPR is the field where the speaker is recognized without consideration of words, that is, who is speaking? Speech signal consists of not only a linguistic message but also speaker-related information in terms of speaking style, emotions, accent, etc. SPR field is also categorized into speaker verification (SV) and speaker identification (SI). SV means to confirm the person who she/he claims to be. SI is to give the identity to a person among all registered persons.

Furthermore, SPR can be text-dependent (TD) and text-independent (TID). TD systems are less flexible but more accurate because the system is trained with the specific dataset, but TID systems are more elastic with complexity because any new word can be spoken into the system. The SPR task is highly dependent upon the cooperation of the user. If SR is used as biometric, then, of course, the user would cooperate, but for forensic applications, the user may or may not be cooperative. Therefore, the vital block of this system lies with the feature extraction stage. The basic building blocks of the SV system are shown in Figure 13.3 [59].

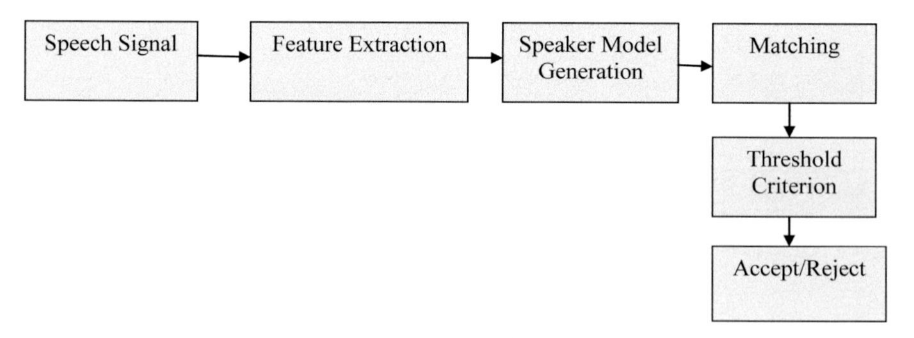

FIGURE 13.3 Block diagram of the SV system.

In the SV system, feature extraction is used to select the speaker-related feature vectors. Speaker models are generated on the basis of training data.

In the testing phase, the claimed user model is matched with the existing models, and according to some threshold criterion, the user model is accepted or rejected. The applications for the SV system include the transaction in banks, access to some restricted area, attendance recording systems, etc. There is little difference between SI systems and SV systems. In SI systems, there may be more than two speaker models. Out of registered speakers, identity is assigned to the claimed speaker. The block diagram for the SI system is shown in Figure 13.4 [16].

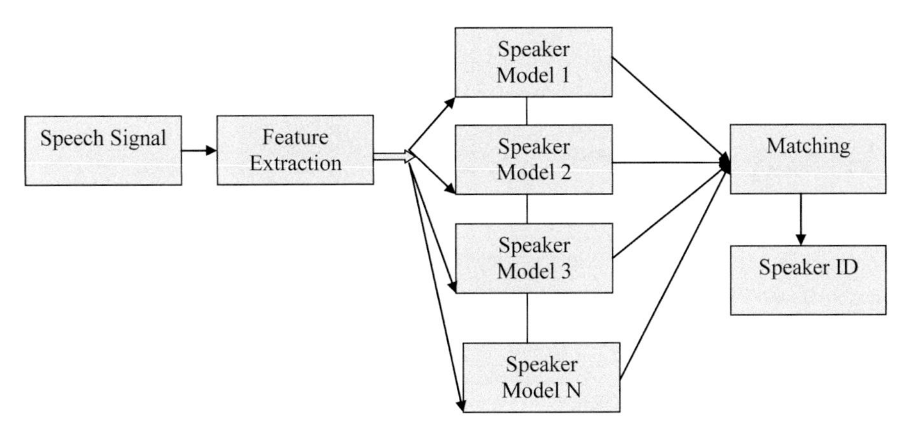

FIGURE 13.4 Block diagram of the SI system.

The SI system is based upon one-to-many comparisons. In this system, the voice pattern of the unknown speaker is compared with all stored speaker models. The best match is reported, and the speaker ID is assigned. The SV and SI systems perform well when they are TD systems. This means that the speaker has to speak all those words/sentences that are known to the system. This type of systems, however, has fewer applications as compared to TID SV/SI systems. TD SV systems are used in home banking services where a specific user is to speak PIN or password. However, because of the security issues, these types of systems are less popular. TID SV systems are used in forensic applications, and they are quite popular.

13.1.3 INTEGRATED SPEAKER AND SPEECH RECOGNITION

SPR and SR are different fields according to the claims. However, some applications require combined SPR and SR systems.

Feature extraction techniques used in SPR and SR are conventional. The combined area of SPR and SR encompasses speaker-dependent SR. The block diagram for ISSR is shown in Figure 13.5. In ISSR, all basic blocks are similar except the feature extraction techniques [48, 49]. Now, the feature vectors should represent both speaker and speech characteristics. This type of system is used in command- and control-related applications, where specific user can only control the devices.

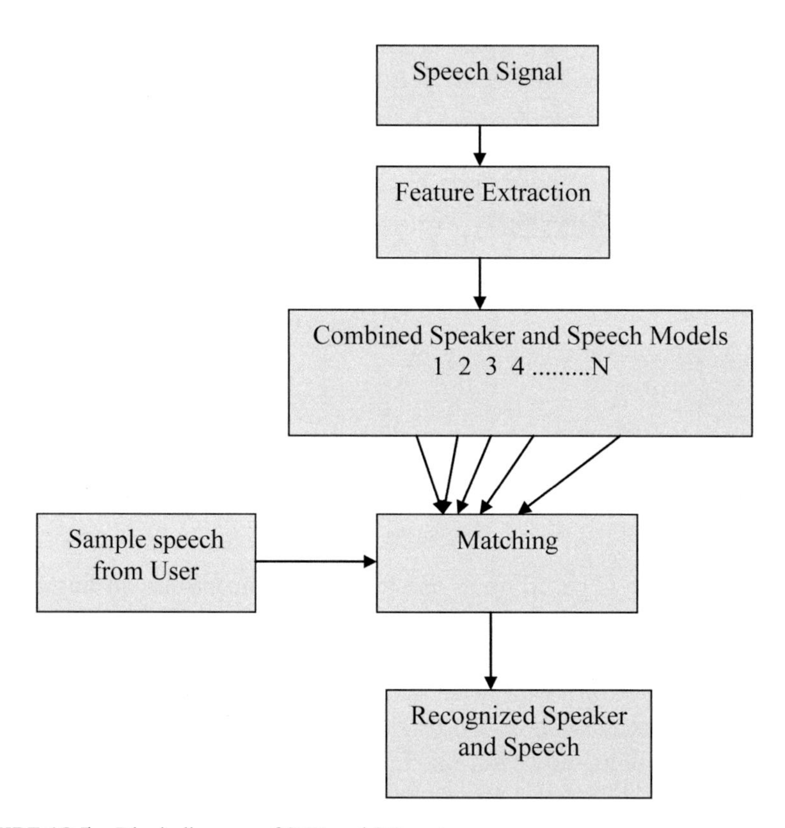

FIGURE 13.5 Block diagram of SPR and SR systems.

13.2 VARIATIONS OF AUTOMATIC SPEECH RECOGNITION SYSTEMS

Kinds of speech mode can be isolated SR (IWSR), connected SR (CWR), and continuous SR (CSR). In IWSR, the speaker has to take a break between the words. In CWR, continuous speaking is there without taking any break.

CSR is like a person talks. In IWSR, the speech is treated as a single phrase, and without any prior knowledge about phonetic, it is recognized. In this, speech is categorized into two parts: starting point and ending point. The isolated word recognition system is implemented using IWSR. Gupta and Sivakumar [15] applied an order for 10 digits of Hindi language. Speech waveform is divided into divisions using a Hamming window. An online recognition system is developed by Venkataramani [55] using 10 digits on a field-programmable gate array programmable chip. The speech signals that are acquired by microphones are converted into phrases using the hidden Markov model (HMM) model. Many authors have done the research using the electromyogram (EMG) signals also. They have shown the dependency of the speech signal with the EMG signal [37].

In CWR, the words are divided by pause. The speech signal strings size may be little, average, or significant. Myers and Rabiner [41] implemented a CWR system using a dynamic warping method. The matching is done between the stored database and the uttered words. Syama and Mary [52] applied a Malayalam LR system using the same method that is used by Myers and Rabiner. In CSR type of operation, it is same like a human speech [14]. Thangarajan et al. [53] implemented a system for the Tamil language. However, it was quite challenging to build the system because of the variations in vowels and consonants. However, the recognition rate can be increased using extensive training. Many authors have implemented the systems on different languages like Abushariah et al. [2] developed a CSR system using the Sphinx and HTK tools. The five stages of the HMM were also deployed, which had three emitting steps corresponding to the triphone acoustic model. The statistical model was comprised of bigram, unigram, and trigram. The simulation was done based on the various amalgamations of speakers and sentences.

Along with this, it included a manual validation and classification process for validating the right pronunciation of the words. This manual validation was done by a human. After getting the results, it was concluded that the system had the best suitability when the speakers were different, but the sentences were similar, and it failed when different speakers and different sentences were considered.

13.3 FUNDAMENTALS OF SR

Speech signal conveys lots of information. At the first stage, the speech signal gives the message; at the second stage, the speech signal tells the idea about the speaker. Speech production is a very complex phenomenon. It comprises many levels of processing. First of all, message planning is done in our mind, and then, language coding is done. After coding, the neural muscular command is generated. After this, the sound is produced through vocal cords. Every human speech is different from each other because of various parameters like linguistics (lexical, syntactic, semantic, and pragmatics), paralinguistic (intentional, attitudinal, and stylistic), and nonlinguistics (physical emotional). Therefore, speech signal contains different segmental and supra segmental features, which can be extracted for the SPR as well as SR [31].

13.3.1 SPEECH PRODUCTION AND PERCEPTION

It is essential to know the individual speech construction model for developing the speech model as one can then extract the speech features more accurately. A physiological model of human speech is shown in Figure 13.6. Glottis is the opening of the vocal tract. The vocal tract is like a cavity, which is having two openings: nostrils and lips. Velum is an articulator organ, which moves up and down, and it is responsible for the radiation of sound either from the nose or through lips. The flow of air is controlled by vocal cords. The vocal cords open and close accordingly. Effectively, it is pulse-like excitation that is given to vocal tract. Vocal tract acts as a filter. Whatever pulses are fed to it, spectral shaping of those pulses is done by the vocal tract. Spectral shaping varies from time to time based upon the utterance of words. The speech signal is quasi-periodic waveform, that is, statistical parameters remain the same within a short time interval of order 10 to 20 ms. Therefore, the short-time analysis of speech signals is required. Different sounds are produced with the help of articulators organs such as velum. When velum moves downward, the air passage is blocked from vocal tract region up to lips, and hence, nasal sounds are emitted. When the movement of velum is in an upward direction, the nostril passage is closed, and voiced sounds are produced through lips. The tongue is another organ that is responsible for different sounds [10].

Voiced speech is produced when the vocal cords tremble through the articulation, for example, /a/, /e/, /i/, /o/, /u/, whereas unvoiced speech is produced when there is no trembling of vocal cords, for example, /t/, /p/ [11]. The human production system is modeled by a source-filter model. The source is excitation generation of the glottis pulses, and vocal tract and effects of radiation are represented by the time-varying system (linear). The actual speech spectrum is not uniform. There are some frequency components, which are more dominating than others. This can be explained with the help of a cylindrical tube model, which is open at one end (lip end). First resonance will occur, creating nodes and antinodes. If the length of the tube is $L = 17$ cm (distance between larynx and lips) and velocity of sound in air is 340 m/s, then first resonance frequency comes out to be 500 Hz, the second resonance frequency is 1500 Hz, and the third resonance frequency is at 2500 Hz. These frequencies called formant frequencies maybe in some range for different persons. Therefore, this is one of the speech features that can be extracted from the speech signal.

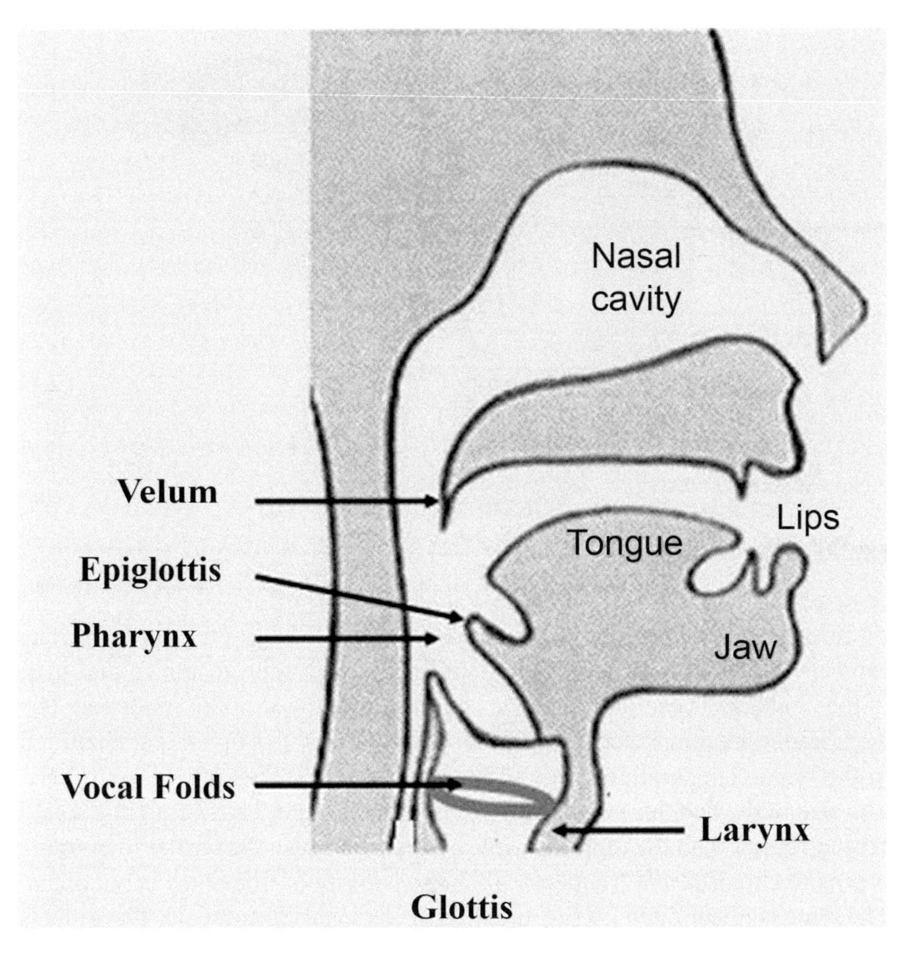

FIGURE 13.6 Human speech production model.

Speech perception involves detecting sounds through ear. The ear consists of an outer part, middle part, and the inner part. The central part of the ear consists of three bones, and inner part consists of a snail-like structure known as the cochlea. Cochlea consists of fluid-filled chambers and is separated by the basilar membrane, as shown in Figure 13.7.

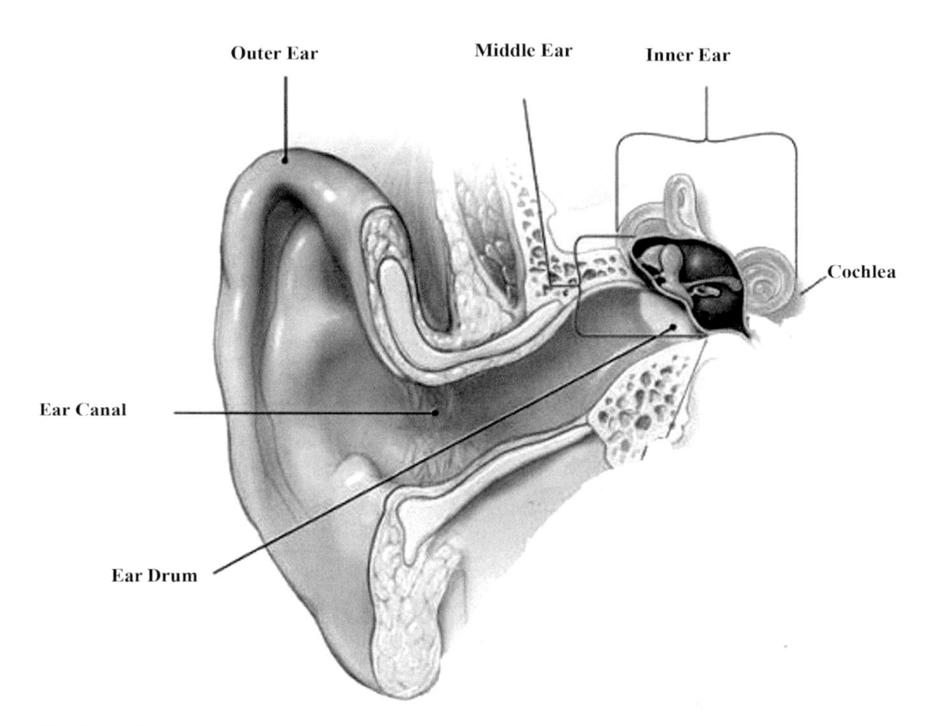

FIGURE 13.7 Human ears.

Sound waves are received by the ear canal and fall on the eardrum. The eardrum vibrates, and these vibrations are passed upon to bones and then to the cochlea. These vibrations set up the resonances in the inner ear. The membranes are connected with the nerves, and thus, the signal is transmitted to the brain. The audible range of human speech is from 20 Hz to 20 kHz. The size of the cochlea is around 30 mm. Resolution of low-frequency sound is much better than the high-frequency sounds because the basilar membrane is sensitive to different frequencies. The sensitivity of frequency is not linear. The scale is linear up to 2 kHz, and then, it is a logarithmic scale. The scale is known as bark or Mel scale. At last, it is concluded that the brain gets much more information about low-frequency components of speech [35].

13.3.2 SPEECH CHARACTERISTICS

Speech waveform is having many characteristics such as loudness, voiced/unvoiced speech, fundamental frequency, pitch, spectral envelope, formants, etc. Speech features can be classified as physical and perceptual. Physical characteristics are dependent upon physical properties of speech signal such as power, fundamental frequency, etc. The amplitude of the speech signal specifies power in it. More power means a louder signal. Silence zones in the speech signal can be discovered using power measurement. Another physical feature is the fundamental frequency. The range of fundamental frequency of females is 165–255 Hz, and that of a male is from 85 to 155 Hz, so male and female voices can be separated using this attribute [26]. Pitch and prosody are perceptual features. Pitch analysis can tell about the emotions of the speaker. Speech features are extracted by mathematical analysis of the signal. Analysis of speech signals can be done in time as well as in the frequency domain. Various speech features, such as the energy of the signal and amplitude of the speech signal, are analyzed in the time domain using zero-crossing rate (ZCR) and autocorrelation (AC) [16]. The simple process is involved in these tasks in the time domain, but in the frequency domain, various methods with extensive calculations are required such as Fourier transform, spectrograms, filter banks, cepstral, etc. Spoken words are not exactly equal to what we write because of background noise, body language, channel variability, speaker variability, speaking style, dialects, etc. All the mentioned factors are responsible for the accuracy of SR.

13.4 DIFFICULTIES IN SR

SR is quite easy for individuals, but it is not easy for devices because of many factors such as style of speech, environment, and speaker features. [26].

- The recognition rate is high when words are uttered indefinite style.
- The environment has a significant role concerning recognition accuracy. Noisy surroundings have less recognition as compared to clean surroundings.
- Emotions of a speaker: how fast or slow he/she speaking, accent, gender, etc. affect the recognition rate. Emotions of a speaker: how fast or slow he/she speaking, accent, gender, etc., have effect on recognition rate.
- There may be many difficulties when SR systems are made with different languages. Quality depends upon the language model and applications where the system is to be used.

13.5 GENERATION OF SR

The work under speech and SPR systems is made into four generations. The first generation (1950–1969) work was mainly on the phonetics methods. Second generation (1970–1990) work was on matching techniques. In the third generation (1990–2000) work was on statistical modeling. Deep learning methods were implemented in the fourth generation (the 2000s onward).

13.5.1 FIRST GENERATION (FROM 1950S TO 1969)

Starting from the year 1950, many researchers implemented the systems using acoustic-phonetic methods. A digit recognizer system was first implemented by Davis et al. [9] in Bell Laboratories formant frequencies. The noise or unintentional data from the audio signal were removed by using the time constant of the sampling model. This was one of the most preferred methods of that era to perform matching of the speaker's word. In this technique, the developed circuit splits the speech spectrum into two different frequencies, that is, below and above 900 Hz, respectively. A set of digits were kept in a table, and then, the speaker's words were matched with the predefined words, and matching was performed based on the correlation among the values. In 1952, an SR system was implemented to recognize numbers from zero to nine. However, its accuracy was depended upon the adjustment of the requirements of the speaker. In 1956, a system was made based upon the phonemes classification by Wiren and Stubbs [57]. The unvoiced and voiced speech was separated using pitch frequency. Zero-crossing levels were used for amplitude measurement of the signal. The authors took the idea from the published book by Jakobson et al. [32] in 1952. In 1962, Fry [11] analyzed the role of acoustics on human sound perception system. Based on that analysis, many authors have done researches on this topic [47].

13.5.2 SECOND GENERATION (FROM 1970S TO 1990S)

In 1970, many authors Itakura and Atal [3,29] defined that vocal tract features could be modeled in terms of time-varying features. The US Department of Defense has given many funds to flourish research in this area from 1971 to 1976. Now, the vocabulary size increased from 10 words. In 1979, Rabiner et al. [46] implemented a recognizing system by taking 26 isolated words. In the training session of the proposed model, the speaker speaks each word of the vocabulary in a direct-dialed up phone line, and then, an AC mechanism is applied to locate the starting and end of the word. The system is divided into two modes: mode one comprises of recording the signals and mode two is clustering the words.. The clustering was performed by using unsupervised without averaging. The simulation was done on four test sets. The proposed work was observed to have a substantial improvement to the alternative recognizer with a similar vocabulary. In 1976, statistical methods were implemented [33]. Many researchers used statistical techniques in their work [4,34]. The vast vocabulary and speaker dependency on the recognition accuracy and mixture density algorithms of HMM techniques were implemented [30]. After the evolution of neural network systems in 1989, this field got the opportunity to enhance the research. Therefore, many researchers have used algorithms such as backpropagation neural network to train the data [7, 23, 42]. Fuzzy-based systems can be combined with neural network systems.

13.5.3 THIRD GENERATION (FROM THE 1990s TO 2000s)

Now, in the era of 1990s, many techniques were implemented for making the recognition systems, but the effect of noise was very prominent. So far, the focus was on to reduce the impact of noise on recognition systems. Therefore, many techniques were combined to reduce the noise such as HMM, neural, and fuzzy techniques [58]. Bourlard et al. [7] extended the work done by Fry [11] by introducing a multilayered neural network. Many advanced neural network-based algorithms such as recurrent neural network were implemented to reduce noise and to increase the accuracy [24].

13.5.4 FOURTH GENERATION (FROM THE 2000s TO TILL DATE)

In fourth generation (the 2000s onward), deep-learning-based algorithms were implemented. Therefore, multilevel processing of features could be used to increase the accuracy. However, the main challenge in this area was how to train the multilevel neural network. Many authors have discovered training methods. In 2006, Hinton et al. [21] found the deep belief network (DBN). In this method, layerwise training was done. Results showed that accuracy rate increased when layerwise training was done as compared to other conventional networks [1, 8, 30, 39]. In 2012, Mohamed et al. [40] implemented the combined techniques using a Gaussian mixture model with the DBN. Deep architecture got famous in this era. Many researchers have explored this area, and many advances have been emerged [13,44,51].

13.6 FEATURE EXTRACTION TECHNIQUES

Feature extraction is a significant step in any recognition process. Unique features are to be selected out of total features, which are responsible for the exact recognition for speaker and speech. Feature extraction is also used to lessen the dimensionality of the feature vector. Although speaker and SR systems are different, their feature extraction methods are overlapped. Various feature extraction algorithms are investigated, including perceptual linear prediction (PLP), Mel-frequency cepstral coefficients (MFCC), relative spectra perceptual linear predictive (RASTA-PLP), and linear predictive coding coefficients (LPCC). The speech signal is not stationary. Therefore, before the feature extraction step, there is a need for the preprocessing of the speech signal. Preprocessing means to get ready for the feature extraction procedure, so that signal is removed, that does not contain any information. After silence removal, the speech signal can be short and requires less memory for its storage and moreover less time for processing also. This can be done with the help of measuring the energy of the whole signal, and some value is set, so the signal is removed according to the selected value. The formula for calculating energy is

$$E = \sum_{n=0}^{i} |x(n)|^2 \tag{13.1}$$

where $x(n)$ is the speech signal.

To perk up the signal-to-noise ratio, the energy level for higher frequencies is boosted for which a high-pass filter is used. This is done because high frequencies have less energy as compared to low frequencies. The speech signal is variable. However, for a little episode of time, it may be treated as a fixed signal because the duration of phonemes is assumed to be stationary for the period of 5–100 ms. Then, windowing is done to deal with the loss of data at the edges of the speech due to framing. This is done with the help of window. The features that are in one frame of size P ms are multiplied by the window size R ms. P should be lesser than R. That is why two frames get overlapped. There are many types of windows available, such as Hanning, Chebyshev, Hamming, Kaiser, etc. There is nothing any weight to choose the type of window except some soft window is used to avoid an abrupt change in signal characteristics.

The effect of noisy surroundings has more impact on SR systems. Therefore, to deal with this problem, many algorithms were proposed [36]. Rabiner and Sambur [45] implemented a system in which the ZCR of the signal is calculated to measure the noise signal. In addition, the energy of the signal is calculated. However, the accuracy decreases when this system is applied for the real-time scenario. The processing time was too slow. Therefore, it was the requirement for an algorithm that should be fast and efficient. Therefore, Hopper and Adhami [25] tried a system using the digital-signal-processing-based algorithm such as fast Fourier transform (FFT). This algorithm achieved excellent results in terms of recognition rate in noisy surroundings. Many authors have tried time-domain features with frequency-domain features. Voiced and unvoiced regions were detected using energy and ZCR that was proposed in [48]. After that, using FFT, elements of speech signal were extracted. Various feature extraction techniques used by multiple investigators are as follows.

13.6.1 LINEAR PREDICTIVE CODING COEFFICIENT

Spectral magnitude can be analyzed using linear predictive coding coefficients. Vocal tract coefficients are generated using this technique, and this algorithm was most popular in early algorithms. In this technique, speech parameters at the current time can be treated as a linear combination of old speech samples. Many researchers have implemented an LPCC feature extraction technique in their systems [27, 28]. The coefficients in this technique can be reduced by other method known as wavelet decomposition [43, 50].

13.6.2 MEL-FREQUENCY CEPSTRAL COEFFICIENTS

It is a well-accepted method for feature extraction. In this method, first of all, the preprocessed speech signal is converted into a frequency domain to know about the frequencies of the speech signal using FFT. The output of FFT contains a lot of data that are not required because at higher frequencies, there is not any difference between the rates. This is based upon the way how human being listens. The scale is linear until 1000 and logarithmic after it. Therefore, to calculate the energy level at each frequency, Mel-scale analysis is done using Mel filters. Then, energy is calculated. After that, logarithmic of filter bank energies is taken. This operation is done to match the features more close to human hearing.

At last, discrete cosine transform of the log filter bank energies is taken to decorrelate the overlapped energies. First, 13 coefficients are selected known as MFCC features. This is because higher features degrade the recognition accuracy of the system. Those features do not carry speaker and speech-related information. Many researchers have implemented the MFCC method for better results in the recognition rate [56, 60]. Experimentation was also done by combining different feature extraction techniques. Trang et al. [54] described a new feature set by combining principal component analysis with MFCC.

13.6.3 PERCEPTUAL LINEAR PREDICTIVE

In the year 1990, Hermansky [18] developed a PLP technique that was based upon the perceptual predictive coefficients. This method was based upon the human hearing system to create specific features for the recognition purpose. This method is like the LPC technique that is about the spectrum of the signal. Transformations are done according to psychophysically, and the frequency of the signal is modified. There is a difference in spectral smoothing of PLP from MFCC. Linear predictions are made for spectral smoothing of PLP. The procedure consists of critical band spectral resolution and then loudness curve and after that intensity loudness power law. All pole models are used for approximation of the auditory system. The PLP technique is the only technique in which speaker-dependent features can be suppressed using a fifth-order all-pole model. Many results showed that PLP is much more useful than LPCC analysis, and for the implementation of speaker-independent systems, PLP is quite popular.

13.6.4 RELATIVE SPECTRAL PROCESSING (RASTA)

Noise has a very severe effect on recognition systems. Therefore, to remove noise, a bandpass filter can be added to the PLP technique. Hermansky et al. [17] implemented this technique. Many researchers have done work in this domain by associating PLP features with RASTA processing [19,20]. The advantage of using this technique is that it gives good results when the conditions are changing unpredictably. The work in this chapter was focused on the convolutional noise in the communication channel in which corrections were made in the log spectral domain. The experimental analysis has been performed on a speech corrupted with convolutional noise. The acquired results concluded an order-of-magnitude improvement in terms of error rate in comparison with LPC or PLP technique. Moreover, the obtained results showed consistency because of different databases as well as different recognition techniques. Hirsch et al. [22] implemented a filtering method in the power spectral domain, and this method has reduced the background noise.

13.7 EXPERIMENTS AND RESULTS

The database is created by four speakers: two females and two males (30–40 years old). Selected words are one, two, three, four, and five. Each speaker was given 100 samples for each word, and thus, a 2000-word database is prepared. MATLAB is used for the implementation of the system. The calculated features from all techniques mentioned above are stored as a reference pattern. Then, in the testing phase, reference patterns are matched with the uttered word using dynamic time algorithm (DTW). This DTW technique is used to calculate the distance between the uttered word and reference patterns. Let the two speech patterns A and B of length n and m be

$$A = a_1, a_2, a_3, a_4, \ldots, a_i, \ldots, a_n \tag{13.2}$$

$$B = b_1, b_2, b_3, b_4, \ldots, b_j, \ldots, b_m \tag{13.3}$$

$$d(a_i, b_j) = \sqrt{(a_i - b_j)^2} \tag{13.4}$$

and the whole distance can be shown as

$$D(i,j) = \min\left[D(i-1,j-1), D(i-1,j), D(i,j-1) + d(i,j)\right]. \tag{13.5}$$

The minimum distance is calculated for the matched pattern for the uttered word. Due to the variations in speech samples, distance has some value but not zero.

White Gaussian noise (WGN) is chosen to corrupt the speech samples. It is a basic noise that is used for many processes occurring in nature, and the best part is that it has a uniform distribution on all frequencies [53]. SR accuracy is measured in different measuring metrics such as precision rate, recall rate, accuracy, sensitivity, and specificity, defined as follows:

$$\text{Precision rate} = \frac{\text{True positive (TP)}}{\left[\text{True positive (TP)} + \text{False positive (FP)}\right]} \tag{13.6}$$

$$\text{Sensitivity} = \frac{\text{True positive (TP)}}{\left[\text{True positive (TP)} + \text{False Negative (FN)}\right]} \tag{13.7}$$

$$\text{Specificity} = \frac{\text{True Negative}(\text{TN})}{[\text{True positive }(\text{TP}) + \text{False Negative }(\text{FN})]} \tag{13.8}$$

$$\text{Accuracy} = \frac{[\text{TP} + \text{TN}]}{[\text{TP} + \text{FP} + \text{TN} + \text{FN}]} \tag{13.9}$$

where true positive (TP) represents accurately selected features, false positive (FP) represents falsely selected features sets, true negative (TN) represents all negative feature those are true, and false negative (FN) represents all negative feature those are false. Table 13.1 shows the results of different implemented techniques.

TABLE 13.1 Average Recognition Accuracy for Different Techniques

Type of Speech Samples	Average Recognition Rate (%) for LPCC Technique	Average Recognition Rate (%) for PLP Technique	Average Recognition Rate (%) for RASTA-PLP Technique	Average Recognition Rate (%) for MFCC Technique
Clean Speech Samples	92.12	93.17	93.16	94.25
Speech Samples with White Gaussian Noise (WGN)	73.54	73.12	73.60	73.98

Results showed that the average recognition rate is 93.12% for clean speech signals using the LPCC technique. For clean speech signals, the average recognition rate is 93.17% using the PLP technique, 93.16% using the RASTA PLP technique, and 94.25% using the MFCC technique. The recognition rate decreases as speech samples are corrupted with WGN. Then, the average accuracy rate decreases to 73.54% using the LPCC technique, to 73.12% using the PLP technique, to 73.98% using the RASTA PLP technique, and to 73.98% by using the MFCC technique. Results conclude that the MFCC technique worked well in clean as well as in a noisy environment.

13.8 CONCLUSION AND FUTURE SCOPE

Speech processing is a very vast field. There are many different fields based upon the applications. Human–computer interface through speech is the best application for SR, for example, a dictation system, customer care services, etc. Similarly, SPR is extensively used for security purposes such as biometrics, forensic, etc. These two fields can be combined for specific applications related to assistive technologies for disabled persons, for example, accessing the computer or the movement of the wheelchair through speech. In the healthcare sector, there is a growing need for this technology. Different feature extraction methods such as MFCC, LPCC, PLP, and PLP RASTA are implemented to the recognition of speech. Results showed that the MFCC technique worked well in clean as well as in a noisy environment. For future scope, in real-time applications, deep neural networks can be used to accommodate variations due to the speaker and background noise. Hence, a more robust system can be implemented.

KEYWORDS

- **speech recognition (SR)**
- **speaker recognition (SPR)**
- **integrated speaker and speech recognition (ISSR)**

REFERENCES

1. Abdel-Hamid O., Mohamed A., Jiang H., and Penn G., (2012), "Applying Convolutional Neural Networks Concepts to Hybrid NN-HMM Model for Speech Recognition," In *IEEE International Conference on Acoustics, Speech and Signal Processing,* pp. 4277–4280.
2. Abushariah, M. A. M., Ainon, R. N., Zainuddin, R., and Elshafei, M., (2010), Natural Speaker-Independent Arabic Speech Recognition System Based on Hidden Markov Models Using Sphinx Tools," In *International Conference on Computer and Communication Engineering*, Kuala Lumpur, Malaysia, pp. 1–6.
3. Atal, B. S., and Hanauer, S. L., (1971), "Speech Analysis and Synthesis by Linear Prediction of the Speech Wave," *Journal of Acoustic Society of America*, *50*(2), pp. 637–655.
4. Baum, E., Petrie, T., Soules, G., and Weiss, N., (1970), "A Maximization Technique Occurring in the Statistical Analysis of Probabilistic Functions of Markov Chains," *The Annals of Mathematical Statistics*, *41*(1), pp. 164–171.
5. Bharti, R., and Bansal, P., (2015), "Real- Time Speaker Recognition System using MFCC and Vector Quantization Technique," *International Journal of Computer Applications*, *117*(1), pp. 25–28.
6. Botros, N., Dieri, M. Z., and Hsu, P., (1989), "Automatic Voice Recognition using Artificial Neural Network Approach," In *IEEE Proceedings of the 32nd Midwest Symposium on Circuits and Systems*, Champaign, IL, USA, pp. 763–765.

7. Bourlard, H., Konig, Y., Morgan, N., and Ris, C., (1996), "A New Training Algorithm for Hybrid HMM/ANN Speech Recognition Systems," In *IEEE European Signal Processing Conference*, Trieste, Italy, pp. 1–4.
8. [8] Dahl, G. E., Ranzato, M., Mohamed, A., and Hinton, G. E., (2010), "Phone Recognition with the Mean-Covariance Restricted Boltzmann Machine," In *International Conference on Neural Information*, pp. 469–477.
9. Davis, K. H., Biddulph, R., and Balashek, S., (1952), "Automatic Recognition of Spoken Digits," *Journal of Acoustic Society of America*, *24*(6), pp. 627–642.
10. Rabiner, L. R., Schafer, R. W., (2007), "Introduction to Digital Speech Processing," *Foundations and Trends in Signal Processing*, *1*, 1–194.
11. Fry, D. B, (1962), *The Role of Acoustics in Phonetic Studies, in Technical Aspects of Sound*, Elsevier Publishing Co. Amsterdam, The Netherlands, pp. 1–69.
12. Furui, S., (2005), "50 Years of Progress in Speech and Speaker Recognition," *ECTI Transactions on Computer and Information Technology*, pp. 64–74.
13. Gavat, I., and Militaru, D., (2015), "Deep Learning in Acoustic Modeling for Automatic Speech Recognition and Understanding—An Overview," In *IEEE International Conference on Speech Technology and Human-Computer Dialogue*, Bucharest, Romania, pp. 1–8.
14. Ghai, W., and Singh, N., (2013), "Continuous Speech Recognition for Punjabi Language," *International Journal of Computer Applications*, *72*(14), pp. 23–28.
15. Gupta, R., and Sivakumar, G., (2006), "Speech Recognition for Hindi Language," Master's Dissertation, IIT Bombay, Mumbai, India.
16. Haton, J. P., (2004), *Automatic Speech Recognition: A Review*. Springer, New York, NY, USA, pp. 1–2.
17. Hermanskey, H., Wan, E. A. and Avendano, C., (1994), "Noise Suppression in Cellular Communications," In *2nd IEEE Workshop on Interactive Voice Technology for Telecommunications Applications*, Kyoto, Japan, pp. 85–88.
18. Hermansky, H., (1990), "Perceptual Linear Predictive (PLP) Analysis of Speech," *Journal of Acoustical Society of America*. *87*(4), pp. 1738–1752.
19. Hermansky, H., and Morgan, N., (1994), "RASTA Processing of Speech," *IEEE Transactions on Speech and Audio Processing*, *2*(4), pp. 578–589.
20. Hermansky, H., Morgan, N., Bayya, A., and Kohn, P., (1992), "RASTA-PLP Speech Analysis Technique," *IEEE International Conference on Acoustics, Speech and Signal Processing*, San Francisco, CA, USA, pp. 121–124.
21. Hinton, G.E., Osindero, S., and Teh, Y. W., (2006), "A Fast Learning Algorithm for Deep Belief Nets," *Neural Computation*, *18*, pp. 1527–1554.
22. Hirsch, H., Meyer, P. and Ruhl, H., (1991), "Improved Speech Recognition using High-Pass Filtering of Sub band Envelopes," In *Second European Conference on Speech Communication and Technology*, Genova, Italy, pp. 413–416.
23. Hochberg, M. M., Silverman, H. F., and Morgan, D. P., (1989), "A Dynamic Programming Neural Network Approach for Connected Speech Recognition," *IEEE International Conference on Acoustics, Speech, and Signal Processing*, Glasgow, UK, pp. 651–654.
24. Hochreiter, S. and Schmidhuber, J., (1997), "Long Short-Term Memory," *Neural Computation*, *9*(8), pp. 1735–1780.
25. Hopper, G., and Adhami, R., (1992), "An FFT-based Speech Recognition System," *Journal of Franklin Institute*, *329*(3), pp. 555–562.
26. Huang, X., and Deng, L., (2010), "An Overview of Modern Speech Recognition," In *Handbook of Natural Language Processing*. CRC Press, Boca Raton, FL, USA, pp. 339–367.

27. Hydari, M., and Karami, M. R., (2009), "Speech Signals Enhancement using LPC Analysis based on Inverse Fourier Methods," *Contemporary Engineering Sciences*, *2*(1), pp. 1–15.
28. Itakura, F., (1975), "Minimum Prediction Residual Principle Applied to Speech Recognition," *IEEE Transactions on Acoustics, Speech, and Signal Processing*, *23*(1), pp. 67–72.
29. Itakura, F., and Saito, S., (1970), "A Statistical Method for Estimation of Speech Spectral Density and Formant Frequencies," *Electronics and Communications in Japan*, *53*, pp. 36–43.
30. Jaitly, N., and Hinton, G., (2011), "Learning a Better Representation of Speech Sound Waves using Restricted Boltzmann Machines," In *IEEE International Conference on Acoustics, Speech and Signal Processing*, Prague, Czech Republic, pp. 5884–5887.
31. Jakobson, R., and Halle, M., (1956), "Fundamentals of Language," *The Library of Technology*, *23*(1), pp. 34–41.
32. Jakobson, R., Fant, C. G. M., and Halle, M., (1952), "Preliminaries to Speech Analysis," Technical Report No. 13, Acoustic Lab., MIT, Cambridge, MA, USA.
33. Jelinek, F., (1976), "Continuous Speech Recognition By Statistical Methods," *Proceedings of IEEE*, *64*(4), pp. 532–556.
34. Juang, B. H., Levinson, S. E. and Sondhi, M. M., (1986), "Maximum Likelihood Estimation for Multivariate Mixture Observations of Markov Chains," *IEEE Transaction on Information Theory*, *32*(2), pp. 307–309.
35. Jurafsky, D., and Martin, J. H., (2000), *Speech and Language Processing*. Prentice-Hall, Englewood Cliffs, NJ, USA, p. 07632.
36. Kaur, G., Srivastava, M., and Kumar A., (2017), "Analysis of feature extraction methods for speaker-dependent speech recognition," *International Journal of Engineering and Technology Innovation*, *7*(2), pp. 78–88.
37. Lee, K.S., (2008), "EMG-Based Speech Recognition using Hidden Markov Models with Global Control Variables," *IEEE Transactions on Biomedical Engineering*, *55*(3), pp. 930–940.
38. Makhoul, J., (1971), "Speaker Adaptation in a Limited Speech Recognition System," *IEEE Transactions on Computers*, *20*(9), pp. 1057–1063.
39. Mohamed A.R., Sainath, T. N., Dahl G., Hinton, G. E., (2011), "Deep Belief Networks using Discriminative Features for Phone Recognition," In *IEEE International Conference on Acoustics, Speech and Signal Processing*, Prague, Czech Republic, pp. 5060–5063.
40. Mohamed, A.R., Dahl, G.E., and Hinton, G., (2012), "Acoustic Modeling using Deep Belief Networks," *IEEE Transactions on Audio, Speech, and Language Processing*, *20*(1), pp. 14–22.
41. Myers, C. S. and Rabiner L. R. (1981), "Connected Word Recognition using a Level Building Dynamic Time Warping Algorithm," In *IEEE International Conference on Acoustics, Speech, and Signal Processing*, pp. 951–955.
42. Nakamura, M. and Shikano, K., (1989), "A Study of English Word Category Prediction Based on Neural Networks," In *IEEE International Conference on Acoustics, Speech, and Signal Processing*, Glasgow, UK, pp. 731–734.
43. Nehe, N. S., and Holambe, R. S., (2012), "DWT and LPC Based Feature Extraction Methods for Isolated Word Recognition," *EURASIP Journal on Audio, Speech, and Music Processing*, *7*, pp. 1–7.
44. Nikoskinen, T., (2015), "From Neural Network to Deep Neural Network," Alto University School of Science, Espoo, Finland, pp. 1–27.

45. Rabiner, L. R., and Sambur, M. R., (1975), "An Algorithm for Determining the Endpoints of Isolated Utterances," *The Bell System Technical Journal*, *54*(2), pp. 297–315.
46. Rabiner, L. R., Levinson, S. E., Rosenberg, A. E. and Wilpon, J. G., (1979), "Speaker Independent Recognition of Isolated Words Using Clustering Techniques," *IEEE Transaction on Acoustics, Speech and Signal Processing*, *27*, pp. 336–349.
47. Rabiner, L. R., (1989), "A Tutorial on Hidden Markov Models and Selected Applications in Speech Recognition," *Proceedings of the IEEE*, *77*(22), pp. 257–286.
48. Rabiner, L., and Juang, B. H., (1993), *Fundamentals of Speech Recognition*. Prentice-hall Publishers, Upper Saddle River, NJ, USA.
49. Rabiner, L., Juang, B. H., and Yegnanarayana, B., (2010), *Fundamentals of Speech Recognition*. Pearson Publishers, London, UK.
50. Sambur, M. R., and Jayant, N. S., (1976), "LPC Analysis/Synthesis from Speech Inputs Containing Quantizing Noise or Additive White Noise," *IEEE Transactions on Acoustics, Speech, and Signal Processing*, *24*(6), pp. 488–494.
51. Sarikaya, R., Hinton, G. E., and Deoras, A., (2014), "Application of Deep Belief Networks for Natural Language Understanding," *IEEE/ACM Transactions on Audio, Speech, and Language Processing*, *22*(4), pp. 778–784.
52. Syama, R. and Mary, S., (2008), "HMM Based Speech Recognition for Malayalam Language," In *The 2008 International Conference on Artificial Intelligence*, Las Vegas, NV, USA.
53. Thangarajan, R., Natarajan, A. M., and Selvam, M., (2008), "Word and Triphone Based Approaches in Continuous Speech Recognition for Tamil Language," *WSEAS Transactions on Signal Processing*, *4*(3), pp. 76–85.
54. Trang, H., Loc, T. H., and Nam, H. B., (2014), "Proposed Combination of PCA and MFCC Feature Extraction in Speech Recognition System," In *IEEE International Conference on Advanced Technologies for Communications*, Hanoi, Vietnam, pp. 697–702.
55. Venkataramani, B., (2006), "SOPC-Based Speech-to-Text Conversion," Nios II Embedded Processor Design Contest, pp. 83–108.
56. Wang, J. C., Lin, C. H., Chen, E. T., and Chang, P. C., (2014), "Spectral Temporal Receptive Fields and MFCC Balanced Feature Extraction for Noisy Speech Recognition," In *Asia Pacific Signal and Information Processing Association Annual Summit and Conference*, Siem Reap, Cambodia, pp. 1–4.
57. Wiren, J., and Stubbs, H. L., (1956), "Electronic Binary Selection System for Phoneme Classification," *The Journal of the Acoustical Society of America*, *28*, pp. 1082–1091.
58. Xydeas, C. S., and Cong, L., (1996), "Robust Speech Recognition using Fuzzy Matrix Quantization, Neural Networks and Hidden Markov Models," In *8th European Signal Processing Conference*, pp. 1–4.
59. Zhang, C., Li, X., Li, W., Lu, P., and Zhang, W., (2016), "A Novel i-Vector Framework using Multiple Features and PCA for Speaker Recognition in Short Speech Condition," In *International Conference on Audio, Language and Image Processing*, Shanghai, China, pp. 499–503.
60. Zhou, X., Fu, Y., Johnson, M. H., and Huang, T. S., (2007), "Robust Analysis and Weighting on MFCC Components for Speech Recognition and Speaker Identification," In *IEEE International Conference on Multimedia and Expo*, Beijing, China, pp. 188–191.

CHAPTER 14

Natural Language Processing

V. VISHNUPRABHA[*], LINO MURALI, and DALEESHA M. VISWANATHAN

School of Engineering, Cochin University of Science and Technology, Kochi, Kerala 682022, India

[*]*Corresponding author. E-mail: vishnuprabha72@gmail.com*

ABSTRACT

Understanding and processing human language has always been a complicated task in computational theory. Natural language processing, a combination of artificial intelligence and computational linguistics, employs computational techniques to understand the structure of human language. The challenges in the field are increasing day by day with the explosive growth of text contents on the Internet, varying text forms used in social media, and handling conversational complexities associated with intelligent devices. Natural language processing ranges from analyzing natural language by tokenizing and parsing techniques to resolving ambiguities and co-references in language. This chapter explains the basics of natural language processing that includes text processing techniques, parsing, semantic analysis, and the latest trend in deep learning models, which promises excellent improvement in natural language processing tasks, with simple examples that give more comprehension about the concepts in brief. The chapter also includes a comparative study of long short-term memory and gated recurrent unit for the sequence to sequence modeling with step-by-step implementation details of opinion summarization.

14.1 INTRODUCTION TO COGNITIVE COMPUTING

Artificial Intelligence (AI) and Neural Network (NN) are two essential technologies in the field of computer science. Both technologies try to create intelligent systems. Human intelligence depends on brain structure, and brain plays a vital role in thinking and formulating decisions. Machine learning, fuzzy logic, etc., can be used to make intelligent machines artificially. However, the most effective technique that creates an intelligent system is a Neural Network that helps machines to think like humans. Therefore, the main difference between AI and NN is that NN is the stepping stone to AI. A Neural Net can solve real-world scenarios by training them in similar situations. The main drawback of NNs is that they cannot handle a new scenario that is not present in the training set. Cognitive computing is a new technology that enables the machines to think like humans, whereas AI tries to create intelligent machines. Cognitive computing can extract information from a large amount of data and can learn continuously and instantly.

For making a machine think like a human, it should have all the capabilities that a human has. Learning from experiences, sensory perceptions, deduction, processing information, and memory are the essential features needed to think like a human.

14.1.1 FEATURES OF COGNITIVE COMPUTING

- The system should be capable of learning new data.
- A cognitive system must have a sensory perception depending on the application, and it should be like a human–human interaction, for example, voice assistants.
- The system should have the ability to handle a new situation efficiently.
- The system should be able to identify the context based on the current environment.

14.1.2 ROLE OF NATURAL LANGUAGE PROCESSING (NLP) IN COGNITIVE COMPUTING

Cognitive computing is a technology that helps in mimicking the human thought process. Most of the people believe that cognitive computing is a standalone technology. However, cognitive computing is a combination of multiple technologies. The key technologies that fuel cognitive computing are machine learning, NLP, machine reasoning, speech recognition, object recognition or computer vision, dialog systems, and human–computer interaction.

NLP plays a vital role in building cognitive systems as natural language understanding and natural language generation (NLG) have been an inevitable human feature. NLP, a combination of AI and computational linguistics, employs computational techniques for understanding the structure of human language. NLP ranges from analyzing natural language by tokenizing and parsing techniques to resolving ambiguities and coreferences in the language. With recent advances in NLP, computers can understand the natural language and respond. NLG plays a vital role in generating responses artificially. NLG by analyzing images, videos, and other nontextual data is still a vast research area. Current trends in NLP include sentiment analysis, dialog systems, machine translation, opinion summarization, and spam detection. Recent advances in deep learning models and methods promise a significant improvement in NLP tasks with a better understanding of cognitive science.

Both NLP and cognitive computing rely on each other. NLP aids cognitive computing and cognitive computing aids NLP. NLP itself can be seen as a cognitive technology because it uses sensory perceptions such as audio and visual perceptions as the primary step for an NLP task. Figure 14.1 shows the cognitive approach to NLP.

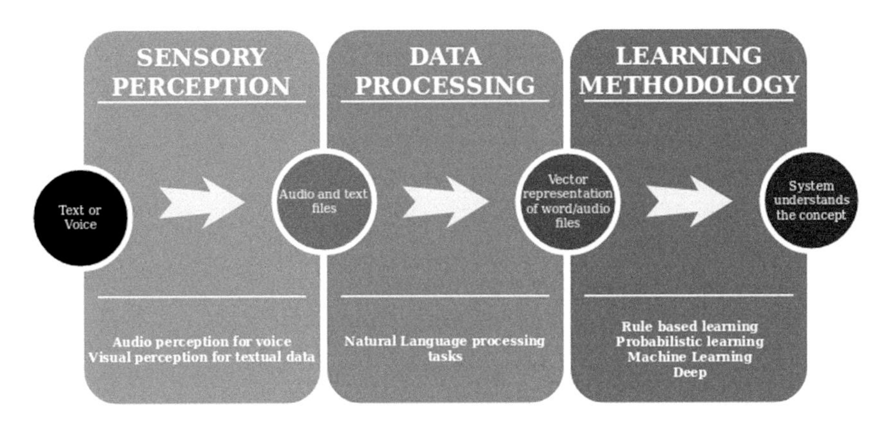

FIGURE 14.1 Cognitive approach to NLP.

14.2 IMPORTANT MILESTONES OF NLP

The journey of NLP began in the 1950s when signal processing scientists started processing speech signals. Machine translations works started during that period. Some of the landmark works in NLP are listed as follows.

1950 Automatic translation from Russian to English using the IBM 701 mainframe computer. They used statistics and grammatical rules for translation.

1956	McCarthy coined the term "Artificial Intelligence" in Dartmouth Conference.
1957	Chomsky's language models made a remarkable change in the field of NLP
1958	McCarthy introduced LISP programming language
1963	Giuliano introduced Automatic language processing concept
1964	ELIZA (NLP computer program) was developed at MIT by Joseph Weizenbaum
1966	Halted research on machine translation as it was very costlier than human translation
1970	SHRDLU (NL understanding computer) project for rearranging blocks by Terry Winograd was able to understand sentences like "Put the blue cube on the top of the red cube."
1975	Parsing program for automatic text to speech
1979	*N*-gram concept
1981	Knowledge-based machine translation
1982	Concept of the chatbot was created, and the project Jabberwacky began
1985–1990	Natural language processing using knowledge base
1990	Speech recognition using HMM
1992	Neural net for knowledge extraction
1995	Use of linguistic patterns for knowledge-based information extraction
1998	Classification of text documents
1999	New methods for syntax and semantic analysis

By the beginning of the 21st century, NLP research becomes more advanced with the evolution of modern computers, increased computation power, and memory. The new technologies such as machine learning, NN, probabilistic methods, and statistics were used to develop NLP applications with cognitive abilities. Apples' SIRI, IBM Watson are examples of the system with cognitive skills. NLP research is continuing to find more efficient methods for NLP, natural language understanding, and NLG.

14.3 CONCEPTS OF NLP

For understanding NLP, one needs to know the basics of natural language, that is, linguistics and the steps involved in processing it.

14.3.1 LINGUISTIC TERMS

14.3.1.1 PHONOLOGY/PHONETICS

Phonology is the study of speech sounds used in a particular language. Every alphabet has a sound associated with it, and a word is pronounced by combining those sounds. Word pronunciation can be explained through phonetics.

For example, in the English language, "read" in the present tense (reed) and "read" in the past tense (red) have different meanings.

14.3.1.2 MORPHOLOGY

Morphology is the study of structures of words/formation of words. A morpheme is the smallest individual unit of language that has a specific meaning. Morphemes can be words, prefixes, or even suffixes. For example, the word "unfairness" means a lack of justice or inequality. Morphemes out of this would be:

Un (not)	–	prefix
fair (treating people equally)	–	the root word
ness (being in a state)	–	suffix

14.3.1.3 SYNTAX

The syntax is nothing but the structure of language. Syntax analysis is the study of the structural relationships among words in a sentence or how words are grouped to form sentences. Every language follows a rule for creating meaningful sentences. Subject, verb, object, parts of speech (POS), etc., help in the formation of meaningful sentences.

In English linguistics, subject–verb–object is a sentence structure where the subject comes first, the verb second, and the object third. The POS categorize words according to their usage. For example, noun is used to represent the name of a place, person, or thing.

14.3.1.4 SEMANTICS

Semantics is the study of the meaning of words in a sentence and how these words are combined to form meaningful sentences. Lexical semantics analyze the relation between words such as synonyms, hypernyms, and others. Semantics try to interpret the meaning of a sentence by combining and finding a relationship between word meanings.

For example, consider a sentence extracted from a paragraph. "That company is facing a huge financial crisis now, and in May it may vary." In this case, "May" represents a month and "may" represent a word. This type of words can confuse the machine in finding the proper meaning of a sentence.

14.3.1.5 PRAGMATICS

Pragmatics studies the situational use of language sentences. It is slightly different from semantics. Semantics try to interpret the meaning of a sentence by combining the meaning of words, whereas pragmatics try to find the meaning of a sentence based on the situation.

Consider the proverb, "Don't judge a book by its cover." Its semantic meaning is the same as the word meaning in the sentence, but its pragmatic meaning is "Don't judge someone or something by appearance alone."

14.3.1.6 DISCOURSE

Discourse is a group of sentences, and it studies or finds the actual meaning of the context by connecting component sentences.

"Radha took a book from the library. Then she went to the coffee shop, and she left the book there."

The above context leads to more than one inferences, and it can answer different questions. Consider the question, "Where is the book now?" The answer is, "Book is in the coffee shop."

14.3.2 NLP TASK FLOW (THEORETICAL AND ENGINEERING APPROACH)

For a given context in natural language, there are various stages of analysis for that sentence/context. The study of natural language for the extraction of useful information is called NLP. NLP tasks are explained in two ways: one is a theoretical approach, and the other is an engineering approach. The theoretical method describes the stages of NLP for a given problem conceptually, whereas the engineering approach describes how a computer accomplish NLP.

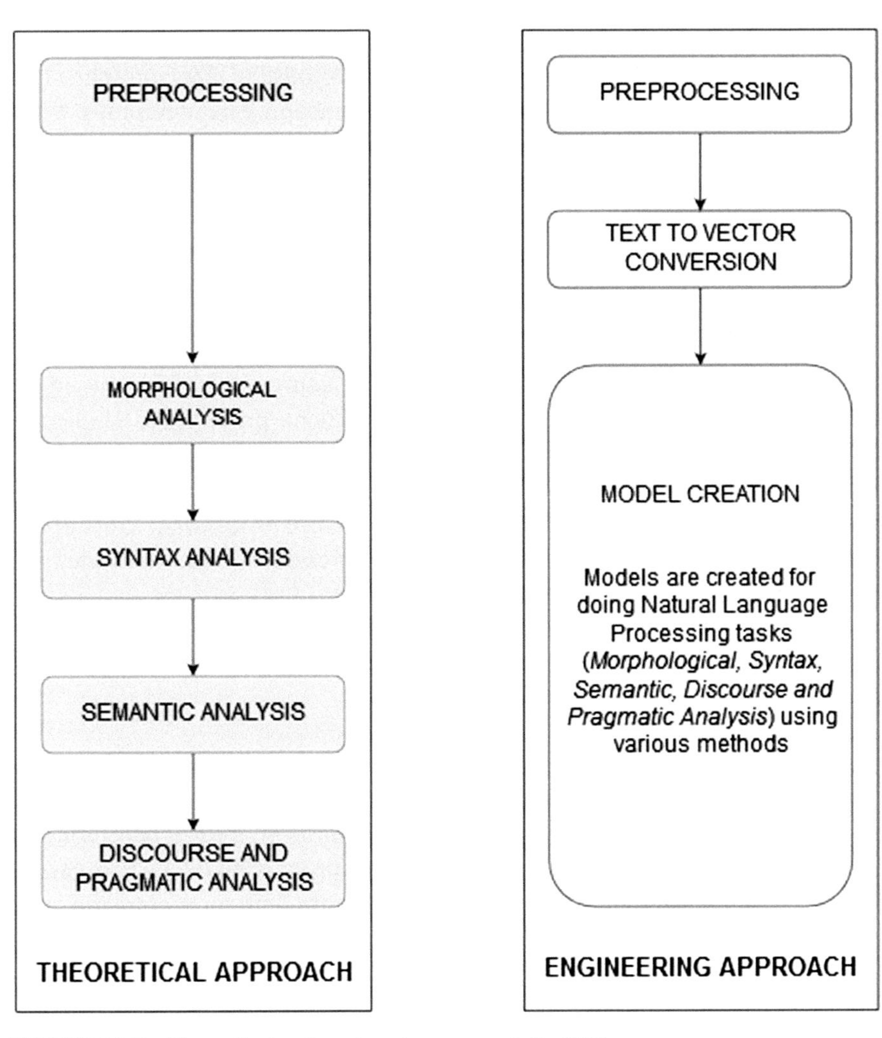

FIGURE 14.2 Theoretical and engineering approach for NLP.

14.4 THEORETICAL APPROACH TO NLP

In the theoretical approach, a given sentence is subjected to initial processing, and words are separated from sentences. The words are in different forms (i.e., past tense and present tense). Therefore, it is required to convert those words into its base form through morphological analysis. For finding out the actual meaning of the sentence, it must be in proper syntax. Syntax analysis enables syntactic checking and tagging the words with POS. POS tagging helps in finding the meaning of a sentence quickly. The semantic study finds out the meaning of a sentence by building relationship charts/graphs. The semantic analysis helps in finding the sentence meaning by combining word meanings. Finally, pragmatic and discourse analysis finds the actual meaning of a sentence by considering the situation.

14.4.1 PREPROCESSING

Real-world data are unstructured and complicated, especially with the emergence of social media. The scattered and unorganized data should be combined and organized adequately for efficient processing. Irrelevant information affects the efficiency of the final output. In addition, the use of abbreviations, smileys, etc., can create confusion. Therefore, the data must be converted into their normalized form before processing. Most of the preprocessing tasks are done using regular expressions. Essential functions in initial processing include the following tasks.

14.4.1.1 SPELLING CORRECTION

Words with incorrect spellings are identified and corrected. Most of the data that are processing today include customer reviews, social media comments, etc. It is more likely to use short word forms in those sites. Therefore, the spellings must be corrected for avoiding the preprocessing delay. Processing text data without spelling correction may take more time for processing, as it is a tedious task in finding a proper word from the dictionary.

Example: "Its ur responsibility to handle dat case" >>> "It is your responsibility to handle that case."

14.4.1.2 CASE CONVERSION

Case conversion is a normalization procedure, in which all data are converted to lower case to make processing easier. Normalization procedure makes processing more manageable, but it may create problems in some situations. For example, "March" represents a month, and in that scenario, the letter "M" is capitalized, but in the case of "march" (a military walk), the letter "m" is small.

Example: How are you? >>>how are you?

14.4.1.3 PUNCTUATION REMOVAL

It is the process of removing punctuations in the text.

Example: Awesome! >>> Awesome.

14.4.1.4 TEXT STANDARDIZATION

It is the process of converting abbreviations to its original for resolving ambiguity while processing.

Example: NLP >>> Natural Language Processing.

14.4.1.5 TOKENIZATION

Breaking of data into smaller units to ease processing. The main tokenization methods include the following.

- *Sentence tokenization:* A piece of text is converted into individual sentences. It is done by using a period(.).
  Example: "Learning is the activity of gaining knowledge. Learning enhances the awareness of the subjects of the study." >>> "Learning is the activity of gaining knowledge." "Learning enhances the awareness of the subjects of the study."
- *Word tokenization:* Breaking up of textual data into individual words. Sentences are converted to words by identifying blank space between words.

Example: "I am fine." >>>"I," "am," "fine."

14.4.1.6 STOP WORDS REMOVAL

A stop word is a word that is commonly used in every article. The words "are," "was," "and," etc., are examples of stop words. The presence of stop words affects feature engineering very severely. Count of these words is high in the document, and there is a chance to take this as a relevant word while classifying documents.The classification process efficiency gets affected by these words. Therefore, stop words are eliminated during preprocessing.

Example: ("The," "book," "is," "on," "the," "table") >>>
("The," "book," "table").

14.4.2 MORPHOLOGICAL ANALYSIS

After tokenization, the word should be converted into their base forms. Word can have suffixes and prefixes that make the sentences express their idea efficiently. However, those words will not be present in the dictionary for further analysis. Therefore, the words must be converted into their base forms by trimming suffixes or prefixes, if any. Morphological analysis is done using two methods.

14.4.2.1 STEMMING

It deals with different forms of the same word. For example, write can appear in many forms, such as write, wrote, writing, written, writer, and so on. Stemming converts words into their base forms by trimming inflectional word forms such as ing, ed, en, etc.

Example: writing >>> write, calves >>>calv.

14.4.2.2 LEMMATIZATION

Lemmatization does the same thing as stemming, but the difference is that lemmatizer checks the presence of the stemmed word in the dictionary.

	Noun Lemmatizer	Verb Lemmatizer
Writing	Writing	Write
Calves	Calves	Calve

14.4.3 SYNTAX ANALYSIS

It is essential to know the syntax and structure of language for efficient processing. Syntax analysis parses the text and annotates the text with POS tags. It helps in understanding the hierarchy of the sentence, and it also makes semantic analysis easier. Standard parsing techniques for understanding text syntax are mentioned below.

14.4.3.1 POS TAGGING

POS tagger labels the words in a sentence with a POS tag that is most suitable for that particular word. POS tagging helps in analyzing the semantic meaning of the sentence without much effort. Different POS include adjective, adverb, conjunction, determiner, noun, number, preposition, pronoun, verb, adjective phrase, adverb phrase, noun phrase, verb phrase, etc.

Example: Ram loves eating pizza >>> Ram\Noun loves\Verb eating\Verb pizza\Noun

14.4.3.2 SHALLOW PARSING OR CHUNKING

Shallow parsing identifies the nonrecursive phrases in a sentence.

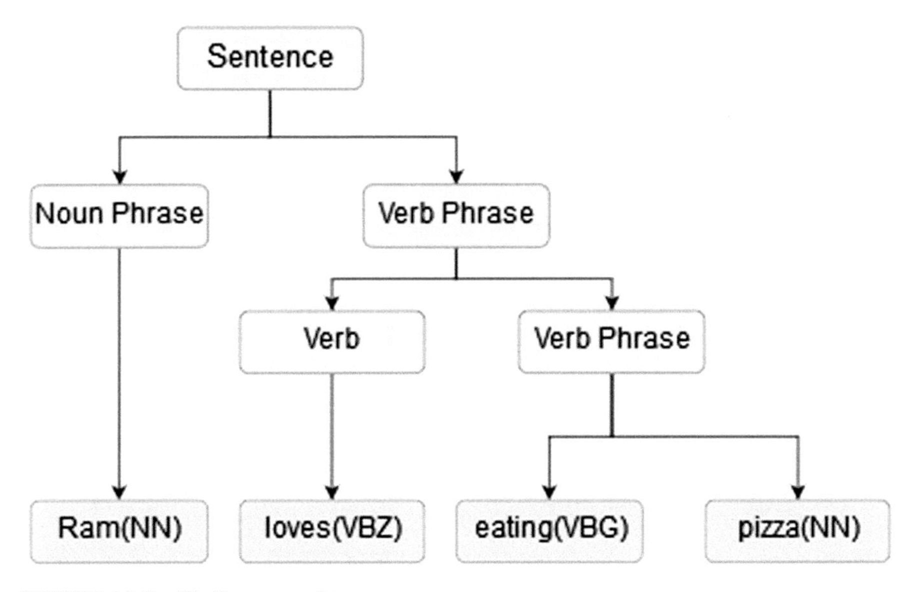

FIGURE 14.3 Shallow parsed tree.

14.4.3.3 CONSTITUENCY PARSING

Constituency parser finds the recursive phrases in a sentence.

14.4.3.4 DEPENDENCY PARSING

Dependency parser focuses on a word in the sentence and tries creating a relationship with other words. A relationship tag labels the edges.

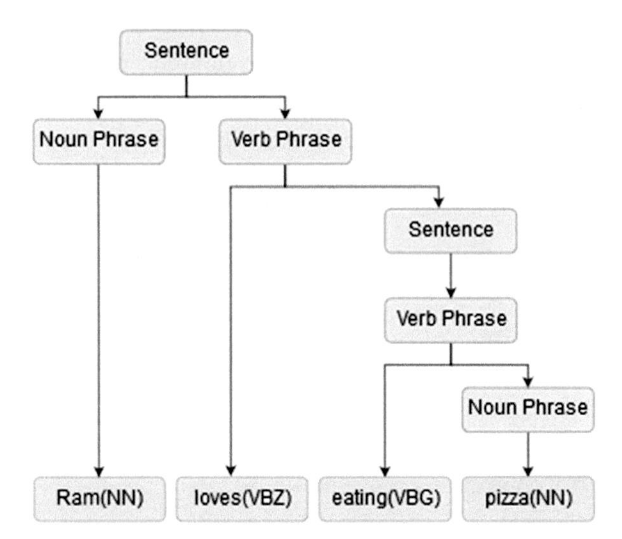

FIGURE 14.4 Constituency parsed tree of a sentence.

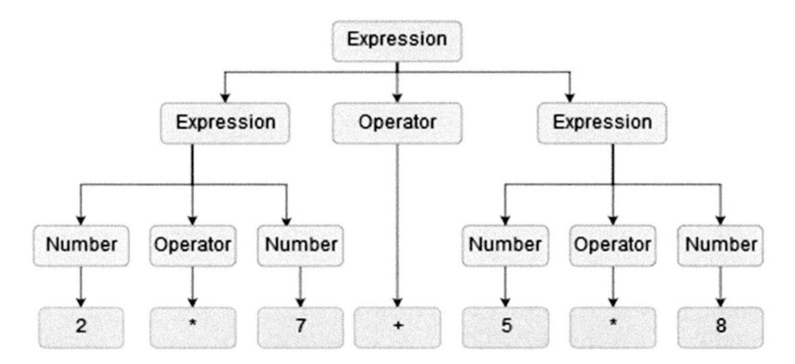

FIGURE 14.5 Constituency parsed tree of the expression ((2*7)+(8*5)).

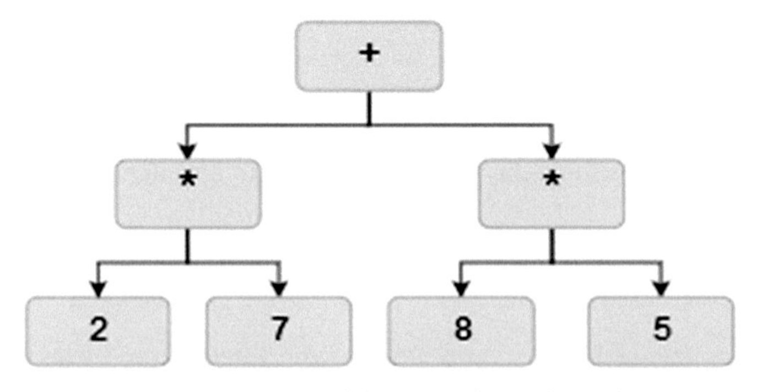

FIGURE 14.6 Dependency parsed tree of the expression ((2*7)+(8*5)).

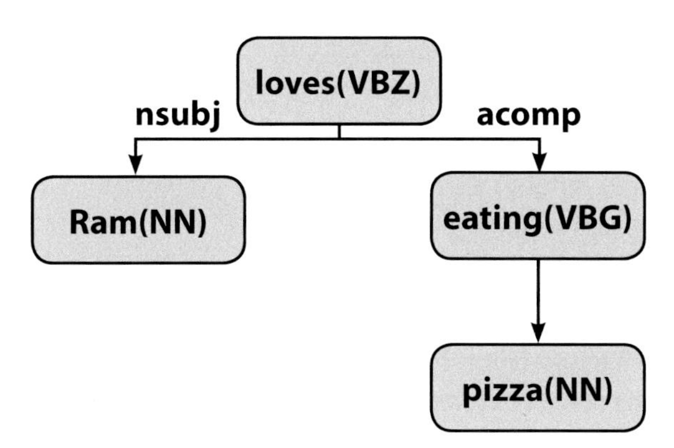

FIGURE 14.7 Dependency parsed tree for a sentence.

14.4.4 SEMANTIC ANALYSIS

Semantic analysis demands contextual awareness and a well-structured knowledge base for extracting an accurate meaning from a sentence. Human–computer interaction systems such as conversational agents generally deal with understanding the human language and generating an accurate response in return. Semantic analysis of human language is the critical technology that assists any cognitive computing system to interpret the meaning of natural language. An efficient cognitive system must have the ability to analyze unstructured data and make connections between related information by refining the learning process continuously. The power of linguistics and advances in semantic analysis facilitates cognitive computing to establish relevant relationships among the words in human interaction by considering the surrounding context and sense of every word.

The grammatical information transferred from the syntactic parser is fed into the semantic analysis phase after assigning a proper part of speech tag to each lexeme. Semantics is the process of understanding the linguistic meaning of each lexeme from the parser. For a sentence or statement, the meaning depends on the meaning of constituent parts as well as the composition of lexical items. Various other factors such as the context of the sentence, senses of words, the inclusion of phrases, utterance, and pronouns mentioned play a significant role in cracking the meaning of a sentence. A rich and versatile knowledge base also supports the power of language understanding and semantic processing to enact better reasoning for comprehending and decision making in cognitive systems.

14.4.4.1 THEORETICAL APPROACHES IN SEMANTIC ANALYSIS

The topic discusses the necessary methods in semantic processing:

- How is the input to a semantic analyzer converted to an intermediate meaning representation?
- How is this intermediate representation used to assign proper meaning to each constituent?

Meaning extraction from the text data can be done based on two different approaches. First is lexical semantics that deals with extracting word meaning of each lexical item, while the later compositional semantics considers the meaning of phrases, the composition of lexical parts, etc. For information extraction (IE), the initial challenge in the semantic analysis is how to represent the meaning of a human utterance in the computer. Hence, first, the output from syntax parser must be converted into proper meaning representation using techniques such as first-order predicate logic, associative networks, frames or scripts, etc. The final transformation is done by approaches such as syntax-directed semantic analysis, semantic grammars, IE, etc.

14.4.4.2 LEXICAL SEMANTICS

The study of meanings of word units in an entire sentence and their interrelation is lexical semantics. Semantic analysis of individual lexemes impose analysis of their structure and labeling their relations to other lexemes, accounting similarities and differences among different lexemes in similar settings and the nature of relations among lexemes in a single setting [1]. Table 14.1 shows a list of lexical relations and their senses significant in computational as well as cognitive semantics. To distinguish between the various senses of a word is one of the major issues identified in lexical semantics. Some of the issues related to lexical semantics are word sense disambiguation, semantic role labeling, and semantic selectional restrictions.

14.4.4.2.1 Word-Sense Disambiguation

The process of distinguishing word senses and choosing the most appropriate sense for a word is called WSD. One common feature is to identify word senses from the different contexts, in which a given word is used. WSD resolves many NLP tasks that address cognitive semantic issues, such as question answering, intelligent document retrieval, and text classification. The way that WSD is exploited in these and other applications varies widely based on the particular needs of the application [1].

One way of WSD is to use dictionaries and ontological relationships to analyze the different sense of words. It is a knowledge-based approach and highly depends on the quality of the knowledge base in use [2]. Another approach used is based on machine learning, which involves supervised and unsupervised learning. Unsupervised learning uses clustering techniques for identifying different contexts and co-occurrence of words. In supervised learning, a selected set of sense tagged words along with corpora (like WordNet) is used to train the algorithm [2].

TABLE 14.1 Common Lexical Relationships in the English Language

Lexical Relations	Description	Example
Homonymy	Words with same spelling and sound but having a different meaning	Light The book is too *light.* *Light* the lamp
Polysemy	Multiple related meaning within a single lexeme	Blood *bank* and question *bank*
Synonymy	Different lexemes with the same meaning	*Big* and *large* *Fare* and *price*
Antonymy	Words that express the opposite meaning	*Fast* and *slow* *Cold* and *hot* *Rise* and *fall*
Hyponymy	The pairing of words where one lexeme denotes a subclass of another	*Car* is a subclass of *Vehicle* *Banana* is a subclass of *fruit*
Meronymy	Words that denote the part of a relation	*An arm* is a part of *the body*
Holonymy	A word that indicates Whole of a part (reverse part-of)	*The building* is the whole part of the *window*

14.4.4.2.2 Semantic Role Labeling

An event represented in a dialog system can be presented in several different syntactical ways. Understanding the idea should not depend on the syntactic arrangement. It is another issue in meaning representation/semantic parsing. The solution is to map the meaning by identifying the verb and its arguments, not by its syntactic order.

Semantic role labeling is to identify a verb or a predicate and its arguments in a sentence [1]. Each argument identified is labeled based on its semantic relation with the predicate. Semantic role labeling within the sentence makes the meaning representation independent of syntactic arrangement. Both supervised and unsupervised methods are used in semantic role labeling. The semantic roles can come variously from resources such as PropBank, FrameNet, or VerbNet. It can also be extended to similar semantic roles that are introduced by other POS, such as nominal or adjectival elements [2]. Semantic role labeling is also called as thematic role labeling, case role assignment, or even shallow semantic parsing [1].

14.4.4.2.3 Semantic Selectional Restriction

Even though disambiguating word senses and attaching role labels to every word argument makes the semantic analysis more accurate, specific environments show semantic violations in a different way.

Example: She likes to eat mountain.

This sentence is syntactically correct, but semantically it is not acceptable. The argument of the verb "eat" will be an edible thing. This cognitive issue in the semantic analysis is solved by semantic selectional restriction. This technique allows predicate words to use semantic constraints on its argument words. This process is known as a semantic selectional restriction. A violation of semantic restriction may produce an anomalous sentence [1].

14.4.4.2.4 Compositional Semantics

The meaning of a phrase or composite expression can be obtained by combining the meaning of each word unit in the phrase and by the syntactic arrangement of these constituent words. The combination of such words often leads to meaning representation issues. Meaning representation is the first task of semantic analysis as every natural language statement should be converted into an equivalent meaning representation for further mapping. In compositional semantics, meaning representation is usually done by a logical approach or a truth-conditional approach, which is based on the principle of compositionality [3].

14.4.4.3 CHALLENGES IN SEMANTICS

Some significant issues in semantics are discussed as follows.

14.4.4.3.1 Meaning Representation

To perform the desired task based on a linguistic input, a cognitive system needs a rich semantic analyzer. It is necessary to bridge the gap between the linguistic input and the knowledge base. Hence, the primary step in any semantic analysis is to construct an equivalent meaning representation of natural language input. An ideal meaning representation should be verifiable, unambiguous, and expressive [1]. A significant issue in meaning representation is understanding sentences that convey the same meaning. Such sentences are structured with different lexical patterns, as in the example given as follows.

Example: Today is sunny.
Today is a bit hot.

The meaning representation for the given examples should be the same. The conventional ways of meaning representation are listed in the following.

14.4.4.3.1.1 Predicate–Argument Structure

The words and phrases in a sentence keep some relationships or dependencies with the underlying concepts and meaning of the input text. Grammatical objects such as verbs in a sentence assert an object argument structure with other constituent words such as nouns, noun phrases, etc., in the sentence. Hence, the predicate–argument structure is a suitable format for semantic representation of human language.

14.4.4.3.1.2 Model-Theoretic Semantics

In this method, a linguistic input is represented as objects, their properties, and object relationships in a model. Here, a model is implemented to represent the state of affairs in the world. If the model accurately captures the facts from the input, then it efficiently expresses the meaning representation. The domain of a model is simply the set of objects that are part of the application, or state of affairs, being represented. Each distinct concept, category, or individual in an application denotes a unique element in the domain.

This approach sometimes initiates a truth conditional, which represents the world as a mathematical abstraction made up of sets and related linguistic expressions to the model. Truth-conditional semantics takes the external aspect of meaning as fundamental. Here, a sentence is true or false depending on the state of affairs that obtain in the world, and the meaning of a proposition is its truth conditions [1].

14.4.4.3.1.3 First-Order Predicate Calculus

First-order logic or first-order predicate calculus (FOPC) is a sound method for representing meaning which satisfies verifiability and expressiveness. The basic building blocks FOPC using are constants, functions, and variables to refer objects and their properties in the real world as modeled in a knowledge base. It uses logical connectives to make composite and complex representations from simple predicates. Quantifiers such as existential quantifiers and universal quantifiers can be applied to represent variables as a collection. One crucial issue of semantics is that meaning representation should support the inference. With the support of inference, FOPC deduces valid conclusions from existing knowledge.

14.4.4.3.1.4 Semantic Networks and Frames

A semantic net is a representation of factual knowledge as entities and their relationships. It uses nodes to represent entities and links to represent relationships. Frames are structures containing all relevant information about the type of entities and their instances. It works like a slot and slot-filler to store fields and values of a record structure and sketches a narrow concept in detail.

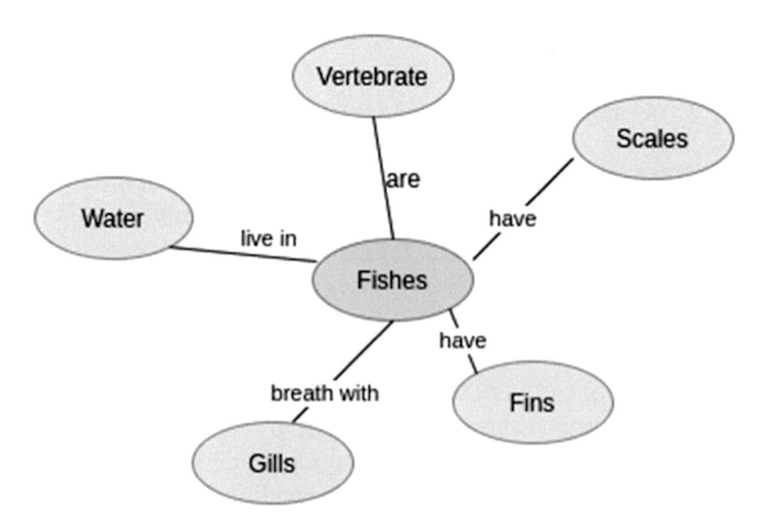

FIGURE 14.8 Semantic net with nodes and links

14.4.4.4 COMPUTATIONAL SEMANTICS

Computational semantics performs the computation and analysis of linguistic meaning for the natural language input. It automatically assigns meanings to the intermediate representation with common-sense reasoning.

14.4.4.4.1 Syntax Directed Semantic Analysis

A syntax analyzed tree is given to the semantic analyzer. It uses grammar rules and lexical semantics to assign the correct semantic value to each lexical item. Semantic grammars encode semantic information into a syntactic grammar. They use context-free rewrite rules with nonterminal semantic constituents. Generally, a semantic error occurs when the meaning of knowledge is not communicated correctly.

14.4.4.4.2 Information Retrieval

The information retrieval system relies on the storage of documents and retrieval of data in response to a textual query. Keyword-based retrieval should apply WSD to ensure proper indexing and error-free document retrieval. In addition, in a multiword query, the relationship between keywords is semantic rather than syntactic. A concept-based document retrieval should adopt a cognitive semantic approach to tackle the issues such as term co-occurrence and relationship between keywords.

TABLE 14.2 Commonly Used Lexical Resources in Semantics for WSD and SRL

Name	Description	Origin
WordNet [17]	Vast computational lexicon or dictionary provides different word senses	Miller, G. A., 1995
FrameNet [18]	Lexical database using a semantic frame, describes a type of event, relation or entity, and its participants	Baker, C. F., Fillmore, C. J., & Lowe, J. B., 1998
VerbNet [19]	Online verb lexicon	Schuler, K. K., 2005
PropBank [20]	Proposition Bank, information about basic semantic proposition	Kingsbury, P. R., & Palmer, M., 2002
ConceptNet [21]	Multilingual semantic net and common sense knowledge base	Liu, H., & Singh, P., 2004
BabelNet [22]	Multilingual encyclopedic dictionary, Integrated with word Net	Navigli, R., & Ponzetto, S. P., 2010
HowNet [23]	Online common sense knowledge base	Dong, Z., & Dong, Q., 2003

14.4.4.4.3 Information extraction

IE is the activity of identifying and extracting data from a document and categorizing it semantically. Various tasks, including IE, are as follows:

- *Name extraction:* The process of identifying the names in a text and classifying them as people, organizations, locations, etc.
- *Entity extraction:* Identifying all phrases which refer to objects of specific semantic classes, and linking phrases which refer to the same object.
- *Relation extraction:* Identifying pairs of entities in a specific semantic relation.
- *Event extraction:* Identifying instances of events of a particular type and the arguments of each event.

The goal of IE is only to capture selected types of relations, types of events, and other semantic distinctions that are specified in advance. IE systems are domain based, and most of them incorporate the semantic structure of that domain. Major domains of IE are news articles, medical records, and biomedical literature, which have large quantities of text, repeated entities, and events of the same type and where there is a need to distill the information to a structured database form [2, 3].

14.4.4.4.4 Named Entity Recognition

Named entity recognition is the process of extracting named entities that are present in a text into pretagged categories such as "individuals," "companies," "places," "organization," "cities," "dates," "product terminologies," etc. It enriches the semantic knowledge of the content and helps promptly understand the subject of any given text. It is useful in applications such as news content analysis, business sentiment analysis, etc. Named entity recognition can provide article scanning based on relevant tags to reveal the significant people, organizations, and places discussed in them. It helps in the automatic classification of articles with fast content discovery. In business sentiment analysis, extracting the identity of people, places, dates, companies, products, jobs, and titles gives an insight into the people's opinion on product and company.

14.4.5 PRAGMATIC AND DISCOURSE ANALYSIS

The study of natural language by the speaker's utterance and its interpretation based on situational context is called pragmatics. It is closely connected to semantics, but the focus is on interpersonal communication. Sentence meaning in semantics may vastly underdetermine speakers meaning. The goal of pragmatics is to bridge the gap between a formal sentence meaning and the speaker's intention based on the context.

14.4.5.1 DISCOURSE

A group of structured related sentences is a discourse [1]. It refers to a sequence of sentences, where each sentence is interpreted in the context of the preceding sentences. Natural language utterances are never disconnected or unrelated sentences; instead, they are structured and continually and coherently related one after another. Such coherent groups of sentences form a discourse.

14.4.5.2 ANAPHORA RESOLUTION

The linguistic action of referring back to a previously mentioned item in the text is termed as anaphora. The word or phrase "referring back" in the text is known as *anaphor,* and the thing which it refers to is its *antecedent*. When the anaphor refers to an antecedent, and when both have the same referent in the real world, they are termed *coreferential*. The interpretation of anaphors, known as *anaphora resolution,* has a vital role in the understanding of discourse.

Example, Santa loves icecream. She works in an icecream shop.

Here, *She* is an anaphor, and *Santa* is the antecedent. Here, *she* and *Santa* are coreferential.

14.4.5.3 COREFERENCE

A natural language sentence in a discourse often containing a reference to a previous word entity is called a referring expression, and the object that is referred to is called the referent. References used in a sentence are often denoted as mentions, and mentions of the same entity are linked as being coreferences, or members of a coreference set. These can include pronouns, nominal entities, named entities, and events [2]. Co-reference resolution intends to find referring expressions in a text that refer to the same entity. The set of expressions that co-refer is known as a co-reference chain. Reference resolution is the task of determining what objects are referred to by which linguistic expressions [1]. Application areas of co-reference resolution are IE, summarization, and conversational agents.

Example, Dr. APJ Abdul Kalam was the President of India. Formerly, he was the chairman of ISRO.

Here, *he* and *Dr. APJ Abdul Kalam* are co-references.

14.4.5.4 TEXT COHERENCE

Text coherence has an influential role in determining the acceptability of a specific discourse. Coherence refers to the meaningful connection between text units. A large structured discourse unit is built by organizing coherent structures in a meaningful way. A coherent paragraph is unified, logical, consistent, and meaningful. However, it makes sense only when reading as a whole.

Example: *One Man's Meat* (by E.B. White) [https://literarydevices.net/coherence/]

"Scientific agriculture, however, sound in principle, often seems strangely unrelated to, and unaware of, the vital, grueling job of making a living by farming. Farmers sense this quality in it as they study their bulletins, just as a poor man senses in a rich man an incomprehension of his own problems. The farmer of today knows, for example, that manure loses some of its value when exposed to the weather…But he knows also that to make hay, he needs settled weather—better weather than you usually get in June."

The above passage is an excellent example of coherent text. The author here described theoretical agriculture, and that topic is intelligently combined with farmer's problems and climate.

14.4.5.5 RHETORICAL STRUCTURE THEORY (RST)

RST explains the coherence of texts. RST is a model of text organization [2]. It can be used for systematic analysis of text using rhetorical relations. A text analysis using RST creates a tree based on the rhetorical relations to represent the content. All relations in RDT are based on the concept of a nucleus and satellite. The nucleus holds a central independent idea, but the satellite is less central, and usually, its interpretation is connected with the nucleus. There is a set of 25 rhetorical relations. These relations include the evidence relation, which applies to two text spans where the satellite provides evidence for the claim contained in the nucleus [2].

14.5 ENGINEERING APPROACH TO NLP

In the engineering approach, a given text is preprocessed, and base forms of words are created. However, one cannot do syntax analysis, etc., in the engineering approach immediately after morphological analysis because the computer needs everything in digital form for analysis. Therefore, the words must be converted into vectors for doing further analysis with the help of machine learning methods.

14.5.1 PREPROCESSING TEXT

The preprocessing step in the engineering approach includes the same steps as in the theoretical approach. The preprocessing tasks are done mainly using regular expressions in computers.

14.5.2 TEXT-TO-VECTOR CONVERSION

Text-to-vector conversion is the process of converting word to vector form for analysis. This process is also known as feature extraction. It is to find relevant features that are required to create a machine learning model.

Word embeddings are nothing but a numerical representation of the text. Embedded word makes the processing faster. There are various techniques for creating word embeddings. Traditional techniques used for numerical representation are One Hot encoding and *N*-gram representation. Most of the new methods based on language modeling use these traditional techniques as the first step. Today word embeddings are created using machine learning techniques. Fast text, Glove, word2vec, etc., are examples of word embeddings created using NNs.

14.5.2.1 ONE HOT ENCODING

Tokenized words are sent to One Hot encoder for mapping.

Example: (“Hai,” “how,” “are,” “you”)

OUTPUT:

Hai – [1, 0, 0, 0]

how – [0, 1, 0, 0]

are – [0, 0, 1, 0]

you – [0, 0, 0, 1]

One hot encoded representation of the word “are” is [0, 0, 1, 0].

14.5.2.2 N-GRAM

N-gram is one of the most efficient techniques used for predicting the next word given the previous word. *N*-gram works based on the chain rule of probability.

Consider a Phone review for sentiment analysis: "Camera quality is not bad" the actual meaning of this sentence is "The phone has a good camera quality." Sentiment analysis is the process of finding sentiment (positive, negative, or neutral) about a product/something. The sentiment is found by tokenizing the words and finding whether that word is present in the list of positive, negative, or neutral category. Here, in this case, both "not" and "bad" fall in the negative category, and that review is classified as a negative one even though it is a positive review. Therefore, for avoiding such situations, the *N*-gram concept is introduced. *N* can be any value above 0.

Example: (Camera quality is not bad)

OUTPUT:

1-gram : ("Camera," "quality," "is," "not," "bad")

2-gram : ("Camera quality," "quality is," "is not," "not bad")

3-gram : ("Camera quality is," "quality is not," "is not bad")

This method eliminates the problem of classifying it as a negative sentence using the bigram technique.

14.5.2.3 WORD EMBEDDINGS

New methods for word embedding helps in finding similarities as well. Words with similar meanings will be having the same representation in word embeddings. There are different types of word embeddings.

Frequency based

- Count Vectorizer
- TF-IDF Vectorizer
- Co-occurrence Vectorizer

Prediction based

- Continuous Bag of words (CBOW)
- Skip-Gram model

14.5.2.3.1 Count Vectorizer

Count vector technique works on the top of one-hot encoding. This method is similar to the one-hot encoding technique, but this can be used for large corpus (collection of documents).

Consider a corpus of $D(=2)$ documents. Unique tokens are extracted out from the corpus excluding stop words.

Doc1: (He is greedy. Raghu is a bad and greedy person)

Doc2: (Ramu is greedy)

TABLE 14.3 Count Vector Representation

	Greedy	bad	Person	Raghu	Ramu
Doc1	2	1	1	1	0
Doc2	1	0	0	0	1

Count vector representation of the word "greedy": [2, 1] (columnwise).

The count vector method counts the occurrence of the word "greedy" in Doc1 and Doc2. It uses either the frequency of the word occurrence or the presence of that word for getting count value. In this case, if using frequency, then "greedy" is represented as [2,1], and if the presence of a word is used, "greedy" is represented as [1,1]

Count vector representation of Doc1 : [2 1 1 1 0] (row-wise).

This technique is not suitable for a large document with millions of words. A matrix of significant size, including the list of words and documents, is to be created for large documents. In addition, if such a matrix is created, it will be sparse.

14.5.2.3.2 Term Frequency-Inverse Document Frequency (TF-IDF)

It is one of the most popular techniques used for finding whether that word is relevant in that specific document that is analyzing. Words with high occurrence count and words that occur less frequently are treated as essential words

$$\mathrm{TF}(T) = \frac{\text{Number of times term } T \text{ appears in a document}}{\text{Total number of terms in the document.}}.$$

Term frequency tells you how frequently a given term appears. For example, If "Learning" appears 20 times in Doc1 and the total number of terms in the document is 200, then $TF(Learning) = 20/200 = 0.1$.

$$\mathrm{DF}(T) = \frac{\text{Number of documents containing a given term } T}{\text{Total number of documents}}.$$

Document frequency finds the importance of that term. Assume that the number of documents containing the term "learning" is 100, and the total number of documents in the corpus is 4,000,000, then $DF(learning) = 100/4{,}000{,}000 = 0.000025$. Taking log() for normalizing this value, $\log(0.000025) = -4.60205999133$. If the total number of documents containing the term T is very less than the total number of documents in the corpus, it gives negative value, and it is challenging to compare. Therefore, the value is inverted, that is,

$$\mathrm{idf}(T) = \frac{\text{Total number of documents in the corpus}}{\text{Number of documents containing a given term } T}$$

$$\mathrm{IDF}(T) = \log\left(\mathrm{idf}(T)\right).$$

IDF can compare the quantities even if there is a considerable difference between the quantities. For example, *if* $idf(learning) = 4{,}000{,}000/100 = 40{,}000$, normalize this value by taking log will give IDF as 4.60205999133

$$TF\_IDF(learning, Doc1) = 0.1 * 4.60205999133 = 4.60205999133.$$

Taking another example, consider the term "this" in Doc1.

$$TF(this) = 60/200 = 0.3$$

$IDF(this) = \log(4{,}000{,}000/4{,}000{,}000) = 0$ (Every document will contain at least one "this")

$TF\text{-}IDF(this) = 0.3 * 0 = 0$ (Indicates that the term "this" is not relevant).

14.5.2.3.3 Co-occurrence Matrix

It is based on the assumption that similar words tend to occur together. The word co-occurrence means the number of times a word pair (W1, W2) has appeared together in a context window. The context window is similar to the N-gram language model. The size of the context window can be varied depending on the application.

Corpus = "I love to learn. I love robotics. I love to read."

Context window size = 2 (takes two words on the left and right of the given word) that is, Consider the word "to." The word "to" occurred with "I" in context three times and "to" occurred with "love" in the context two times. The phrases "love to" and "I love" tend to occur together.

I	love	to	Learn	I	love	robotics	I	love	to	read

TABLE 14.4 Co-occurence Matrix Representation

	I	Love	To	Learn	Robotics	Read
I	0	4	3	1	2	0
love	4	0	2	1	2	1
to	3	2	0	1	0	1
learn	1	2	1	0	0	0
robotics	2	2	0	0	0	0
read	0	1	1	0	0	0

This model is rarely used as this needs a large matrix if the number of terms is high. In addition, the co-occurrence matrix is not a word vector representation, and it needs to be converted into word vector using principal component analysis. This model helps in building relationships, i.e., (Boy, Girl), (Husband, Wife) pairs.

14.5.2.3.4 Continuous Bag of Words

CBOW is a machine learning model created for predicting the probability of a word in a given context. A context is formed using the *N*-gram concept. It may be a single word or a collection of words. One-hot encoded vector is given as input to an NN, and the probability of the given word is predicted. This method consumes less memory and is more accurate. However, training an NN needs more time as it formulates rules from examples. However, this is an efficient and widely used model that can predict the probability of word being in a given context. The word "bat" can be a bird or a cricket bat. CBOW places the word 'bat' in between the cluster of birds and a cluster of sports. CBOW takes a context as input and predicts the probability of the current word based on the context. The bag of words matrix is calculated as follows.

TABLE 14.5 Bag of Words Representation

	Not	Bad	Good	Story	Casting
Not bad	1	1	0	0	0
Good story	0	0	1	1	0
Good casting	0	0	1	0	1

14.5.2.3.5 Skip-Gram Model

It is a machine learning model with a context window on both sides. This model predicts the word surrounding the target word. Consider an example, a skip-gram model with context window size = 1.

Example: Corpus = Hello! How are you?

TABLE 14.6 Skip-Gram Model Representation

INPUT (Given Word)	OUTPUT 1 (Next Word)	OUTPUT 2 (Previous Word)
Hello	How	–
How	Are	Hello
Are	You	How
You	–	Are

The skip-gram model with negative subsampling has a better accuracy over all other methods. In addition, this can identify the semantics of a word quickly. That is, the word "bank" can be a financial institution or riverbank.

- Google's word2vec uses CBOW and skip-gram model.
- Facebook's Fast Text is an extension of the word2vec model, but it feeds *N*-grams of words instead of feeding *N*-grams of sentences. It helps in identifying the context of the given word even if that word is not present in the dictionary. Consider a legal term; it can be present in the standard dictionary, but some of the legal terms may not be present in the standard dictionary that can be found out with the help of this method.

  Example: Trigram of the word 'crane'>>> {cra,ran,ane}.

14.5.3 NLP METHODOLOGIES

Features are used to implement different applications with the help of machine learning or rule-based methods. Every application need not use the other processing steps such as syntax, semantics, and pragmatic analysis. It varies depending on the application. There are different approaches to processing natural language.

Symbolic approach: This approach to NLP handles the problem of using human developed rules. Regular expressions and context-free grammars are examples of symbolic approaches. Morphological analysis and syntactic parsing are mainly done using symbolic methods.

Statistical approach: In this approach, NLP is done with the help of mathematical analysis. A large corpus is analyzed, and trends are found out. After that, linguistic rules are formulated using the analyzed trend. Machine learning models for classification, semantic processing, discourse processing, etc. are based on statistical approaches. Finite automata (FA), Hidden Markov model, etc., are examples of statistical methods.

Connectionist approach: The connectionist approach to NLP combines symbolic and statistical approaches. This approach uses formulated rules for processing text and combines them with inference obtained from the statistical approach for efficiently making specific applications. WSD, NLG, etc., can be done through the connectionist approach.

Another classification for NLP is based on the methods used for processing syntax, semantics, pragmatics, and discourse. It includes the following:

- Rule-based methods.
- Probabilistic and machine learning techniques.
- Deep learning methods.

14.5.3.1 RULE-BASED METHODS

Rule-based approaches are very efficient for rule-based tasks. Natural language has a specific rule for syntax and semantics, and hence, those tasks are rule based. Therefore, the analysis of syntax and semantics can be quickly done using rule-based approaches. The main disadvantage of a rule-based system is that the efficiency of the system depends on the programmer who creates rules. If he is efficient, the system will also be efficient. A skilled programmer can build the system within a limited period. Rule-based methods used in computers are regular expressions and FA. Regular expressions are very fast and easy to use. Most of the preprocessing and morphological tasks are done using regular expressions. One of the everyday NLP tasks called string matching can be done quickly using regular expressions.

For example: The set of strings over {0,1} that end in three consecutive 1's.

(0 | 1)* 111.

The above expression matches a string that ends with 111. The matched strings for the above regular expression include 0000001111, 0001110010111, 10101010111, etc.

- For every regular expression R, there is a corresponding FA that accepts the set of strings generated by R.
- For every FA, there is a corresponding regular expression that generates the set of strings accepted by that FA.

14.5.3.2 PROBABILISTIC AND MACHINE LEARNING METHODS

Probabilistic models for NLP are used in many NLP tasks. However, the most exciting application of the probabilistic model in NLP is predicting the next word, given a word. Likelihood maximization, conditional random fields, etc., are the most commonly used probabilistic methods in the area of NLP.

Machine learning models learn from examples. They formulate rules based on given examples, and similar data can be handled by a machine learning model very quickly. Many machine learning algorithms can be used for different purposes. There are two types of learning.

14.5.3.2.1 Supervised Learning

In this learning method, the programmer feeds the system with a set of example data about which he already knows the answer. It is used for the structured dataset. Supervised NLP machine learning algorithms are:

- Support vector machines.
- Bayesian networks.
- Maximum entropy.
- NNs.

14.5.3.2.2 Unsupervised Learning

This method of learning is used for unstructured data whose labels are unknown. Unsupervised machine learning algorithms include clustering and dimensionality reduction.

Machine learning methods cannot handle new situations efficiently. For handling such problems, deep learning methods are introduced. Sentiment analysis, dialog systems, etc., are modeled using deep learning models as it comes up with new data every time. NLP tasks such as named entity recognition, text or document classification, etc., are done using machine learning models.

14.5.3.3 DEEP LEARNING METHODS

An NN with more than one hidden layer is called as a deep learning network. In machine learning, classification is done by feeding examples along with essential features, whereas, in deep learning, there is no need to supply features for classification. The deep learning network learns the features for classification automatically from given examples. However, the main disadvantage of a deep learning network is that it needs a massive dataset of examples since it learns features from examples. Consider a dataset of vehicle images; it is required to predict whether the input image is a car or not, and for that, the network should be fed with the features of a car (i.e., it has four wheels, four doors, etc.). However, in the case of deep learning, no need to specify the features of a car.

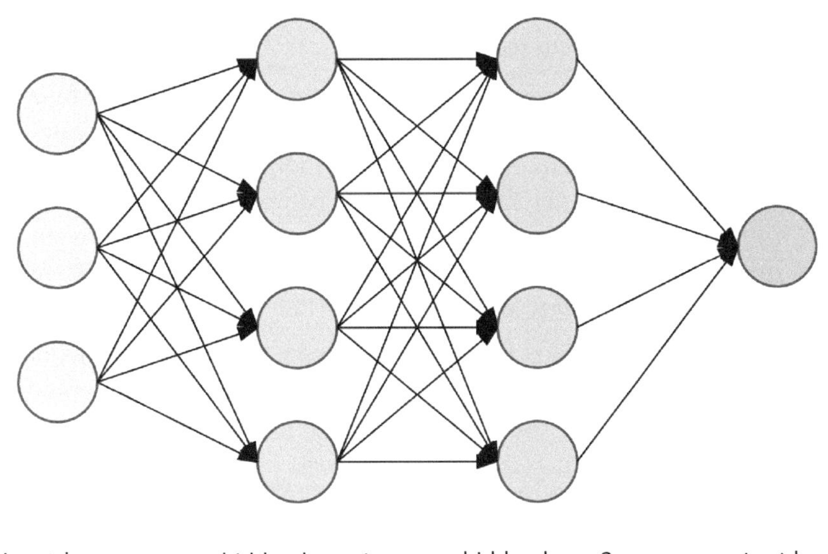

FIGURE 14.9 Architecture of a feedforward deep neural network.

14.5.3.3.1 Convolutional Neural Network (CNN)

CNN is a deep neural network (DNN) model mainly used for image classification.

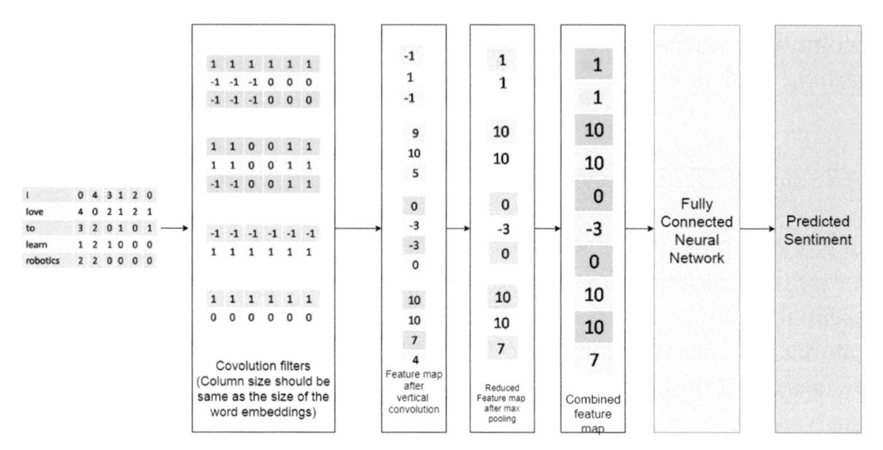

FIGURE 14.10 Convolutional neural network.

Most of the computer vision applications, such as photo tagging and age prediction that exist today, are based on the CNN. It uses convolutional filters for analyzing different features simultaneously. The main advantages of the CNN include the following:

- *Ability to learn features from unformatted data:* It can find relevant data from raw data. The CNN can easily find the linguistic feature in a sentence. It is used to process diverse morphology, syntax analysis, and things that do not follow a specific pattern.
- *Less training time as the weights are shared:* The same weight for inputs helps in reducing the training time. Since the weights are the same for every neuron, there is no need to train every neuron separately.

The CNN with pooling is better than general CNN. It helps in reducing the feature maps. Pooling can reduce the number of locations to be examined by the following layers. It also helps in reducing the training time.

14.5.3.3.2 Recursive Neural Network (RcNN)

RcNNs are created by applying the same set of weights recursively over a structured input to generate a structured prediction. It can be used to learn the sequential and hierarchical structure. Language is seen as a recursive structure where words become the leaf nodes, and subphrases become child nodes that compose the parent node, sentence. It is easy to model parse trees using RcNN since it is used for building hierarchical structures. The CNN shares weights between nodes, whereas RcNN shares weights between layers.

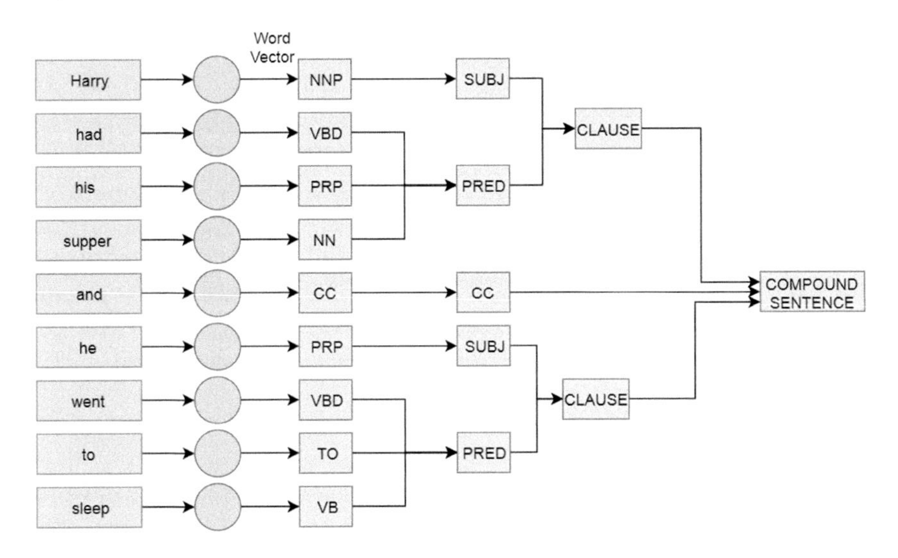

FIGURE 14.11 Recursive neural network.

14.5.3.3.3 Recurrent Neural Network (RNN)

RNN is a feedforward NN with a feedback loop, and it is used mainly for sequential and time-series data. The RNN can be used for classifying, clustering, and making predictions. While writing an email in Google, one will get predictions for the next word based on the previous words. It is an application of RNN. The absence of memory makes it unable to process long sequences efficiently. As there is a vanishing gradient problem in RNN, it is challenging to train such a network.

14.5.3.3.4 Long Short-Term Memory (LSTM)

LSTM is a modified version of RNN that eliminates the shortcomings of RNN. RNN cannot retain memory and fails in processing sequences. RNN with a memory cell inside the neurons is called an LSTM. The presence of memory in LSTM helps in capturing syntax and semantic dependencies very quickly. It has memory capacity and can store data over time. LSTM has input, output, and forget gates. Learning LSTM cell sends its hidden state and memory cell state to the adjacent LSTM cells. There are different types of LSTM networks.

- One to One model
  It is used for predicting common phrases.
  Example: ("Pretty")>>>("Good")
- One to many model
  It can be used for question generation.
  Example: ("Fine")>>>("How," "are," "you," "?")
- Many to one model
  Many to one model can be used for predicting the next word.
  Example: ("How," "are")>>>("you")
- Many to many model
  Many to many models can be used for generating a sequence of words, given a sequence of words.
  Example: ("How," "are," "you," "?")>>>("I," "am," "fine")

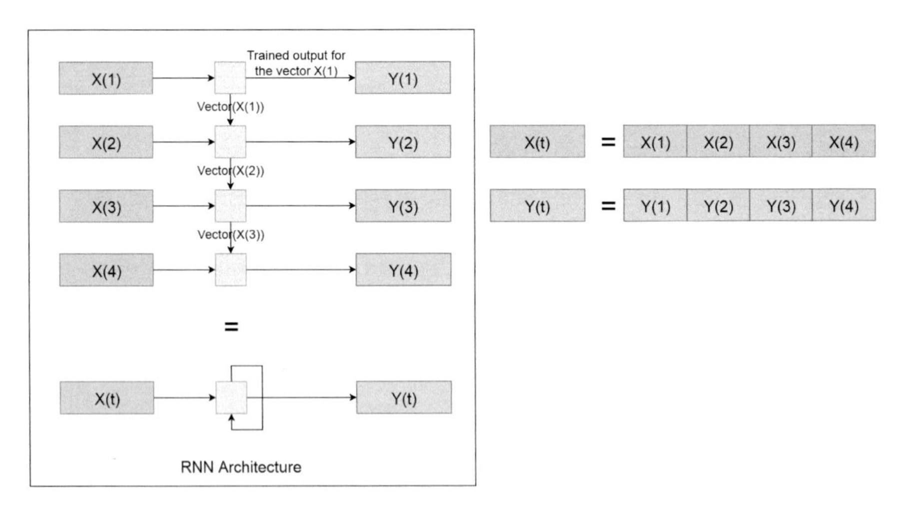

FIGURE 14.12 Recurrent neural network.

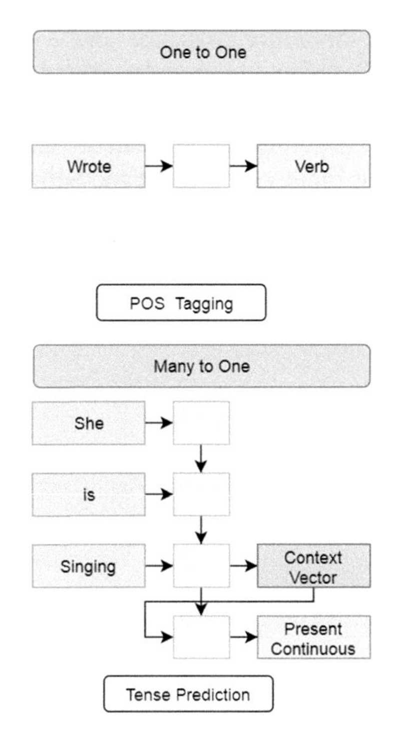

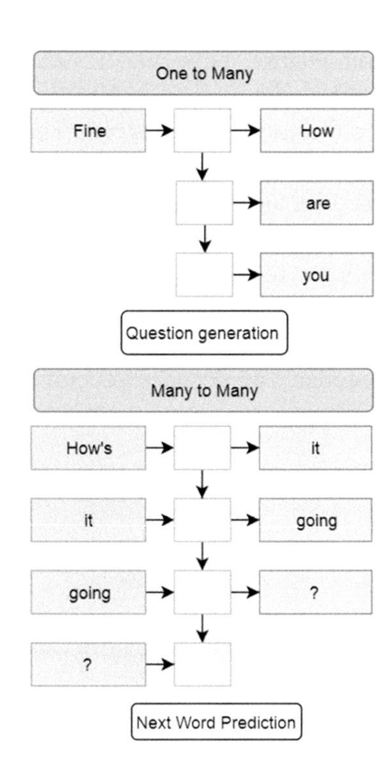

FIGURE 14.13 Types of LSTM.

14.5.3.3.5 Gated Recurrent Unit (GRU)

GRUs is introduced to overcome the vanishing gradient problem of RNN and is similar to LSTM. However, the GRU has only two gates, namely, reset and update gates. The GRU can perform well, like LSTM, without a memory unit. GRU has more control over hidden states and can eliminate the vanishing gradient problem without using memory. The learning procedure for GRU is the same as LSTM, but the GRU uses hidden states as an internal vector.

14.5.3.3.6 Encoder–Decoder network

LSTM cannot handle variable-length input and output. LSTM cannot read the entire input sequence and are not able to retain the entire sequence in its memory. For example, if "How are you?" is to be converted into any other language, then it is required to feed the entire sequence to the decoding architecture. However, LSTM cannot retain the entire sequence and will not be able to feed it directly. To solve this problem, an encoder–decoder architecture is introduced. The encoder–decoder network is a sequence to sequence architecture in which both input and output are sequences. The encoder uses an LSTM network for receiving input and convert the entire sequence into a fixed-length vector. Then, the fixed-length vector is given as input to the decoder LSTM. This method can increase output efficiency. Figure 14.14 shows the encoder–decoder network, and "How are you?" is converted to a fixed-length vector "W." That "W" is given as input to the decoder. The decoder generates a reply to the input given.

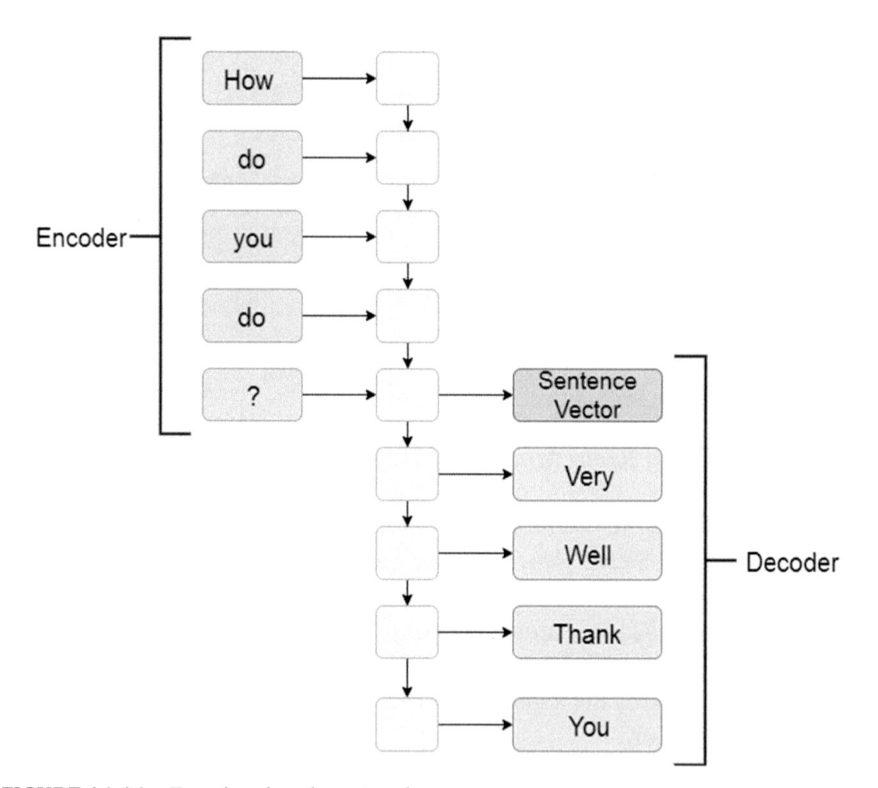

FIGURE 14.14 Encoder–decoder network.

14.6 NATURAL LANGUAGE GENERATION

With advances in technologies such as cognitive computing and intelligent systems, NLP needs to satisfy the demands of spoken dialog systems and intelligent digital assistants. Such systems should mainly converse with humans in natural language as if they were talking to a native speaker. NLG technology strives to express information stored and modeled in software to speak with humans.

NLG is the process of automatic generation of human language by a computer program to render a thought or stored data. As the primary intention of such systems is to communicate a piece of the required information, the core functionality is to decide on what to convey and then to organize the theme by selecting proper sentences and organizing them rhetorically to fit it into a grammar [3].

There are two main tasks of NLG. (i) The first is to decide and choose the content to be transferred as text. This task is done by a phase generally known as a document planner, which finally delivers chunks of information required to structure the output text. (ii) The second is to realize the actual text to be generated from the output of the document planner as a discourse in a coherent way. The primary function of the realizer is to determine the syntactic structure of the content to be generated and present it in a grammatically and syntactically correct way.

14.7 NLP CHALLENGES

- Word sense disambiguation
- Coreference resolution
- Discourse coherence
- Semantic role labeling
- Named entity recognition
- Co-occurrence, that is, how words occurring together form a relationship
- Paraphrase generation, that is, automatic paraphrase generation of a text.

14.7.1 FUTURE DIRECTIONS

Cognitive approaches can be applied in solving several research issues in NLP. A few of them are listed as follows.

Paraphrase generation—rewording a passage to get a concise idea

Example:

Original: *A total of 3000 kg of solid waste has been collected from world's highest peak, Mt. Everest since April 14 when Nepal launched an*

ambitious clean-up campaign aimed at bringing back tonnes of thrash left behind by climbers.

Paraphrase: *Since April 14, 3000 kg of solid waste left by climbers has been removed from Mt. Everest.*

Text entailment—It can be thought of as capturing relationships of the form $t \Rightarrow h$, where

t is some natural language text, and h is some hypothesis, also expressed in natural

language. Need to infer from the given text.

Example: *He is snoring*

Entailed text: He is sleeping

Example:

Premise: *A large grey elephant walked beside a herd of zebras*

Hypothesis: *The elephant was lost*

Metonymy resolution:

Metaphor-verbal figurative expressions

A metaphor often describes a word or phrase in a different style or symbolic way.

Example:

Life is a roller-coaster means *life has ups and down*

Metonymy—figurative expressions

Example:

Wall Street, Silicon Valley, White House, Hollywood, Silverfox

Metonymy resolution, text entailment, and paraphrase generation are complex research issues in NLP where human common sense thinking is highly required to extract the exact semantics. In future, cognitive self-learning algorithms with a rich context-aware knowledge base can be used to simulate human thought processes. Such cognitive systems can rely on deep learning or statistical approach based on NLP issues addressed to expand the growth of NLP.

14.8 NLP APPLICATIONS

The typical applications of NLP include the following.

- *Question answering:* It is a system that answers to the question automatically. It is an information retrieval system.
- *Machine translation:* It is one of the oldest but beneficial applications of NLP. It automatically translates text from one language to another by considering the syntax, semantics, etc., of both languages.
- *Text summarization:* It takes a document/piece of text as input and generates a compressed form of text packing essential content without any change in meaning.
- *Optical character recognition:* Extracts text out from an image embedded with text.
- *Text similarity and clustering of documents:* Finding similar texts helps in building relationships quickly. Consider (Man, Woman), (Boy, Girl) pairs; these words are not the same, but they have some similarities. These types of relationships are identified by finding text similarity between two words.
  - *Levenshtein distance:* The similarity of two strings is calculated by accounting the total number of editing operations (insertion, deletion, replace) required to convert one string into another. The Levenshtein distance of the pair (Hook, Hack) is 2. The second letter "o" should be replaced with "a" and the next "o" is replaced with "c" to get the string "Hack." Here, two substitution operations were required to convert "hook" to "hack." Therefore, the Levenshtein distance is 2.
  - *Cosine similarity:* After converting the text to vectors, the similarity of vectors can be found out using the cosine similarity measure.
  - *Phonetic similarity:* The voice-to-text converter applications use a phonetic matching concept. It tries to find a matched word from the dictionary that is phonetically similar.

14.9 TEXT CORPUS

For most of the NLP tasks, large bodies of text are used. Such text bodies are known as text corpus. It will be in a structured format. For social media analytics, a text corpus is created using text from social media and the web. For handling new problems, one can build their corpus. Some of the available text corpora include the following:

- Reuters—It is a collection of news documents.

- Gutenberg—Collection of books.
- IMDB Movie Review—A collection of movie reviews from the website imdb.com.
- Brown's corpus—Large sample of English words.
- Stanford Question Answering Dataset (SQuAD).

14.10 IMPLEMENTATION PROCEDURE FOR SEQUENCE-TO-SEQUENCE MODELING

14.10.1 COMPARATIVE STUDY OF LSTM AND GRU FOR SEQUENCE-TO-SEQUENCE MODELING

14.10.1.1 INTRODUCTION

The emergence of DNN revolutionized the entire field of NLP. The use of DNN for NLP gave rise to high-performance language models. The DNN learns features automatically from the dataset. The automatic feature learning capacity of DNN makes the language model efficient and independent of the programmer. The existing techniques for NLP mainly depend on the programmer's ability to create a language model.

The DNN works well for most of the NLP applications. Text summarization, NLG, next word prediction, question answering, Chatbots, etc., can be done efficiently using the DNN. Text summarization has a wide range of applications in this busy world due to time constraints. Text summarization is the process of extracting meaningful concepts from a large text and creating a smaller version of the given text without losing its meaning. A summary can reduce the reading time and convey the document content effectively. Automatic summarization is an active research topic in NLP. Extractive summarization creates the summary by extracting essential sentences or words from the document and combining the extracted sentences to form a summary text, whereas abstractive summarization understands the main concepts in the piece of text and then creates a summary in natural language using NLG techniques.

Various approaches are there for performing the summarization task. However, abstractive summarization is not easy to create using existing methods since it involves NLG. Abstractive summarization can be done effectively using deep learning techniques. Many architectures are there for implementing DNNs. The main architectures include CNN and RNN. The CNN works well with image data, and RNNs are suitable for processing sequential data. Simple RNNs are rarely used since it does not have memory, and they cannot remember long sequences. LSTM and GRUs are an improvised version of RNN specially designed for sequence processing. LSTM and GRU can capture long-term dependencies with the help of memory units.

14.10.1.2 ABSTRACTIVE TEXT SUMMARIZATION USING LSTM AND GRU

The text summarization approach is used for creating summaries of reviews about products in shopping websites such as Amazon. The problem of product review summarization can be modeled as a sequence to sequence problem. A text sequence is converted to another text sequence with the help of NLG. Typically, sequence to sequence language models is created using encoder–decoder architecture. An LSTM/GRU encoder encodes the text sequence to a context vector. The decoder takes context vector and internal states of LSTM/GRU encoder as an input and generates the summary of the review word by word. The decoder is trained to create the next word prediction. The decoder starts producing the summary when it receives a "start" token and stops on getting an "end" token.

Here, a summary of product reviews from Amazon is created by building an LSTM model and a GRU model and compares the performance of both models using sparse categorical cross-entropy function as the loss function. The workflow is given as follows.

14.10.1.2.1 Load Dataset

The Amazon product review dataset can be downloaded from http://jmcauley.ucsd.edu/data/amazon/. The reviews of books, Electronics, etc. are available on the website. The reviews of "Cell Phones and Accessories" that contain 194,439 reviews are used for performing text summarization. The metadata of the dataset used was given as follows.

- reviewerID—ID of the reviewer.
- asin—ID of the product.
- reviewerName—the name of the reviewer.
- helpful—helpfulness rating of the review.
- reviewText—text of the review.
- overall—the rating of the product.
- summary—summary of the review.

	reviewerID	asin	reviewerName	helpful	reviewText	overall	summary	unixReviewTime	reviewTime
194434	A1YMNTFLNDYQ1F	B00LORXVUE	eyeused2loveher	[0, 0]	works great just like my original one i really...	5.0	this works just perfect	1405900800	07 21, 2014
194435	A15TX8B2L8B20S	B00LORXVUE	Jon Davidson	[0, 0]	great product great packaging high quality and...	5.0	great replacement cable apple certified	1405900800	07 21, 2014
194436	A3JI7QRZO1QG8X	B00LORXVUE	Joyce M. Davidson	[0, 0]	this is a great cable just as good as the more...	5.0	real quality	1405900800	07 21, 2014
194437	A1NHB2VC68YQNM	B00LORXVUE	Nurse Farrugia	[0, 0]	i really like it becasue it works well with my...	5.0	i really like it becasue it works well with my...	1405814400	07 20, 2014
194438	A1AG6U022WHXBF	B00LORXVUE	Trisha Crocker	[0, 0]	product as described i have wasted a lot of mo...	5.0	i have wasted a lot of money on cords	1405900800	07 21, 2014

FIGURE 14.15 Sample preview of Amazon review dataset.

- unixReviewTime—the time of the review (unix time).
- reviewTime—the time of the review (raw).

14.10.1.2.2 Preprocess the Dataset

The dataset contains many columns that are not relevant to our problem. The dataset is to be preprocessed in such a way that the dataset only contains information relevant to our problem. In addition, the contractions such as [can't, aren't] should be expanded to [can not, are not]. Then, the special characters and stop words should be removed from the dataset. Apply preprocessing, separately to the summary, and review text. After preprocessing, save cleaned_text and cleaned_summary as separate columns in the data frame. Append "start" and "end" tokens at the beginning and end of the summary.

14.10.1.2.3 Prepare the Dataset for Modeling the Language Model Based on Our Problem

After preprocessing, prepare the dataset for feeding it into an NN. The dataset is split into training data and test data. Random split is done using train_test_split function in sklearn. About 20% of the data is used for testing and the remaining for training.

After splitting, a corpus specific dictionary is built. The dictionary is created from the corpus with the help of a tokenizer. The tokenizer creates a dictionary of unique words in the corpus. The unique word is kept as key and the occurrence of that word as its count. Then, it sorts the entire dictionary based on the occurrence count. The value after sorting is kept as word index. For example, the word "start" has the highest occurrence count in the corpus; then, it is at index 1 in the word_index dictionary.

Then, each review that contains a sequence of words is converted into integer list of integer indices. For example, consider the cleaned_summary "good quality" with "start" token at the beginning and "end" token at the end, that is, [start good quality end]. It is converted to [1, 7, 26, 2].

Next, find the review sequence having a maximum length, and convert every sequence to that maximum length by padding 0s. Before feeding into the network, standardize the length of review and summary to decide the number of neurons needed in the embedding layer of the encoder and the decoder.

	reviewText	summary	cleaned_text	cleaned_summary
0	they look good and stick good i just don't lik...	looks good	look good stick good like rounded shape always...	_START_ looks good _END_
1	these stickers work like the review says they ...	really great product	stickers work like review says stick great sta...	_START_ really great product _END_
2	these are awesome and make my phone look so st...	love love love	awesome make phone look stylish used one far a...	_START_ love love love _END_
3	item arrived in great time and was in perfect ...	cute	item arrived great time perfect condition howe...	_START_ cute _END_
4	awesome stays on and looks great can be used o...	leopard home button sticker for iphone 4s	awesome stays looks great used multiple apple ...	_START_ leopard home button sticker for iphone...

FIGURE 14.16 Data representation after cleaning.

```
{'end': 1, 'start': 2, 'phone': 3, 'case': 4, 'great': 5, 'one': 6, 'good': 7, 'like': 8, 'would': 9, 'screen
': 10, 'use': 11, 'battery': 12, 'well': 13, 'iphone': 14, 'get': 15, 'product': 16, 'charge': 17, 'charger':
18, 'works': 19, 'really': 20, 'nice': 21, 'love': 22, 'time': 23, 'also': 24, 'price': 25, 'quality': 26, 'm
uch': 27, 'fit': 28, 'little': 29, 'back': 30, 'work': 31, 'usb': 32, 'charging': 33, 'easy': 34, 'protector
': 35, 'device': 36, 'even': 37, 'power': 38, 'better': 39, 'used': 40, 'still': 41, 'using': 42, 'got': 43,
'bought': 44, 'two': 45, 'fits': 46, 'cable': 47, 'cover': 48, 'buy': 49, 'recommend': 50, 'protection': 51,
'first': 52, 'need': 53, 'perfect': 54, 'could': 55, 'looks': 56, 'best': 57, 'for': 58, 'the': 59, 'car': 6
0, 'sound': 61, 'galaxy': 62, 'put': 63, 'without': 64, 'new': 65, 'it': 66, 'cases': 67, 'way': 68, 'bluetoo
th': 69, 'pretty': 70, 'samsung': 71, 'around': 72, 'hard': 73, 'headset': 74, 'thing': 75, 'want': 76, 'long
': 77, 'devices': 78, 'make': 79, 'cannot': 80, 'enough': 81, 'made': 82, 'look': 83, 'bit': 84, 'phones': 8
5, 'right': 86, 'feel': 87, 'plastic': 88, 'looking': 89, 'not': 90, 'small': 91, 'take': 92, 'since': 93, 'b
utton': 94, 'far': 95, 'another': 96, 'see': 97, 'keep': 98, 'day': 99, 'easily': 100, 'color': 101, 'lot': 1
02, 'perfectly': 103, 'note': 104, 'however': 105, 'problem': 106, 'think': 107, 'came': 108, 'cheap': 109, '
makes': 110, 'ear': 111, 'and': 112, 'port': 113, 'ipad': 114, 'review': 115, 'fine': 116, 'comes': 117, 'lig
ht': 118, 'many': 119, 'know': 120, 'though': 121, 'say': 122, 'sure': 123, 'design': 124, 'seems': 125, 'fin
```

FIGURE 14.17 Word_index dictionary.

14.10.1.2.4 Build the Model

An encoder–decoder architecture is used for building sequence to sequence models where input and output are sequences of different lengths. Here, it is required a word-level sequence to sequence model for comparison. The encoder–decoder architecture can be build using SimpleRNN, LSTM, GRU, etc., but SimpleRNN fails to remember the lengthy sequences. LSTM and GRU are introduced to capture the dependencies in lengthy sequences. An encoder–decoder architecture is modeled using both LSTM and GRU.

The encoder encodes the sequence and generates a context vector. The output generated by the encoder, along with its internal states, is given as input to the decoder. The decoder decodes the context vector and uses the internal states for generating the output sequence word by word. The encoder and the decoder are separate NNs and are trained by considering the problem.

Note: RNN can be LSTM/GRU. If it is LSTM, the internal state vector consists of hidden states and cell states in each time step. If it is GRU, internal state vector consists of hidden states only. V indicates the list of vocabulary.

For solving the text summarization problem, the encoder is trained as a classifier network. The encoder classifies each word in the review as an important word and unimportant word. A context vector is created using the essential words and is generated as an encoder output.

On the decoder side, it takes summary as input and generates the same summary with one offset. The decoder is trained in such a way that it predicts the next word given the previous word.

In addition to this, the attention mechanism can be used for predicting a more accurate summary. Generally, the encoder gives equal importance to all words in the sequence. However, an encoder that gives more attention to specific parts of the sequence is to be designed. Consider the Amazon reviews; mostly, the starting of the review and end of the review conveys more information. Therefore, those parts need more attention. It can be achieved using the attention mechanism. A weight is assigned to all parts, and parts that need more importance are assigned with an increased weight.

14.10.1.2.5 Train the Model

An encoder–decoder architecture is created with an attention mechanism. The Amazon reviews dataset was fit into the encoder–decoder network. Each review from the dataset is processed, and output was generated. About 80% of the data was used for training, and 20% was used for validation or

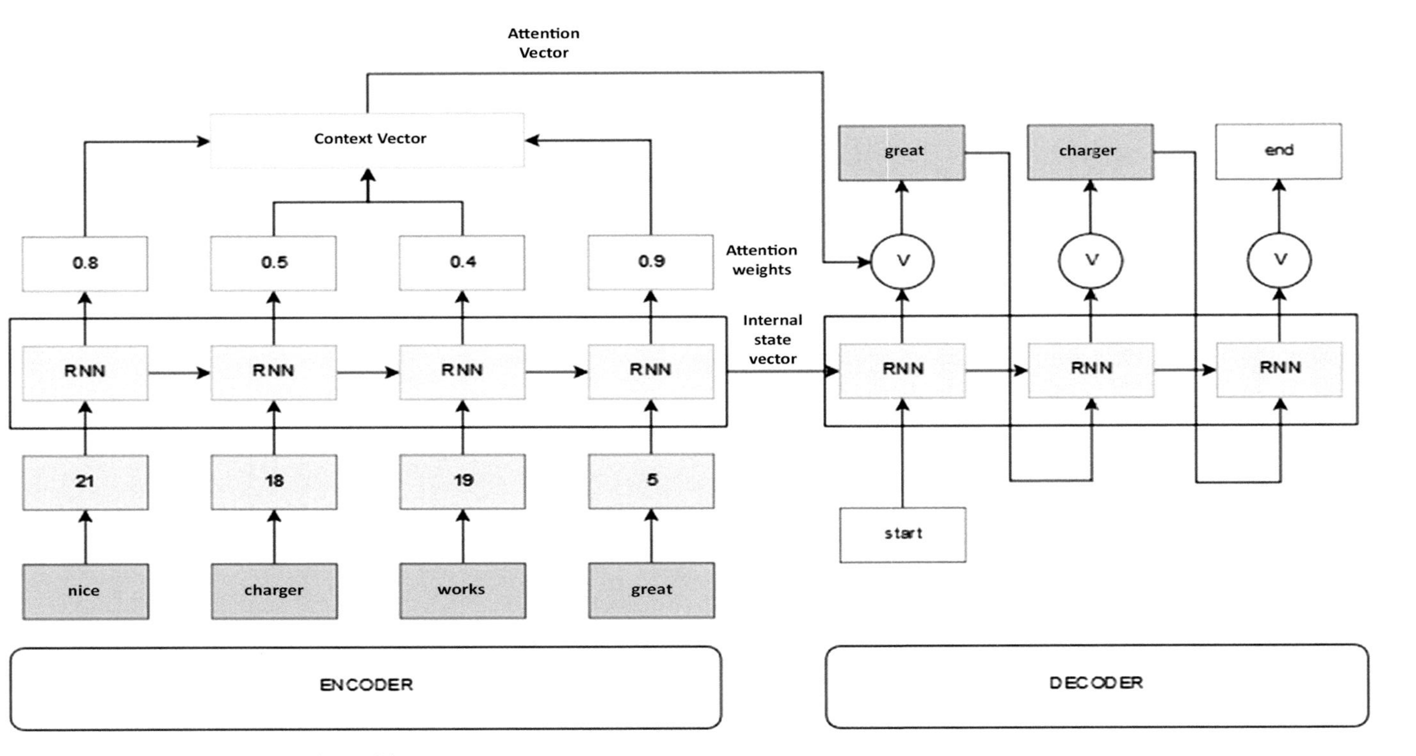

FIGURE 14.18 Encoder–decoder architecture

testing.ie. Train on 155,240 samples and validate them on 38,810 samples. An early stopping mechanism was used to stop the training when there is no significant improvement in the parameter used for evaluation. Here, the training stops when the validation loss increases or there is no significant decrease in the validation loss value.

14.10.1.2.6 Make Predictions Using the Trained Model

The final step is to make predictions using the trained model. A new review is sent to the encoder as input. The encoder generates a context vector for that input sequence. The output of the encoder, along with the internal states, is sent as input to the decoder. The decoder starts predicting the target sequence word by word when it receives a "start" token along with encoder output and internal states. The decoder predicts the next word until it encounters an "end" token or maximum summary length is attained. This predicted sequence will be in the form of word vectors like [1,20,21,2]. Decode the sequence using the dictionary that has already created. The word vector [1,20,21,2] will be converted to [start really nice end].

14.10.1.2.7 Results

Dataset used: Amazon Cell Phone and Accessories review.

No of reviews: 194,439.

Optimizer: rmsprop.

Loss: sparse_categorical_crossentropy.

Activation Function: Softmax.

Batch_size :512.

TABLE 14.7 LSTM and GRU Performance Comparison

	LSTM	GRU
Early stopping occurred on	18th Epoch	16th Epoch
Loss on last epoch	1.2436	1.4488
Validation loss on last epoch	1.3864	1.6191
Average training time per sample on each epoch	9 ms/sample	7 ms/sample

TABLE 14.8 LSTM and GRU Output Comparison

		PREDICTED OUTPUT		
47	this portable usb port car charger is a must h...	a must have for people on the go	5.0	not bad
48	i love the way it can charge 2 devices at once...	works best without location /gps on device	4.0	not bad for the price
49	bought so we could charge to phones at the sam...	works	4.0	great case
		LSTM		
47	this portable usb port car charger is a must h...	a must have for people on the go	5.0	good sound
48	i love the way it can charge 2 devices at once...	works best without location /gps on device	4.0	great phone but not the best
49	bought so we could charge to phones at the sam...	works	4.0	fits well

14.10.1.3 CONCLUSION

For the comparison of LSTM and GRU, a model of both LSTM and GRU is created using Keras for Amazon review summarization under similar constraints. The performance of both models is evaluated using sparse categorical cross-entropy loss function and validation loss. LSTM had less loss than the GRU network. However, GRU reached almost the same loss value with less time when compared to LSTM. In addition, the words predicted using LSTM and GRU are different for the same review set.

14.11 CURRENT AND FUTURE TRENDS

NLP is still in its infancy when compared to other fields of research. One of the main reasons for that is changing the structure of texts. The structure of the text is changing day by day with the emergence of social media. Therefore, there is a need for developing new methods for processing natural language. In addition, the use of NLP in cognitive systems made NLP more complex. Cognitive computing is a field where active research is going on, and they are focusing on human–human-like interaction between human and computer. Enabling a computer think like a human is not easy, and incorporating one of the main features of humans, that is, NLP, is a tedious task. Natural language understanding, NLG, and thereby creating a natural language interaction between humans and the computer is one of the main aims of cognitive computing.

NLP relies on cognitive computing for doing NLP tasks. Introducing a cognitive approach to NLP tasks makes the NLP efficient. The influence of cognitive computing in NLP tasks. Human-like processing of natural language is made possible through bringing a cognitive approach in NLP tasks such as syntax analysis, semantic analysis, etc. One of the main areas where less research is going on is syntactic processing using the cognitive approach.

There are some areas in syntax analysis where the cognitive approach can bring a change. Human-like thinking is introduced with various strategies such as machine learning, deep learning, etc. However, it takes more time for training. Since the syntax of the language has some rules, the human thought process in syntax analysis can be mimicked easily using fuzzy logic. One of the fuzzy-based approaches for syntax analysis is predicting the kinds of sentences. That is whether the sentence is assertive, imperative, interrogative, or exclamatory. It can be done using fuzzy logic. Given a sentence as input, the probability of sentence being an assertive, declarative, interrogative, or exclamatory can be predicted using fuzzy logic combined with a knowledge base.

Another is to predict the POS tag associated with a word. There are many methods for predicting the POS tag associated with the word, but there are some words that may belong to different POS according to how they are used. Consider the word "after"; it can be an adverb, preposition, adjective, or conjunction depending on the sentence in which that word is used. The fuzzy system can predict the probability of the input word being an adverb, verb, noun, preposition, adjective, pronoun, or conjunction based on the sentence.

Cognitive computing can bring a significant change in the future by making Artificial Intelligent agents more like a human (e.g., voice assistants). However, the key to achieving that is the development of more efficient NLP methodologies and tasks. Therefore, the NLP is still serving as a hot research area where researchers can contribute.

KEYWORDS

- **cognitive computing**
- **neural network**
- **encoder decoder network**
- **RNN**
- **LSTM**
- **GRU**

REFERENCES

1. Martin, J. H., & Jurafsky, D. (2009). *Speech and Language Processing: An Introduction to Natural Language Processing, Computational Linguistics, and Speech Recognition.* Upper Saddle River, NJ, USA: Pearson/Prentice-Hall.
2. Clark, A. S., Fox, C., & Lappin, S. (Eds.). (2010). *The Handbook of Computational Linguistics and Natural Language Processing* (p. 800). West Sussex, U.K.: Wiley-Blackwell.
3. Indurkhya, N., & Damerau, F. J. (Eds.). (2010). *Handbook of Natural Language Processing* (Vol. 2). Boca Raton, FL, USA: CRC Press.
4. Raghunathan, K., Lee, H., Rangarajan, S., Chambers, N., Surdeanu, M., Jurafsky, D., & Manning, C. (2010, October). A multi-pass sieve for coreference resolution. In *Proceedings of the 2010 Conference on Empirical Methods in Natural Language Processing* (pp. 492–501).

5. Clark, P., & Harrison, P. (2008, September). Boeing's NLP system and the challenges of semantic representation. In *Proceedings of the 2008 Conference on Semantics in Text Processing* (pp. 263–276).
6. Manning, C., Surdeanu, M., Bauer, J., Finkel, J., Bethard, S., & McClosky, D. (2014). The Stanford CoreNLP natural language processing toolkit. In *Proceedings of 52nd Annual Meeting of the Association for Computational Linguistics: System Demonstrations* (pp. 55–60).
7. Berant, J., & Liang, P. (2014). Semantic parsing via paraphrasing. In *Proceedings of the 52nd Annual Meeting of the Association for Computational Linguistics (Volume 1: Long Papers)* (Vol. 1, pp. 1415–1425).
8. Al-Rfou, R., & Skiena, S. (2013). SpeedRead: A fast named entity recognition Pipeline. *arXiv preprint arXiv:1301.2857.*
9. McShane, M. (2017). Natural language understanding (NLU, not NLP) in cognitive systems. *AI Magazine, 38*(4), 43–56.
10. Sarkar, D. (2016). *Text Analytics with Python: A Practical Real-World Approach to Gaining Actionable Insights from your Data.* New York, NY, USA: Apress.
11. Khurana, D., Koli, A., Khatter, K., & Singh, S. (2017). Natural language processing: State of the art, current trends and challenges. *arXiv preprint arXiv:1708.05148.*
12. Liddy, E. D. (2001). Natural language processing.
13. Young, T., Hazarika, D., Poria, S., & Cambria, E. (2018). Recent trends in deep learning-based natural language processing. *IEEE Computational Intelligence Magazine, 13*(3), 55–75.
14. Otter, D. W., Medina, J. R., & Kalita, J. K. (2018). A survey of the usages of deep learning in natural language processing. *arXiv preprint arXiv:1807.10854.*
15. Goyal, P., Pandey, S., & Jain, K. (2018). *Deep Learning for Natural Language Processing: Creating Neural Networks with Python.* New York, NY, USA: Apress.
16. Marlabs. (2018). Cognitive Computing |A Primer (Whitepaper)
17. Miller, G. A. (1995). WordNet: a lexical database for English. *Communications of the ACM, 38*(11), 39–41.
18. Baker, C. F., Fillmore, C. J., & Lowe, J. B. (1998, August). The berkeley framenet project. In *36th Annual Meeting of the Association for Computational Linguistics and 17th International Conference on Computational Linguistics, Volume 1* (pp. 86–90).
19. Schuler, K. K. (2005). VerbNet: A broad-coverage, comprehensive verb lexicon.
20. Kingsbury, P. R., & Palmer, M. (2002, May). From TreeBank to PropBank. In *LREC* (pp. 1989–1993).
21. Liu, H., & Singh, P. (2004). ConceptNet—a practical commonsense reasoning tool-kit. *BT Technology Journal, 22*(4), 211–226.
22. Navigli, R., & Ponzetto, S. P. (2010, July). BabelNet: Building a very large multilingual semantic network. In *Proceedings of the 48th Annual Meeting of the Association for Computational Linguistics* (pp. 216–225).
23. Dong, Z., & Dong, Q. (2003, October). HowNet—a hybrid language and knowledge resource. In *International Conference on Natural Language Processing and Knowledge Engineering, 2003. Proceedings. 2003* (pp. 820-824). IEEE.
24. [Online]. Available: https://literarydevices.net/coherence/ on Date: 28 April 2019
25. [Online]. Avialable: https://www.analyticsvidhya.com

Index

B

C

E

F

G

H

I

O

P

R